FIRST EDITION

SAFETY, HEALTH, AND ENVIRONMENT

Center for the Advancement of Process Technology

Prentice Hall

Boston   Columbus   Indianapolis   New York   San Francisco   Upper Saddle River   Amsterdam
Cape Town   Dubai   London   Madrid   Milan   Munich   Paris   Montreal   Toronto   Delhi
Mexico City   Sao Paulo   Sydney   Hong Kong   Seoul   Singapore   Taipei   Tokyo

Editor in Chief: Vernon Anthony
Acquisitions Editor: David Ploskonka
Editorial Assistant: Nancy Kesterson
Director of Marketing: David Gesell
Senior Marketing Coordinator: Alicia Wozniak
Marketing Assistant: Les Roberts
Associate Managing Editor: Alexandrina Benedicto Wolf
Project Manager: Alicia Ritchey
Senior Operations Supervisor: Pat Tonneman
Operations Specialist: Laura Weaver
Art Director: Diane Ernsberger

Cover Designer: Jayne Conte
Cover Art: Center for the Advancement of Process Technology and Corbis
Lead Media Project Manager: Karen Bretz
Full-Service Project Management: Lisa Garboski, bookworks
Composition: TexTech
Printer/Binder: LSC Communications
Cover Printer: LSC Communications
Text Font: Times Ten Roman

Credits and acknowledgments borrowed from other sources and reproduced, with permission, in this textbook appear on appropriate page within text.

Library of Congress Cataloging-in-Publication Data
Safety, health, and environment / CAPT.
    p. cm.
  Includes index.
  ISBN 0-13-700401-X
1. Manufacturing processes—Safety measures. 2. Manufacturing processes—Health aspects. 3. Manufacturing processes—Environmental aspects.   I. Center for the Advancement of Process Technology.
  TS183.S24   2010
  570.28'9—dc22   2009003592

Prentice Hall
is an imprint of

www.pearsonhighered.com

ISBN-10: 0-13-700401-X
ISBN-13: 978-0-13-700401-0

Contents

Preface

The Process Industries Challenge

In the early 1990s, the process industries recognized that they would face a major manpower shortage due to the large number of employees retiring. Industry partnered with community colleges, technical colleges, and universities to provide training for their process technicians, recognizing that substantial savings on training and traditional hiring costs could be realized. In addition, the consistency of curriculum content and exit competencies of process technology graduates could be ensured if industry collaborated with education.

To achieve this consistency of graduates' exit competencies, the Center for the Advancement of Process Technology and its partner alliances identified a core technical curriculum for the Associate Degree in Process Technology. This core, consisting of eight technical courses, is taught in partner member institutions throughout the United States. This textbook provides a common standard reference for the *Safety, Health, and Environment* course that serves as part of the core technical courses in the degree program.

Purpose of the Textbook

Instructors who teach the process technology core curriculum, and who are recognized in the industry for their years of experience and their depth of subject-matter expertise, requested that a textbook be developed to match the standardized curriculum. Reviewers from a broad array of process industries and education institutions participated in the production of these materials so that the widest audience possible would be represented in the presentation of the content.

The textbook is intended for use in high schools, community colleges, technical colleges, universities, and corporate settings in which process technology is taught. However, educators in many disciplines will find these materials useful as a complete reference for both theory and practical application. Students will find this textbook to be a valuable resource throughout their process technology career.

Organization of the Textbook

This textbook has been organized into 24 chapters. Chapter 1 provides an overview of the safety, health, and environmental issues and practices that may be found within the process industries. Chapters 2–17 detail the various types of hazards a process technician might encounter. Chapters 18–24 discuss the various controls and equipment used to reduce the risks associated with each of these hazards.

Each chapter is organized in the following way:

- Learning Objectives
- Key Terms
- Introduction

- Key Topics
- Summary
- Checking Your Knowledge
- Student Activities

The **Learning Objectives** for a chapter may cover one or more sessions in a course. For example, Chapter 2 may take two weeks (or two sessions) to complete in the classroom setting.

The **Key Terms** are a listing of important terms and their respective definitions that students should know and understand before proceeding to the next chapter.

The **Introduction** may be a simple introductory paragraph, or may introduce concepts necessary to the development of the content of the chapter itself.

Any of the **Key Topics** can have several subtopics. Although these topics and subtopics do not always follow the flow of the learning objectives, all learning objectives are addressed in the chapter.

The **Summary** is a restatement of the learning outcomes of the chapter.

The **Checking Your Knowledge** questions are designed to help students self-test on potential learning points from the chapter.

The **Student Activities** section contains activities that can be performed independently or with other students in small groups, and activities that should be performed with instructor involvement.

Chapter Summaries

CHAPTER 1: INTRODUCTION TO SAFETY, HEALTH, AND ENVIRONMENT

This chapter describes industrial accidents and other incidents in the process industries that have impacted safety, health, and the environment (SHE). It describes government agencies and regulations, along with industry organizations and voluntary standards. The role of the process technician in safety, health, and environmental issues is detailed.

CHAPTER 2: TYPES OF HAZARDS AND THEIR EFFECTS

This chapter identifies hazards found in the process industries, including chemical, biological, physical, and ergonomic. The routes of entry for chemical and biological hazards into the body are discussed, along with short and long-term effects these hazards can have on health. Air, water, and soil pollution are described as hazards to the environment.

CHAPTER 3: RECOGNIZING CHEMICAL HAZARDS

This chapter describes chemical hazards in the process industries, along with their impact on safety, health, and the environment. The purpose of chemical labeling systems is discussed, along with how to read and understand a Material Safety Data Sheet (MSDS). Government regulations relating to chemical hazards are also outlined.

CHAPTER 4: RECOGNIZING BIOLOGICAL HAZARDS

This chapter describes biological hazards in the process industries, along with their impact on safety, health, and the environment. Bloodborne pathogens and their effect on the human body are discussed. Government regulations relating to biological hazards are also outlined.

CHAPTER 5: EQUIPMENT AND ENERGY HAZARDS

This chapter identifies the physical hazards posed by equipment, including rotating equipment, fired equipment, and energized equipment. Electrical hazards are described, along with the hazards of automation and robotics. Government regulations relating to equipment and energy hazards are also outlined, including machine guarding and electrical work.

CHAPTER 6: FIRE AND EXPLOSION HAZARDS

This chapter discusses the physical hazards of fire, explosions, and detonation. The fire triangle and tetrahedron are defined, along with classes of fires. Government regulations relating to fire and explosions are also described.

CHAPTER 7: PRESSURE, TEMPERATURE, AND RADIATION HAZARDS

This chapter identifies physical hazards associated with pressure, including vacuum, high pressure, compressed gases, and pressure vessels. The hazards of extremely hot and extremely cold temperatures in a working environment are described, as well as ionizing and non-ionizing radiation. Government regulations relating to pressure, temperature, and radiation are also outlined.

CHAPTER 8: HAZARDOUS ATMOSPHERES AND RESPIRATION HAZARDS

This chapter describes hazardous atmospheres and respiratory-related issues. The dangers of hazardous atmospheres to respiration are detailed, along with how ventilation and respirators are used. Confined-space issues relating to hazardous atmospheres are discussed. Government regulations relating to hazardous atmospheres and respiratory protection are also outlined.

CHAPTER 9: WORKING AREA AND HEIGHT HAZARDS

This chapter discusses hazards associated with work areas, including working surfaces, means of egress (exit), heights, and confined spaces. Fall protection and confined-space entry are described. Government regulations relating to work areas, heights, and confined spaces are also outlined.

CHAPTER 10: HEARING AND NOISE HAZARDS

This chapter describes what noise in the workplace is, and how it can affect hearing. Hearing and the human ear are discussed, along with variables that can impact hearing such as volume of noise and length of exposure. Types of hearing protection are listed. Government regulations relating to noise and hearing protection are also outlined.

CHAPTER 11: CONSTRUCTION, MAINTENANCE, AND TOOL HAZARDS

This chapter identifies hazards associated with construction around a process facility and how to maintain a safe work environment. Maintenance-related hazards are described. Working with tools such as hand, power, pneumatic, and electric is discussed, as well as welding hazards. Government regulations relating to construction, maintenance, and tool use are also outlined.

CHAPTER 12: VEHICLE AND TRANSPORTATION HAZARDS

This chapter identifies the hazards associated with forklifts and powered industrial trucks, boats, trucks, railcars, aircraft, personal vehicles, pipelines, and bicycles/carts. Government regulations relating to vehicle hazards are also outlined.

CHAPTER 13: NATURAL DISASTERS AND INCLEMENT WEATHER

This chapter discusses a variety of natural disasters and related hazards, and how the can affect the process industries, including hurricanes, tornadoes, floods, lightning/storms, extreme temperatures, and earthquakes. Emergency preparedness plans and government agencies/organizations related to natural disasters are discussed.

CHAPTER 14: PHYSICAL SECURITY AND CYBERSECURITY

This chapter identifies hazards posed by physical threats and computer/information (cybersecurity) threats. Terrorists, insiders, and criminal elements are described, along with the risks they present. Terrorist threats and acts, workplace violence, criminal acts, and industrial espionage are discussed. Physical security controls and cybersecurity practices are detailed, along with government regulations.

CHAPTER 15: RECOGNIZING ERGONOMIC HAZARDS

This chapter discusses the hazards associated with ergonomic stress, including lifting and handling materials, working at heights, working in confined space, using tools, and similar situations. Proper lifting and ergonomic techniques are described, along with government and industry guidelines related to ergonomics.

CHAPTER 16: RECOGNIZING ENVIRONMENTAL HAZARDS

This chapter describes specific classifications of hazardous chemicals and the hazard they pose to the environment, along with EPA regulations. The factors that can lead to spills and the potential dangers they present are also discussed.

CHAPTER 17: INTRODUCTION TO HAZARD CONTROLS

This chapter introduces the three types of hazard controls: engineering, administrative, and personal protective equipment (PPE). This chapter also describes why, when, and how these controls are applied.

CHAPTER 18: ENGINEERING CONTROLS: ALARMS AND INDICATOR SYSTEMS

This chapter identifies various engineering controls relating to alarms and indicator systems, and how these systems minimize or eliminate hazards. Systems described include fire alarms and detection, toxic gas alarms and detection, redundant alarms and shutdown devices, automatic shutdown devices, and interlocks.

CHAPTER 19: ENGINEERING CONTROLS: PROCESS CONTAINMENT AND PROCESS UPSET CONTROLS

This chapter describes engineering controls relating to process containment and process upset systems. Process containment systems described include closed systems, closed-loop sampling, tanks, HVAC, effluent control, waste treatment, and other similar systems and equipment. Process upset systems discussed include flares, pressure-relief valves, deluge systems, explosion-suppression systems, and other similar systems and equipment.

CHAPTER 20: ADMINISTRATIVE CONTROLS: PROGRAMS AND PRACTICES

This chapter identifies administrative controls, including programs and practices. Programs discussed include written programs, policies, procedures, mutual aid agreements, and plans, along with the impact of government regulations. Practices described include documentation/shipping papers, training, housekeeping, permits, inspections, investigations, HAZOPs, monitoring, safe work observations, and community awareness.

CHAPTER 21: PERMITTING SYSTEMS

This chapter identifies types of permits, permitting procedures, placement of lockout devices, and job-safety analyses. Permits discussed include confined space permits, lockout/tagout, hot work, opening and blinding, radiation, critical lifting, and scaffolding tags.

CHAPTER 22: PERSONAL PROTECTIVE EQUIPMENT AND FIRST AID

This chapter describes potential injuries, along with the personal protective equipment (PPE) that can prevent them. PPE described includes respiratory protection, hearing protection, foot and legwear, hand and arm wear, body protection, and eye and face protection. PPE information includes descriptions, selection, fit, use, care, and maintenance.

CHAPTER 23: MONITORING EQUIPMENT

This chapter identifies different types of monitoring equipment, process upset controls, and industrial hygiene monitoring. The types of monitoring equipment addressed include LEL shutdown equipment, O_2 meters, various types of gas detection equipment, personal monitoring devices, alarm systems and indicators, redundant alarm and shutdown systems, automatic shutdown devices, and interlocks. The process upset controls include flares, pressure-relief devices, deluge systems, explosion suppression systems, explosive gas alarms and detectors, spill containment, and industrial hygiene monitoring.

CHAPTER 24: FIRE, RESCUE, AND EMERGENCY RESPONSE EQUIPMENT

This chapter identifies fire, rescue and emergency response techniques, procedures, and equipment. The fire and rescue techniques discussed include fire brigades, fire hoses and hydrants, emergency vehicles, fire extinguishers and extinguishing agents. The rescue equipment addressed in this chapter includes ropes and mechanical retrieval devices, stretchers and backboards, breathing apparatus, eyewash stations and safety showers, and chemical protective clothing.

APPENDIX A: GOVERNMENT/REGULATORY AND INDUSTRY RESOURCES

This appendix lists various government agencies and industry organizations that relate to safety, health, and environmental issues in the process industries. Web sites are listed, along with contact information.

Acknowledgements

The following organizations and their dedicated personnel voluntarily participated in the production of this textbook. Their contributions to making this a successful project are greatly appreciated. Perhaps our gratitude for their involvement can best be expressed by this sentiment:

> The credit belongs to those people who are actually in the arena . . . who know the great enthusiasms, the great devotions to a worthy cause; who at best, know the triumph of high achievement; and who, at worst, fail while daring greatly . . . so that their place shall never be with those cold and timid souls who know neither victory nor defeat. – Theodore Roosevelt

INDUSTRY CONTENT DEVELOPERS AND REVIEWERS

Candy Carrigan, ConocoPhillips, New Jersey
Christine Archer, TAP Safety Services, Texas
Diane McGinn, Innovene, Texas
Don Parsley, Valero Refining, Texas
F.D. (Bubba) Diaz, Mississippi Power, Mississippi
Gary Allison, Valero Refining, Louisiana
Gerry Swieringa, ConocoPhillips, Washington
James Barnes, Boeing, California
James Turlington, Lyondell-Citgo Refining, Texas
Jim Duplantis, Valero Refining, Louisiana
John Engelman, S.C. Johnson & Sons, Inc., Wisconsin
Jon Leacroy, The Dow Chemical Company, Texas
Larry Hensley, Innovene, Texas
Linda Brown, Pasadena Refining System, Inc., Texas
Lisa Arnold Diederich, Independent Reviewer, Pennsylvania
Mark Bowers, Compliance Assurance Associates, Inc., Texas
Michael P. Gerrity, ConocoPhillips, New Jersey

Pamelyn G. Lindsey, DuPont, New Jersey
Paul Dietrich, ConocoPhillips, New Jersey
Paul Summers, The Dow Chemical Company, Texas
Randy Armstrong, Shell Oil Company, Texas
Ray Player, Eastman Chemical Company, Texas Operations, Texas
Richard Honea, The Dow Chemical Company, Texas
Robert Walls, Sherwin Alumina, Texas
Roy J. Murdock, Shell Oil Products, Washington
Russell Karins, Chevron Phillips, Texas
Susanne Kolodzy, Troubleshooting Resources, Texas
Ted Borel, Lyondell, Texas
Terry Richey, Chemturn Corporation, Louisiana
William (Billy) Joiner, Equistar Chemical, Texas

EDUCATION CONTENT DEVELOPERS AND REVIEWERS

Anita Brunsting, Victoria College, Texas
Barb Bessette-Henderson, Independent Reviewer, Texas
Barbara Foster, West Virginia University, West Virginia
Bennett Willis, Brazosport College, Texas
Dan Schmidt, Bismarck State College, North Dakota
David Corona, College of the Mainland, Texas
David Gilfillan, College of the Mainland, Texas
Diane Trainor, Middlesex County College, New Jersey
Dorothy Ortego, McNeese University, Louisiana
Douglas Detman, Independent Reviewer, Texas
Ernest Duhon, Sowela Technical Community College, Louisiana
Gary Hicks, Brazosport College, Texas
Jerry Duncan, College of the Mainland, Texas
Jerry Layne, Baton Rouge Community College, Louisiana
John Galiotos, Houston Community College-Northeast, Texas
Kathy T. Brossette, Baton Rouge Community College, Louisiana
Keith James Tolleson, Louisiana Technical College-River Parishes, Louisiana
Louis Babin, ITI Technical College, Louisiana
Lyndon Pousson, Louisiana Technical College, Louisiana
Mark Demark, Alvin Community College, Texas
Martha McKinley, McKinley Consulting, Texas
Michael Gunter, Independent Reviewer, Louisiana
Michael High, Baton Rouge Community College, Louisiana
Paul Rodriguez, Lamar Institute of Technology, Texas
Pete Rygaard, College of the Mainland, Texas
Robert (Bob) Weis, Delaware Technical and Community College, Delaware
Robert (Bobby) Smith, Texas State Technical College - Marshall, Texas
Ryan Caya, Bismarck State College, North Dakota
Tommie Ann Broome, Mississippi Gulf Coast Community College, Mississippi
Vicki Rowlett, Lamar Institute of Technology, Texas
Walter Tucker, Lamar Institute of Technology, Texas

CENTER FOR THE ADVANCEMENT OF PROCESS TECHNOLOGY STAFF

Bill Raley, Principal Investigator
Jerry Duncan, Director
Melissa Collins, Associate Director
Angelica Toupard, Instructional Designer
Scott Turnbough, Graphic Artist
Cindy Cobb, Program Assistant

Joanna Perkins, Outreach Coordinator
Chris Carpenter, Web Application Developer

This material is based upon work supported, in part, by the National Science Foundation under Grant No. DUE-0532652. Any opinions, findings, and conclusions or recommendations expressed in this material are those of the author(s) and do not necessarily reflect the views of the National Science Foundation.

CHAPTER 1

Introduction to Safety, Health, and Environment

Objectives

This chapter provides an overview of safety, health, and environmental issues and practices in the process industries.

After completing this chapter, you will be able to do the following:

- Recall industrial accidents and other events in the process industries that have impacted safety, health, and the environment.
- Understand the necessity of occupational safety regulations.
- Describe governmental agencies and regulations that address safety, health, and environmental issues:

  OSHA

  Environmental Protection Agency (EPA)

  Department of Transportation (DOT)

  Nuclear Regulatory Commission (NRC)
- Discuss how you can impact safety, health, and environmental issues.
- Describe good safety habits.
- Understand safe work practices.

  Follow all procedures.

  Attend training and use documentation.

  Perform housekeeping and sanitation.

  Handle materials properly.

Key Terms

Administrative controls—policies, procedures, programs, training, and supervision to establish rules and guidelines for workers to follow in order to reduce the risk of exposure to a hazard.

Air pollution—the contamination of the atmosphere, especially by industrial waste gases, fuel exhausts, smoke, or particulate matter (finely divided solids).

Attitude—a state of mind or feeling with regard to some issue or event.

Behavior—an observable action or reaction of a person under certain circumstances.

Biological hazard—a living or once-living organism, such as a virus, a mosquito, or a snake, that poses a threat to human health.

Chemical hazard—any hazard that comes from a solid, liquid, or gas element, compound, or mixture that could cause health problems or pollution.

Cybersecurity—security measures intended to protect information and information technology from unauthorized access or use.

DOT—U.S. Department of Transportation; a U.S. government agency with a mission of developing and coordinating policies to provide an efficient and economical national transportation system, taking into account need, the environment, and national defense.

Engineering controls—controls that use technological and engineering improvements to isolate, diminish, or remove a hazard from the workplace.

EPA—Environmental Protection Agency; a Federal agency charged with the authority to make and enforce national environmental policy.

Ergonomic hazard—hazards that can create physical and psychological stresses because of forceful or repetitive work, improper work techniques, or poorly designed tools and workspaces.

Facility—also called a plant. A building or place that is used for a particular industry.

Hazardous agent—a substance, method, or action by which damage or destruction can happen to personnel, equipment, or the environment.

Hazards—substances, methods, or actions by which damage or destruction can happen to personnel, equipment, facilities, and/or the environment.

Heat—the transfer of energy from one object to another as a result of a temperature difference between the two objects.

ISO 14000—an international standard that addresses how to incorporate environmental aspects into operations and product standards.

Material Safety Data Sheet (MSDS)—a document that provides key safety, health, and environmental information about a chemical.

NRC—Nuclear Regulatory Commission; a U.S. government agency that protects public health and safety through regulation of nuclear power and the civilian use of nuclear materials.

OSHA—Occupational Safety and Health Administration; a U.S. government agency created to establish and enforce workplace safety and health standards, conduct workplace inspections and propose penalties for noncompliance, and investigate serious workplace incidents.

Personal Protective Equipment (PPE)—specialized gear that provides a barrier between hazards and the worker using the PPE.

Physical hazard—any hazard that comes from environmental factors such as excessive levels of noise, temperature, pressure, vibration, radiation, electricity, and rotating equipment.

Physical security—security measures intended to counter physical threats from a person or group seeking to intentionally harm other people or vital assets.

Process industries—a broad term for industries that convert raw materials, using a series of actions or operations, into products for consumers.

Process technician—a worker in a process facility who monitors and controls mechanical, physical, and/or chemical changes, throughout many processes, to produce either a final product or an intermediate product, made from raw materials.

Process technology—a controlled and monitored series of operations, steps, or tasks that converts raw materials into a product.

Security hazard—a hazard or threat from a person or group seeking to intentionally harm people, computer resources, or other vital assets.

SHE—Safety, Health, and the Environment. Also referred to as HSE or EHS.

Soil pollution—the accidental or intentional discharge of any harmful substance into the soil.

Unit—an integrated group of process equipment used to produce a specific product or products. All equipment contained in a department.

Voluntary Protection Program (VPP)—an OSHA program designed to recognize and promote effective safety and health management.

Water pollution—the introduction, into a body of water or the water table, of any EPA-listed potential pollutant that affects the water's chemical, physical, or biological integrity.

Introduction

This chapter provides an overview of various hazards that process technicians might encounter in the workplace, government agencies and regulations that address **Safety, Health, and the Environment (SHE)**, the cost of non compliance, voluntary programs and industry standards that promote workplace safety, and how process technicians can practice good safety habits.

Throughout this textbook, the term *process industries* is used. **Process industries** is a broad term for industries that convert raw materials, using a series of actions or operations, into products for use by consumers.

Generally speaking, the process industries involve **process technology**—processes that take quantities of raw materials and transform them into other products. The result might be an end product for a consumer, or an intermediate product that is used to make an end product. Each company in the process industries uses a system of people, methods, equipment, and structures to create products. A **process technician** monitors and controls mechanical, physical, and/or chemical changes, throughout many processes, to produce either the final or intermediate product.

The process industries are some of the largest industries in the world, employing hundreds of thousands of people in almost every country. These industries, either directly or indirectly, create and distribute thousands of products that affect the daily lives of almost everyone on the planet.

There are a variety of industries classified as process industries. These include:

• Oil and gas The exploration and production segment locates oil and gas, then extracts them from the ground using drilling equipment and production facilities.

The transportation segment moves petroleum from where it is found to the refineries and petrochemical facilities, then takes finished products to markets.

The refining segment of the oil and gas process industries takes quantities of hydrocarbons and transforms them into finished products, such as gasoline and jet fuel, or into feedstocks (a component used to make something else, like plastics).

• Chemical manufacturing Chemical manufacturing plays a vital role in our economy, resulting in products such as plastic, fertilizers, dyes, detergent, explosives, film, paints, food preservatives and flavors, and synthetic lubricants.

• Mining Mining is a complex process which involves the extraction and processing of rocks and minerals from the ground. Mining products are integral to a wide range of industries, serving as base materials for utilities and power generation, construction, transportation, agriculture, electronics, food production, pharmaceuticals, personal hygiene, consumer products, and more.

• Power generation Power generation involves the production and distribution of electrical energy in large quantities to industries, businesses, residences, and schools. The role electricity plays in everyday life is enormous, supplying lighting, heating and cooling, and power to everything from coffee pots to refineries. The three main segments within the power generation industry are generation, transmission, and distribution. Power can be generated in a variety of ways, including burning fuels, splitting atoms, and using water.

• Water and wastewater treatment Clean water is essential for life and many industrial processes. It is through water treatment facilities that process technicians are able to process and treat water so it is safe to drink, and safe to return to the environment.

• Food and beverage The food manufacturing industry links farmers to consumers through the production of finished food products. The products created by this industry can vary dramatically, and can range anywhere from fresh meats and vegetables to processed foods that need only be heated in the microwave.

• Pharmaceuticals Modern drug manufacturing establishments produce a variety of products, including finished drugs, biological products, bulk chemicals and botanicals used in making finished drugs, and diagnostic substances such as pregnancy and blood glucose kits.

Modern drugs save lives and improve the well being of countless patients, while improving health and quality of life, and reducing healthcare costs.

• Pulp and paper Paper plays a huge role in everyday life. If, along with paper, you include items made from natural wood chemicals, then the pulp and paper industry creates and distributes thousands of products used daily around the world. The products include items such as packaging, documents, bandages, insulation, textbooks, playing cards, and money.

This textbook generally applies to all of these industry segments, but focuses specifically on the oil and gas and chemical process industries.

In the process industries, workers routinely encounter hazardous substances and face potential hazards that can cause injury, illness, or death. Some of these hazards can also impact the environment in the short and long term.

Safety, Health, and Environmental Hazards in the Process Industries

Hazards are substances, methods, or actions by which damage or destruction can happen to personnel, equipment (all the equipment contained in a department or an integrated group of process equipment used to produce a specific product is called a **unit**), **facilities** (or plant), and/or the environment.

Different government agencies, industry groups, and individuals have created various ways of classifying and describing hazardous agents. Many companies and their safety professionals use the following classification system to categorize **hazardous agents**, dividing these agents into these major types: chemical, physical, ergonomic, biological, security, and environmental.

- Chemical hazard—any hazard that comes from a solid, liquid, or gas element, compound, or mixture that could cause health problems or pollution.

- Physical hazard—any hazard that comes from environmental factors such as excessive levels of noise, temperature, pressure, vibration, radiation, electricity, or mechanical hazards. (Note: This is not the OSHA definition of a physical hazard.)

- Ergonomic hazard—any hazard that can create physical and psychological stresses because of forceful or repetitive work, improper work techniques, or poorly designed tools and workspaces.

- Biological hazard—any living or once-living organism, such as a virus, a mosquito, or a snake, that poses a threat to human health.

- Security hazard—a hazard or threat from a person or group seeking to intentionally harm people, computer resources, or other vital assets.
  - Physical security—security measures intended to prevent physical threats from a person or group seeking to intentionally harm other people or vital assets.
  - Cybersecurity—security measures intended to protect electronic assets from illegal access and sabotage.

- Environmental hazard—any hazard that results in air pollution, water pollution, and/or soil pollution.

AIR POLLUTION

The contamination of the atmosphere, especially by industrial waste gases, fuel exhausts, smoke or particulate matter (finely divided solids).

WATER POLLUTION

The introduction of any EPA-listed potential pollutant that affects the chemical, physical, or biological integrity of water.

SOIL POLLUTION

The accidental or intentional discharge of any harmful substance into the soil.

In the process industries, you might be exposed to any of these hazards. It is vital that you understand the potential causes and their effects, along with how you can minimize or control these hazards.

Incidents in the Process Industries

Companies in the process industries work diligently to provide safe work environments and protect workers as well as the surrounding community. Millions of dollars and thousands of hours are spent annually to minimize or control chemical, physical, ergonomic, biological, security, and environmental hazards.

Accidents can still happen, though, just as in any workplace, home, or public setting. Whenever any accident occurs, an investigation is launched to study the following elements:

- The accident itself (e.g., the vessel leaked)
- The immediate causes (e.g., the alarm was faulty)
- The contributing causes (e.g., the worker did not understand the procedure)
- The results of the accident (e.g., a vapor cloud was released and a person died as a result)

In some cases, accidents result from not just one cause, but from multiple causes that were not identified and corrected in time (cascading failures). By studying accidents, process technicians can better understand the causes and use this knowledge to identify and correct hazards before they become accidents.

Most often, accidents are minor. However, in rare situations major accidents occur with catastrophic results. Following are some examples of industrial accidents and incidents that caused the loss of life, extensive property damage, and damage to the environment:

- Texas City, Texas, 1947: The freighter *Grandcamp,* anchored at the docks near town, caught fire and its cargo of ammonium nitrate exploded. Later a second ship, the *High Flyer,* also exploded. More than 500 people lost their lives, and millions of dollars of damage was done to the nearby process facilities and homes. It was one of the worst industrial disasters in U.S. history.

- Bhopal, India, 1984: A large vapor cloud of methyl isocyanate (MIC, a component in pesticides) was released from the Union Carbide India Limited plant. The cloud reached the nearby town of Bhopal, killing 1,400 people. Long-term effects have killed or disabled thousands of others.

- Chernobyl, USSR (now Ukraine), 1986: The chain reaction in one of the power plant's nuclear reactors went out of control, causing an explosion and fireball that did extensive damage to the facility and released radiation into the environment. Thirty people were killed immediately, and more than 135,000 people were evacuated from the region. Figures indicate that long-term effects have resulted in at least 2,500 deaths. The effects of the radiation have been felt all over the northern hemisphere.

- *Exxon Valdez,* Alaska, 1989: An oil tanker struck a reef in Prince William Sound, Alaska. The tanker spilled more than 11 million gallons of crude oil into the sensitive environmental region that was home to thousands of species including migratory birds, shore birds, bald eagles, otters, porpoises, sea lions, and whales. A commercial fishing industry also worked the waters in the sound. The accident was one of the most devastating environmental disasters ever to occur at sea.

- Pasadena, Texas; Channelview, Texas; Sterlington, Louisiana: In a 20-month span (October 1989–May 1991) three major accidents at process facilities in Texas and Louisiana resulted in numerous fatalities and injuries, along with multi-millions of

---

Did You Know?

DANGER

RADIOACTIVE MATERIAL

Many scientists believe that the Chernobyl nuclear disaster, which happened in Ukraine in 1986, occurred because the plant was improperly designed and plant operators ignored important safety measures.

As a result, large amounts of radioactive materials were emitted into the environment. This has led to serious health problems and/or death for many of those who were exposed.

dollars in damages. In October 1989, an explosion at the Phillips Chemical Company plant in Pasadena, Texas resulted in 23 deaths, more than 100 injuries, and hundreds of millions in damage as the facility was leveled. The explosion was caused by a chemical release of ethylene, hexane, isobutane, and hydrogen; this flammable mix formed a vapor cloud that ignited. Then in July 1990, a treatment tank filled with hundreds of thousands of gallons of wastewater and chemicals exploded at the ARCO chemical plant in Channelview, Texas. Seventeen people were killed. In May 1991, an explosion at the International Minerals and Chemicals Fertilizer Group (IMC) facility in Sterlington, Louisiana caused the deaths of eight people and more than 100 injuries. A fire in a tube array heated an adjacent tube filled with nitromethane, which caused the explosion.

These accidents and other similar ones resulted in sweeping safety, health, and environmental changes. Government agencies created new regulations to address the accident causes, industry organizations implemented voluntary standards and programs, and companies spent time and money improving safety programs, equipment, training, and documentation.

Regulatory Agencies and Their Responsibilities

Several different U.S. government agencies enforce regulations that protect workers, the public, and the environment from safety, health, and environmental hazards. Among these agencies are the **Occupational Safety and Health Administration (OSHA)**, the **Environmental Protection Agency (EPA)**, the **Department of Transportation (DOT)**, and the **Nuclear Regulatory Commission (NRC)**.

Refer to *Appendix A* for contact information on these and many other agencies and organizations.

OCCUPATIONAL SAFETY AND HEALTH ADMINISTRATION (OSHA)

On December 29, 1970, President Richard Nixon signed the Occupational Safety and Health (OSH) Act of 1970. The purpose of the OSH Act was, and continues to be, "to assure so far as possible every working man and woman in the nation safe and healthful working conditions and to preserve our human resources."

The OSH Act established several agencies to oversee the protection of the American worker. These included the following:

• The Occupational Safety and Health Administration (OSHA)—created to establish and enforce workplace safety and health standards, conduct workplace inspections and propose penalties for noncompliance, and investigate serious workplace incidents.

• The Occupational Safety and Health Review Commission (OSHRC)—formed to conduct hearings with employers who were cited for violation of OSHA standards and contested their penalties.

• The National Institute for Occupational Safety and Health (NIOSH)—established to conduct research on workplace safety and health problems, specifically injuries and illnesses that may be attributed to exposure to toxic substances.

Some important OSHA regulations are covered in other sections of this chapter, while others are described in later chapters.

ENVIRONMENTAL PROTECTION AGENCY (EPA)

On January 1, 1970, President Richard Nixon signed the National Environmental Policy Act (NEPA). NEPA was enacted to set national policy regarding the protection of the environment, to promote efforts to prevent or eliminate pollution of the environment, to advocate knowledge of ecological systems and natural resources, and to establish a Council on Environmental Quality (CEQ) to oversee NEPA policy.

President Nixon soon realized that the CEQ, as structured, did not have the resources or the power to fulfill its mission. In Reorganization Order No. 3, issued on

Did You Know?

The Department of Transportation (DOT), the governmental institution responsible for regulating our highways, is also responsible for regulating the transportation of natural gas, petroleum, and other hazardous materials through pipelines.

July 9, 1970, the president stated: "It also has become increasingly clear that only by reorganizing our federal efforts can we develop that knowledge, and effectively ensure the protection, development, and enhancement of the total environment itself."

Through Reorganization Order No. 3, the **Environmental Protection Agency (EPA)** and the National Oceanic and Atmospheric Administration (NOAA) were formed by transferring control of many environmentally related functions from other governmental offices and agencies to the EPA and NOAA.

The EPA's mission is "to protect human health and the environment." The EPA works for a cleaner, healthier environment for Americans. The NOAA seeks to "observe, predict, and protect our environment."

Some important EPA regulations are covered in other sections of this chapter, while others are described in later chapters.

DEPARTMENT OF TRANSPORTATION (DOT)

On October 15, 1966, President Lyndon Johnson signed Public Law 89-670, which established the **Department of Transportation (DOT)**. DOT's mission is "to develop and coordinate policies that will provide an efficient and economical national transportation system, with due regard for need, the environment, and the national defense."

On September 23, 1977, Secretary of Transportation Brock Adams established the Research and Special Programs Administration (RSPA) within the Department of Transportation, consolidating various diverse functions that dealt with intermodal activities.

Eventually the RSPA came to oversee the Office of Pipeline Safety (OPS) and the Office of Hazardous Materials Safety (OHMS), two entities with considerable jurisdiction over the petrochemical industry.

NUCLEAR REGULATORY COMMISSION (NRC)

Congress established the Atomic Energy Commission (AEC) in the Atomic Energy Act of 1946. The AEC's mission was regulation of the nuclear industry. Eight years later, Congress replaced that act with the Atomic Energy Act of 1954, which enabled the development of commercial nuclear power. The AEC's new mission became twofold: encouraging the use of nuclear power and regulating its safety.

In the 1960s, critics charged that the AEC's regulations were not rigorous enough in several important areas, including radiation protection standards, reactor safety, plant location, and environmental protection. The AEC was disbanded in 1974 under the Energy Reorganization Act. This act created the **Nuclear Regulatory Commission (NRC)**, which started operations in 1975. Today, the NRC's regulatory activities focus on reactor safety oversight, materials safety oversight, materials licensing, and management of both high- and low-level radioactive waste.

The NRC also regulates instruments in the process industries that use radioactive materials, such as testing devices (e.g., gas chromatographs) and inspection equipment (e.g., X-ray machines).

OTHER GOVERNMENT AGENCIES RELATED TO THE PROCESS INDUSTRIES

The table below shows some other government agencies that can directly or indirectly affect safety, health, and environmental issues in the process industries.

TABLE 1-1   Government Agencies and How They Relate to the Process Industries

SHE Related Issues	*Agencies*
General workplace safety	• Department of Energy • Department of Labor (which includes OSHA) • National Institute of Standards and Technology (NIST) • Mine Safety and Health Administration (MSHA)
Production of food, beverages, and pharmaceuticals	• Food and Drug Administration (FDA)
Health, safety, chemical and biological hazards	• Department of Health and Human Services, including the Centers for Disease Control and Prevention (CDC) and the National Institutes of Health (NIH)
Transportation	• Coast Guard • Federal Aviation Administration (FAA) • Federal Railroad Administration (FRA) • National Transportation Safety Board (NTSB) • Pipeline and Hazardous Materials Safety Administration (PHMSA)
Physical security and cybersecurity	• Federal Bureau of Investigation (FBI) • Department of Homeland Security (DHS) • Coast Guard • FAA
Weather	• National Weather Service • National Oceanic & Atmospheric Association (NOAA)

While the table covers many common government agencies, it is not a comprehensive list. State and local agencies also regulate certain SHE elements within their jurisdiction. Your company will inform you of the government regulations and agencies with which you must comply.

Regulations Impacting the Process Industries

> Please note that the terms and definitions used in this textbook can vary from one government agency to another. Also, note that government regulations and industry standards can change. Refer to *Appendix A* for more information on government and industry Web sites that can provide specific definitions to terms along with the most current regulations and standards.

The U.S. government has enacted numerous regulations to minimize workplace hazards. These regulations are administered through various federal agencies such as OSHA, the EPA, and the DOT. Some regulations are generic in scope and affect a variety of industries. Other regulations were created specifically to oversee a certain industry and even the handling of certain hazardous substances.

OSHA administers many of the government regulations that significantly impact the day-to-day operations of the process industries. Four of the most important regulations described in this section are Process Safety Management (PSM), Personal

Protective Equipment (PPE), Hazard Communication (HAZCOM), and Hazardous Waste Operations and Emergency Response (HAZWOPER).

Two other major regulations administered by other agencies, the EPA and DOT, are also described in this section. These regulations address hazardous materials and their shipment.

OSHA 1910.119—PROCESS SAFETY MANAGEMENT (PSM)

The OSHA Process Safety Management of Highly Hazardous Materials (PSM)—29 CFR 1910.119 standard seeks to prevent or minimize the consequences of catastrophic releases of toxic, reactive, flammable, or explosive chemicals.

This standard establishes 14 elements aimed at improving worker safety:

- Employee involvement—employees must be involved in the process hazard analysis effort, including identification, information gathering, and information communication. The employer must create a written program detailing employee involvement.

- Process safety information—the hazards posed by processes involving highly hazardous chemicals must be identified and understood. Process safety information includes the following:
  - Hazards of the chemicals used or produced by the process (including toxicity, permissible exposure limits, physical data, reactivity data, corrosivity data, thermal and chemical stability data, and the hazardous effects of accidentally mixing different materials that could possibly occur).
  - Technology of the process (drawings representing the process, such as a Block Flow Diagram or a simplified Process Flow Diagram; process chemistry; safe upper and lower limits for variables such as temperature, pressure, and flow; and an evaluation of the consequences of deviations).
  - Equipment in the process (materials of construction; Piping and Instrument Diagrams, or P&IDs; electrical classification; relief system design and basis; ventilation system design; material and energy balances; and safety systems).

- Process Hazard Analysis (PHA)—a hazard evaluation is conducted on processes covered by the PSM standard. The PHA is used to identify, evaluate, and control hazards in a process. PHA information includes the following:
  - The hazard
  - Identification of any previous incident that could have caused catastrophic results in the workplace
  - Engineering and administrative controls applicable to the hazard and their relationships
  - Consequences if engineering and administrative controls fail
  - Location within the facility
  - Human factors
  - An evaluation of the range of possible safety and health effects on employees, if controls fail

- Operating procedures—employers must provide clear instructions for safely conducting activities within the covered process area. The procedures must include steps for each operating phase, the operating limits, safety and health considerations, and safety systems and their functions. Written procedures, which may also be maintained electronically, must be accessible by all employees who work on or maintain a covered process. The procedures must be reviewed as often as necessary to ensure they reflect current operating practices. They must also include safe work practices, where needed, to provide for special circumstances, such as lockout/tagout and confined-space entry.

- Training—employers must provide training on all covered processes to ensure employees are trained on an overview of the process, all required operating procedures, safety and health hazards, emergency operations, and safe work practices. Employees receive initial training and certification and then must receive refresher training periodically.

- Contractors—contract employees must meet the same PSM requirements as company employees. Contract employers are also required to train their employees to safely perform their jobs around highly hazardous chemicals and to document that employees received and understood training. In addition, they are responsible for ensuring that contract employees know about potential process hazards and the work-site emergency action plan.

- Pre-startup safety review—employers are required to perform a safety review of new or modified equipment or facilities prior to starting up operations. This helps ensure that equipment is constructed to meet design specifications, procedures are developed and in place, training is completed, and all required PHAs are performed and changes implemented.

- Mechanical integrity—employers are required to establish and implement written procedures to ensure the ongoing integrity of process equipment that contains and/or controls a process covered under the PSM standard. This section of the standard does not apply to contract employers; however, contract employees are required to follow the written procedures.

- Hot work permit system—hot work permits must be issued for hot work operations conducted on or near a process covered under the PSM standard. Hot work operations include electric or gas cutting, welding, brazing, solder, grinding, hot tar projects, any portable gas procedures, and steam-generating work.

- Management of change process—this ensures that when changes are made to a process, those changes will provide employees with the same protection from highly hazardous chemicals as the original equipment or process. Employers are required to provide training to all on-site employees prior to starting up the renovated process or equipment. Contract employers must train their own contract personnel on the same procedures. Management of change also ensures that process safety information and operating procedures are updated correctly as needed. Typically, all changes to procedures are reviewed by supervisors, engineers, and other key personnel to ensure accuracy. These changes are then documented and implemented into the existing or new procedures and employees are trained on the changes.

- Incident investigations—the PSM standard requires employers to investigate as soon as possible following an incident but no later than 48 hours after an incident. The incidents that are covered include those that either resulted in or could have resulted in a catastrophic release of covered chemicals. The standard requires that an investigation team, which includes at least one person knowledgeable in the process, and others with knowledge and experience in investigations and analysis of incidents, work together to develop a written incident report. The employer must keep these reports for five years.

- Emergency planning and response—employers must develop and implement an emergency action plan. This plan must include procedures for handling small releases of highly hazardous chemicals. Employees must be trained to follow these procedures and the procedures must be accessible to all employees who may be affected.

- Compliance audits—internal audits are required every three years in facilities with covered processes. These audits must certify that employers have evaluated their compliance with process safety requirements. Employers are required to respond promptly to audit findings and must document how deficiencies were corrected. Employers must retain the most recent internal audits.

- Trade secrets—companies can protect their trade secrets from competitors, by having employees sign confidentiality agreements.

OSHA 1910.1200—HAZARD COMMUNICATION (HAZCOM)

The OSHA Hazard Communication (HAZCOM) - 29 CFR 1910.1200, or "Right to Know," standard seeks to ensure that the hazards of all produced or imported chemicals are evaluated and that information relating to the hazards is provided to employees.

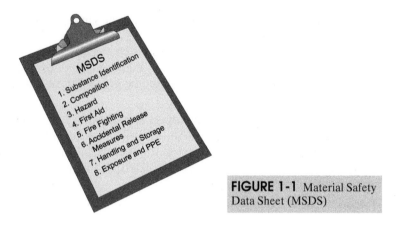

FIGURE 1-1 Material Safety Data Sheet (MSDS)

This standard requires that chemical information is communicated through comprehensive hazard communication programs. Information covered includes physical hazards (e.g., flammability) and health hazards (e.g., irritation, cancer). This information is communicated using a written program, container labeling and other forms of warning, a **Material Safety Data Sheet (MSDS)**, employee training, and other information (Figure 1-1).

The MSDS provides key safety, health, and environmental information about a chemical. This information includes physical properties, proper storage and handling, toxicological data, established exposure limits, fire fighting information, and other useful data. This information is provided in a standardized format. MSDS information must be made available for any material manufactured, used, stored, or repackaged by an organization.

OSHA 1910.120—HAZARDOUS WASTE OPERATIONS AND EMERGENCY RESPONSE (HAZWOPER)

The OSHA Hazardous Waste Operations and Emergency Response (HAZWOPER)—29 CFR 1910.120 standard outlines the establishment of safety and health programs and level of training for employees in hazardous waste operations and emergency response. This standard applies to any facility that has employees involved in the following:

- Cleanup operations involving hazardous substances
- Cleanup operations at sites covered by the Resource Conservation and Recovery Act (RCRA)
- Voluntary cleanup operations at sites recognized by governmental agencies as uncontrolled hazardous waste sites
- Operations involving hazardous wastes that are conducted at treatment, storage, and disposal facilities licensed under the RCRA
- Emergency response operations for release of, or substantial threats of release of hazardous substances

Employers must take the following steps to identify, evaluate, and control safety and health hazards in operations involving hazardous waste or emergency response:

Safety and health program—the program must be designed to identify, evaluate, and control safety and health hazards. It must also provide a documented plan for emergency response in the event of a release of the hazardous materials.

Site control program—this program must include a site map, site work zones, site communications, safe work practices, and identification of the nearest medical assistance facility. Employers are also required to implement a buddy system as a protective measure in particularly hazardous situations. This would allow one employee to keep watch on another to ensure that quick aid could be provided if needed.

Employee training—there are two different levels of HAZWOPER training. For example, employees involved in hazardous waste cleanup are required to have more intensive training than an equipment operator with little potential for hazardous waste exposure. Most sites require initial training and periodic refresher training on hazardous waste operations or emergency responses.

Medical surveillance—this is required for all employees exposed to any hazardous substance at or above established exposure levels. It is also required for those who wear approved respirators for more than 30 days on site in a year and for workers exposed to unexpected or emergency releases. In addition to annual medical checks, the same employees must have a medical check upon termination of their employment.

Reduction of exposure levels—engineering controls, work practices, personal protective equipment, or a combination of all three must be implemented to reduce exposure below established levels for any hazardous substances. In other words, employers must make all possible efforts to reduce exposure levels to acceptable levels to protect their employees.

Air monitoring—on-site air monitoring is required to identify and quantify levels of hazardous substances. This monitoring must be performed periodically to ensure that the proper protective equipment is used on site.

Information program—employers must supply personnel with a program that provides the names of key personnel responsible for site safety and health, names of alternate personnel responsible for site safety and health, and a listing of the HAZWOPER standard requirements.

Decontaminating procedures—employees and equipment must be decontaminated before leaving an area where they may have been exposed to hazardous materials. These operating procedures must minimize exposure through contact with exposed equipment, other employees, or used clothing. Showers and changing rooms must be provided where needed.

Emergency response plans—plans are required to handle possible on-site emergencies as well as off-site emergencies. These plans are often performed as drills and may involve non-employees, such as emergency response technicians, fire fighters, and medical personnel.

OSHA 1910.1000 AIR CONTAMINANTS

The OSHA Air Contaminants standard (29 CFR 1910.1000) establishes the Permissible Exposure Limits (PELs) for a variety of toxic and hazardous substances. A PEL describes the amount of an airborne toxic or hazardous substance to which an employee can be exposed over a specified amount time. OSHA 1910.1000 through 1910.1500 list specific toxic or hazardous substances and the assigned PEL for each.

OSHA VOLUNTARY PROTECTION PROGRAM (VPP)

OSHA established the **Voluntary Protection Program (VPP)** to recognize and promote effective safety and health management. The following list outlines the cooperative relationship between management, employees, and OSHA to implement a VPP:

1. Management agrees to operate an effective program that meets an established set of criteria.
2. Employees agree to participate in the program and work with management to ensure a safe and healthful workplace.
3. OSHA initially verifies that a site's program meets the VPP criteria.
4. OSHA publicly recognizes the site's exemplary program and removes the site from routine scheduled inspection lists. (Note: OSHA may still investigate major accidents, valid formal employee complaints, and chemical spills.)

There are two OSHA VPP Ratings: Star and Merit.

- **Star** participants meet all VPP requirements.
- **Merit** participants have demonstrated the potential and willingness to achieve Star program status and are implementing planned seps to fully meet all Star requirements.

Periodically (every three years for the Star program and every year for the Merit program), OSHA reassesses the site to confirm that it continues to meet VPP criteria.

DOT CFR 49.173.1—HAZARDOUS MATERIALS—GENERAL REQUIREMENTS FOR SHIPMENTS AND PACKAGING

The DOT Hazardous Materials - General Requirements for Shipments and Packaging - 49 CFR 173.1 standard establishes requirements for preparing hazardous materials to ship by air, highway, rail, or water.

This standard establishes requirements for preparing hazardous materials to be shipped by air, highway, rail, water, or any combination of these. It also covers the inspection, testing, and retesting responsibilities for persons who retest, recondition, maintain, repair, and rebuild containers used or intended for transporting hazardous materials.

EPA CFR 264.16—RESOURCE CONSERVATION AND RECOVERY ACT (RCRA)

The EPA Resource Conservation and Recovery Act (RCRA)—40 CFR 264.16 standard promotes "cradle-to-grave" management of hazardous wastes.

This standard classifies and defines requirements for hazardous waste generation, transportation and treatment, storage, and disposal facilities. Additionally, it requires industries to identify, quantify, and characterize their hazardous wastes prior to disposal. It holds the generator of the hazardous waste responsible for management from the point of inception to the final disposal of materials.

EPA CLEAN AIR AND CLEAN WATER ACTS

The 1990 Clean Air Act "sets limits on how much of a pollutant can be in the air anywhere in the United States." The act ensures that all Americans are covered using the same basic health and environmental protections. Each state must carry out its own implementation plan to meet the standards of the act. For example, it would be up to a state air pollution agency to grant permits to power plants or chemical facilities and fine companies for violating the air pollution limit

In 1972, the Federal Water Pollution Control Act Amendments were enacted, reflecting growing public concern for controlling water pollution. When amended in 1977, this law became commonly known as the Clean Water Act. This act regulates the discharges of pollutants into U.S. waters. The act gives the EPA the authority to implement pollution control programs (e.g., setting wastewater standards for industry). The act also sets water quality standards for all contaminants in surface waters, making it illegal to discharge any pollutant from a source into navigable waters without a permit.

MINE SAFETY AND HEALTH ACT

The Federal Mine Safety and Health Act (MSHA) of 1977, also referred to as the "Mine Act," shifted the oversight of coal and non-coal mining safety and health from the Department of the Interior to the Department of Labor. This helped consolidate all safety and health regulations governing the mining industry under one single piece of legislation. This legislation strengthened and expanded the rights of miners, enhanced the protection of miners from retaliation on the part of mine owners, and established the Federal Mine Safety and Health Review Commission that provides independent review of enforcement actions.

United Nations Standards

The United Nations uses a globally recognized system of classifying and labeling dangerous goods, which the U.S. Department of Transportation (DOT) recognizes. This system sets the standard for the following procedures in relation to hazardous materials:

- Identification and classification
- Labeling, marking, and packing
- Documentation
- Emergency response

Consequences of Non-Compliance with Regulations

If a process technician fails to comply with regulations, this can cause many consequences: legal, moral, and ethical; or safety, health, and environment. These consequences can be imposed as a result of a minor accident, a major accident, or from an on-site inspection by a government agency representative.

LEGAL

Legal consequences fall into one of two major types:

- Fines and/or citations levied by federal, state, or local regulatory agencies (and possibly even criminal charges)
- Lawsuits filed by affected parties, such as injured workers or local citizens

MORAL AND ETHICAL

Moral and ethical consequences can manifest as:

- Burden of contributing to injuries or deaths
- Responsibility for causing damaged equipment, lost production, and associated costs
- Guilt for not complying with regulations, policies, and procedures

SAFETY, HEALTH, AND THE ENVIRONMENT

Numerous safety, health and environmental consequences can result from non-compliance. These include:

- Exposed or injured workers
- Exposed or injured citizens
- Air pollution
- Water pollution
- Soil pollution

Industry Groups and Voluntary Standards for the Process Industries

The following are some groups and voluntary standards that can affect SHE.

THE ISO 14000 STANDARD

The International Organization of Standardization (ISO), headquartered in Geneva, Switzerland, consists of a network of national standards institutes from more than 140 countries.

ISO has published more than 13,700 International Standards. ISO standards are voluntary, since the organization is non-governmental and has no legal authority to enforce the standards.

ISO 14000 addresses how organizations can voluntarily incorporate environmental aspects into operations and product standards. It requires a site to implement an Environmental Management System (EMS) using defined, internationally recognized standards as described in the ISO 14000 specification.

ISO 14000 specifies requirements for the following:

- Establishing an environmental policy
- Determining environmental aspects and impacts of products, activities, and services
- Planning environmental objectives and measurable targets
- Implementing and operating programs to meet objectives and targets
- Checking against the standard and making corrective actions
- Performing management review

ISO 14001, one of the sub-classifications of ISO 14000, addresses the following:

- Sites must document and make available to the public their environmental policy.
- Procedures must be established for ongoing review of environmental aspects and impacts of products, activities, and services.
- Environmental goals and objectives must be established that are consistent with the environmental policy, and programs must be set in place to implement goals and objectives.
- Internal audits of the EMS must be conducted routinely to ensure that non-conformances to the system are identified and addressed.
- Management review must ensure top management involvement in the assessment of the EMS and, as necessary, address the need for change.

The Environmental Management System (EMS) document is the central document that describes the interaction of the core elements of the system.

The Environmental Policy and Environmental Aspects and Impacts provide the following:

- Analysis, including legal and other requirements
- Direction for the environmental program by influencing the selection of specific, measurable environmental goals, objectives, and targets
- Recommendations for specific programs and/or projects that must be developed to achieve environmental goals, objectives, and targets
- Ongoing management review of the EMS and its elements to help ensure continuing suitability, adequacy, and effectiveness of the program

NFPA HAZARDOUS MATERIALS STORAGE STANDARDS

The National Fire Protection Association (NFPA) publishes voluntary standards about how hazardous materials can be stored, including the following:

- Ensuring the compatibility of materials stored in the same area
- Ensuring proper ventilation within the storage container as well as in the immediate area
- Allowing adequate traffic routes and escape routes
- Ensuring heat and ignition sources are not in the area. **Heat** is added energy that can cause an increase in the temperature of a material (sensible heat) or a phase change (latent heat - e.g., when fuel generates enough vapors to ignite)
- Ensuring proper labeling of hazardous materials
- Maintaining MSDS for all chemicals at the facility

Other NFPA voluntary standards are described in later chapters of this book.

RESPONSIBLE CARE® GUIDING PRINCIPLES

The American Chemical Council (ACC) provides Responsible Care guiding principles to respond to public concerns about the manufacture and use of chemicals. Through

Responsible Care, member chemical companies are committed to support a continuing effort to improve the industry's responsible management of chemicals.

Following are some guiding principles of Responsible Care:

- Recognize and respond to community concerns about chemicals and operations.
- Develop and produce chemicals that can be manufactured, transported, used, and disposed of safely.
- Make health, safety, and environmental considerations a priority in planning for all existing and new products and processes.
- Promptly report information on chemical-related health or environmental hazards and recommended protective measures to officials, employees, customers, and the public.
- Counsel customers on the safe use, transportation, and disposal of chemical products.
- Operate plants and facilities in a manner that protects the environment and the health and safety of employees and the public.

Other American Chemical Council voluntary standards are described in later chapters of this book.

ADDITIONAL GROUPS AND STANDARDS

- American Conference of Governmental Industrial Hygienists (ACGIH®)—an organization that addresses worker health and safety through education and the development and distribution of scientific and technical knowledge. Each year, the organization publishes guidelines which can be use to evaluate and control workplace exposure to chemical and physical substances. Threshold Limit Values (TLVs), exposure limits for employees, are provided for more than 700 chemical and physical substances. More details are provided in Chapter 3, *Recognizing Chemical Hazards.*
- American National Standards Institute (ANSI)—an organization that develops voluntary technical, industrial, and manufacturing standards in the United States, including Personal Protective Equipment. ANSI is the U.S. representative of the International Standards Organization.
- American Society of Mechanical Engineers (ASME)—an organization that focuses on technical, educational, and research issues relating to engineering and technology communities, setting voluntary standards for industrial and manufacturing codes.
- American Society for Testing & Materials (ASTM)—an organization that establishes and publishes voluntary test standards, methods, and practices for materials, products, systems, and services for various industries.
- National Safety Council (NSC)—an organization originally chartered by the U.S. government that is dedicated to helping protect the safety of Americans at home, on the roads, and at work.

Engineering Controls, Administrative Controls, and Personal Protective Equipment

The process industries use three methods to minimize or eliminate worker exposure to hazards. These methods are listed below, beginning with the highest priority:

- Engineering controls—controls that use technological and engineering improvements to isolate, diminish, or remove a hazard from the workplace. Following are examples of some engineering controls:
  - Using a non-hazardous material in a process that will work just as well as a hazardous material
  - Placing a sound reducing housing around a pump to muffle the noise it makes
  - Adding guards to rotating equipment

- Administrative controls—if an engineering control cannot be used to address a hazard, an administrative control is used. Administrative controls involve implementing programs and activities to address a hazard.

  Programs consist of written documentation such as policies and procedures. Activities involve putting a program into action.

  Administrative control is also called a work practice control or a managerial control. Following are examples of administrative controls:

  - Writing a procedure to describe the safe handling of a hazardous material
  - Limiting the amount of time a worker is exposed to loud noises
  - Training a worker on how to safely perform a potentially dangerous activity
  - Documenting how workers should select and properly wear Personal Protective Equipment suited to a specific task

  Administrative programs can use a variety of methods to promote safety and employee participation: training, teams and safety committees, incentives, suggestion and feedback, competition, and notices and reminders (signs, mugs, pens).

- **Personal Protective Equipment (PPE)**—when engineering and administrative controls are not adequate enough to protect workers, PPE is used. PPE is specialized gear that provides a barrier between hazards and the body and its extremities. Following are examples of PPE:

  - Hearing protection
  - Hard hats
  - Flame-retardant clothing (FRC)
  - Gloves and shoes

The Impact of the Process Technician on Safety, Health, and Environment

Along with regulations, engineering controls, administrative controls, and PPE, process technicians must share in the responsibility to avoid safety, health, and environmental hazards.

The duties of a process technician include maintaining a safe work environment. Process technicians must be trained and able to recognize hazards and security threats, and understand the impact they have on the plant or facility where they work. They must do the following:

- Keep safety, health, environmental, and security regulations in mind at all times.
- Follow all safe operating policies and procedures.
- Wear the appropriate personal protective equipment for the tasks being performed.
- Practice good safety habits.
- Look for unsafe acts.
- Watch for signs of potentially hazardous situations.

A major expectation of companies in the process industries is that employees have a proactive attitude regarding safety. Two crucial safety points that process technicians should remember are that accidents are preventable and that they must be proactive (thinking ahead) before a problem occurs. To achieve this, process technicians must first understand unsafe conditions and unsafe acts:

- An unsafe condition is a situation within a work environment that increases a worker's chances of having an accident or experiencing an injury.
- An unsafe act is any behavior that will increase the likelihood of a worker experiencing an accident or injury.

Unsafe conditions and unsafe acts are the root cause of accidents, injuries, and illnesses. Although many people think that these three terms are synonymous, each has its own distinct definition.

- **Accident**—an unplanned, uncontrolled situation that results in injury to a worker or damage to equipment or facilities.

- Injury—a wound or other condition of the body caused by external force, including stress or strain. The injury is identifiable as to time and place of occurrence and member or function of the body affected, and is caused by a specific event or incident or series of events or incidents within a single day or work shift. The back is the most frequently injured body part, followed by the legs and fingers. Injuries at work are commonly caused by the following:
  - Vehicle accidents
  - Falls
  - Electricity
  - Fire, smoke, and explosions
  - Exposure to chemical and biological hazards
  - Impacts with objects
  - Compression
  - Temperature extremes
  - Overexertion
- Illness—a physiological harm or loss of capacity produced by systematic infection; continued or repeated stress or strain; exposure to toxins, poisons, and fumes; or other continued and repeated exposures to conditions of the work environment over a period of time.

ATTITUDES AND BEHAVIORS THAT HELP PREVENT ACCIDENTS

Safety studies show that, historically, human errors play a significant role in almost every accident at a plant.

One study (Figure 1-2) examined almost 200 accidents in various chemical plants (Nimmo, 1995), reporting the most frequent causes as the following:

- Insufficient knowledge (34%)
- Procedural errors (24%)
- Operator errors (16%)

Another study in the petrochemical and refining industry (Figure 1-3) found the following accident causes:

- Equipment and design failures (41%)
- Operator and maintenance errors (41%)
- Inadequate or improper procedures (11%)
- Inadequate or improper inspections (5%)
- Miscellaneous causes (2%)

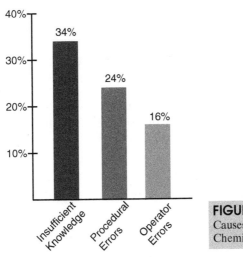

FIGURE 1-2 Most Frequent Causes of Accidents in Chemical Plants

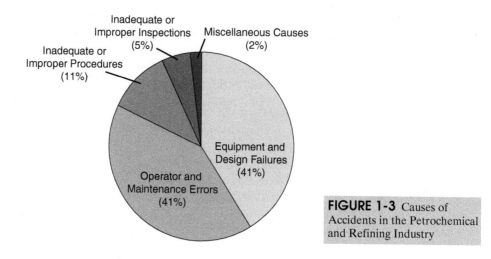

Inadequate or
Improper Inspections
(5%)

Miscellaneous Causes
(2%)

Inadequate or
Improper Procedures
(11%)

Equipment and
Design Failures
(41%)

Operator and
Maintenance Errors
(41%)

FIGURE 1-3 Causes of Accidents in the Petrochemical and Refining Industry

Although human error did not account for all of these accidents and incidents, it did account for many of them.

Personal attitudes and behaviors toward safety can play a significant part in preventing accidents or incidents.

An **attitude** is defined as a state of mind or feeling with regard to some issue or event. Process technicians who maintain a safety mind-set, always thinking about safety, experience fewer (if any) accidents than process technicians who are not safety-oriented.

A **behavior** can be defined as an action or reaction of a person under certain circumstances. Process technicians must respond immediately and appropriately to potential hazards, do a job right the first time, and perform housekeeping duties in a timely manner.

Process technicians must understand and follow not only governmental regulations on safety, health, and the environment, but also plant policies and procedures, general safety principles, and common sense. The technician must obey safety, health, and environment rules and report unsafe conditions or unsafe behaviors of co-workers.

Many employers try to determine during a job interview if a candidate will exhibit a safe attitude and behave safely on the job.

FACTORS THAT CAN AFFECT SAFETY

Process technicians might face the following factors, which can potentially affect safety:

- Encountering a variety of hazards, and combinations of hazards, including chemical, physical, biological, and/or ergonomic
- Working in all types of weather, including extreme conditions
- Participating in a team environment, where everyone is responsible for safety
- Using tools and lifting some heavy objects
- Working shift work in a facility that operates 24 hours a day, 7 days per week

IMPACT OF SHIFT WORK

Since most process facilities operate 24 hours a day, seven days a week, 365 days a year, process technicians are typically shift workers. Each plant or process facility is unique in the way shifts and workdays are arranged.

Most shift work involves 8 or 12-hour rotations, with numerous variations on the way non-working days are arranged. Following are some examples of work day arrangements:

August						
	1	2	3	4	5	6
7	8	9	10	11	12	13
14	15	16	17	18	19	20
21	22	23	24	25	26	27
28	29	30	31			

- Four days on, four days off - the process technician works four consecutive days and then is off four consecutive days.
- EOWO (every other weekend off) - the process technician works one weekend and is off the next weekend.

Process technicians must be able to adjust to the work schedules within a plant or process facility. Because of this, it is vital that process technicians understand the impact of shift work.

A person operates on a natural time clock that is different from work hours. Two mental low points can occur every 24 hours, typically between 2-6 a.m. and 2-6 p.m. The "sunup effect" may also occur, where a person wakes up when the sun rises, no matter how sleep deprived the person is.

Shift work has been compared to having permanent jetlag. People who work long or irregular hours tend to experience the following:

- Fatigue
- Reduced attention span
- Slowed reaction time
- Conflicting body clock and work schedule
- Less attentive
- Difficulty thinking and remembering clearly
- Accident prone

Shift work can affect several areas, including the following:

- Physical health - resulting in high rates of alcohol, drug, and tobacco use, overeating, lack of exercise, and long-term sleep disturbances
- Emotional health - resulting in increased irritability and a tendency toward depression and a lack of social life or healthy leisure activities
- Family life - resulting in higher divorce rates; little time with children and spouse, few shared family activities, and missing out on social outings

Process technicians can reduce the impact of shift work by taking care of themselves. To maintain physical and mental health, a process technician should take the following steps:

- Establish as regular a schedule as possible.
- Create a day-sleeping environment.
- Take naps when possible.
- Avoid stimulants, alcohol, and caffeine.
- Eat only light snacks in the 2-6 a.m./p.m. period.
- Compensate for lower awareness.

Workplace Stress

Stress is a reaction to various situations, perceived by a person as threatening, which can be caused by the following:

- Psychological factors
- Social interactions (e.g., friends, co-workers, family)
- Work (see the list below)
- Environment (e.g., temperature, pollution)

Stress can produce health problems (e.g., fatigue, increased blood pressure, and illness), absenteeism, tardiness, and emotional states such as anger, fear, aggression, guilt, and anxiety. It can lead to poor decision making, confusion, lack of concentration, and sluggishness. It can also make people prone to accidents and lead to unsafe attitudes and behaviors. On the job, a number of factors can result in work-related stress:

- Job security
- Repetitive tasks
- Boredom
- New responsibilities
- Technology
- Task complexity
- Deadlines
- Quality demands
- Workload
- Supervisor and/or co-worker negativity
- Perceived danger
- Workplace safety

Although this list is not definitive and not intended to be, there are various ways to handle stress:

- Meet with a professional counselor
- Recognize the source of the stress and its symptoms
- Analyze stressful situations clearly and decide what is worth worrying about
- Take stress-reduction training
- Speak with a supervisor
- Talk with an impartial third party (e.g., clergy, friend)
- Listen to music
- Exercise
- Spend time with family
- Enjoy a hobby
- Take a vacation
- Change job duties or the work situation

Companies can provide resources to help employees better handle stress. Check with your company for what resources are available.

PRACTICING GOOD SAFETY HABITS

The following are some good personal safety habits that process technicians can develop:

- Take care of yourself. Get plenty of sleep. Eat properly. Exercise regularly.
- Get regular medical, eye, hearing, and other health checkups. Stay current on immunizations and other shots.
- Avoid caffeine, tobacco products, alcohol, and drugs (process facilities are drug and alcohol-free work environments).
- Learn how to relax and handle stress. Participate in family time and activities and hobbies that you enjoy.
- Wear appropriate clothing to work. Some types of clothing and jewelry are prohibited in process facilities (e.g., loose clothing, open-toe footwear, rings).
- Practice good hygiene. Shower after your shift. Wash your hands thoroughly before eating or drinking.
- Do not eat in areas where hazardous substances might be present. Drink water only from potable sources (approved drinking water).
- Beards are not permitted in facilities where respirator use might be necessary, as facial hair interferes with the respirator's effectiveness.
- Properly secure long hair.
- Be on time and ready to work at the start of your shift.

FOLLOWING SAFE WORK PRACTICES

The following are some work-specific safety habits:

- Have a safe attitude and exhibit safe behavior.
- Stay focused and alert.
- Be prepared and keep a clear head in emergency situations.
- Familiarize yourself with applicable government regulations.
- Follow all plant policies and procedures, since many of these are written to ensure compliance with regulations.
- Attend training to stay current with applicable regulations, and pay attention.
- Read all relevant documentation, including policies, Standard Operating Procedures (SOPs), emergency plans, MSDSs, and other important information. Know where to find the most current information.
- Learn to recognize hazards and report and handle them appropriately.
- Know how to use safety equipment and protective gear.
- Perform all job tasks in a timely and accurate way while following safe work practices.
- Maintain a safe work environment by performing good housekeeping tasks and sanitation, as required by your job.
- Recognize all alarms and know the corresponding response procedures.
- Understand and properly use the equipment with which you work.
- Handle and store materials properly.
- Watch for suspended loads (e.g., forklifts, cranes).
- Stay clear of accident scenes, unless you are part of an emergency response team.
- Report injuries and incidents immediately to appropriate personnel.
- Stay in your assigned area. If you must go to another area, make sure to tell appropriate personnel.
- Obey traffic regulations in the plant and never park in fire lanes.
- Use the proper tool for the job.
- Follow your facility's polices on the use of cameras, cell phones, and electronic devices. Also, facilities do not permit workers to bring firearms on the premises.
- Smoke only in designated areas.

Summary

The process industries involve processes that take quantities of raw materials and transform them into other products. Process industries include these segments: oil and gas, chemical, mining, power generation, water and waste water treatment, food and beverage, pharmaceutical, and pulp and paper. The process industries pose various safety, health, and environmental hazards. These hazards can be chemical, physical, ergonomic, biological, security, and environmental in nature.

Previous accidents in the workplace, such as those in Texas City, Bhopal, on the *Exxon Valdez,* and at Chernobyl resulted in sweeping safety, health, and environmental changes. These changes included government regulations, voluntary industry standards, and company safety changes.

To minimize these hazards, U.S. government agencies enforce regulations that protect workers, the public, and the environment from safety, health, and environmental hazards.

The Occupational Safety and Health Administration (OSHA) establishes workplace safety and health standards, conducts workplace inspections, and proposes penalties for noncompliance. The Environmental Protection Agency (EPA) protects human health and the environment. The Department of Transportation (DOT) develops and coordinates policies for an efficient and economical national transportation system. The Nuclear Regulatory Commission (NRC) protects public health and safety through regulation of nuclear power and nuclear materials.

In an attempt to protect workers and the environment, OSHA, DOT, and the EPA have created many regulations. These regulations help minimize the consequences of catastrophic releases of toxic, reactive, flammable, or explosive chemicals; prevent worker exposure to potentially hazardous substances; ensure that the hazards of all produced or imported chemicals are evaluated, and that information relating to the hazards is provided to employers and employees; establish emergency response operations for the releases of hazardous substances; and set requirements for handling hazardous materials. Failure to comply with regulations can have legal, moral, ethical, safety, health, and environmental consequences.

Various industry groups have developed voluntary standards, including ISO 14000, the NFPA Hazardous Materials Storage standards, Responsible Care guiding principles, and other SHE-related guidelines.

To minimize or control hazards, process industries often employ engineering controls (technology and engineering improvements), administrative controls (programs, procedures, and activities) and Personal Protective Equipment, or PPE (specialized gear used to protect the body), to make a safer workplace.

Process technicians play a vital role when it comes to safety and health. They must always maintain a safety-conscious attitude and behave in a safe, responsible, and appropriate manner. They must be familiar with government regulations, follow plant policies and procedures, attend training, practice good safety habits, and follow safe work practices.

Checking Your Knowledge

1. Define the following terms:
   a. Administrative controls
   b. Air pollution
   c. Attitude
   d. Engineering controls
   e. Ergonomic hazard
   f. Facility
   g. Hazardous agent
   h. Process industries
   i. Process technician
   j. Process technology
   k. Unit

2. A(n) _____ hazard comes from environmental factors such as excessive levels of noise, temperature, or pressure.

3. Which type of hazard comes from living, or once-living, organisms such as viruses, mosquitoes, or snakes?
   a. Ergonomic
   b. Physical
   c. Environmental
   d. Biological

4. Which type of pollution occurs when there is an emission of a potentially harmful substance or pollutant into the atmosphere?

5. Name the four elements of an accident.

6. Which government agency was created to "establish and enforce workplace safety and health standards, conduct workplace inspections, and propose penalties for noncompliance, and investigate serious workplace incidents"?
   a. OSHA
   b. NIOSH
   c. EPA
   d. DOT

7. Which OSHA standard seeks to prevent or minimize the consequences of catastrophic releases of toxic, reactive, flammable, or explosive chemicals?
   a. Process Safety Management
   b. HAZCOM
   c. HAZWOPER
   d. HAZOP

8. Name at least four types of information included in a PHA.

9. Describe OSHA's PSM Management of Change process.

10. *(True or False)* An MSDS provides key safety, health and environmental information about a material.

11. The intent of OSHA's Voluntary Protection Program is to:
    a. Regulate all environmental activities of the process industries.
    b. Recognize and promote effective safety and health management.

    c. Gauge the impact that the process industries have on the safety and health of process technicians.

    d. Help companies implement operation programs that are within EPA guidelines.

12. The _____ _____ Act "sets limits on how much of a pollutant can be in the air anywhere in the United States."

13. Which ISO standard addresses environmental management?

    a. 9000

    b. 9001

    c. 1910.119

    d. 14000

14. _____controls involve implementing programs and activities to address a hazard.

15. *(True or False)* An illness is an unplanned, uncontrolled situation that results in injury to a worker or damage to equipment or facilities.

16. According to the Nimmo (1995) study, what is the most frequent cause of accidents in chemical plants?

    a. Operator errors

    b. Procedural errors

    c. Insufficient knowledge

    d. Improper inspects

17. A(n) _____can be defined as an action or reaction of a person under certain circumstances.

18. Name at least four ways to reduce the effects of shift work.

19. Performing good _____ tasks and sanitation is one way to maintain a safe work environment.

Student Activities

1. Select a process industry (e.g., oil and gas, food and beverage, etc.) and research potential hazards to which workers could be exposed. Categorize the hazard as either chemical, physical, ergonomic, biological, physical security, cybersecurity, or environmental. Then, write a three-page paper relating your findings.

2. Walk around your home and make a list of potential hazards that you encounter. Describe the hazard, then label the type of hazard it is. Did you find any situations where multiple hazards occur? If so, describe them. Then, take seps to correct the hazards.

3. Keep a journal for a week, describing how you use a safe attitude and behaviors on a daily basis (e.g., checking the blind spot before changing lanes, or being alert to your surroundings).

CHAPTER

2

Types of Hazards and Their Effects

Objectives

This chapter provides you with an overview of hazards in the process industries, which can affect people and the environment.

After completing this chapter, you will be able to do the following:

■ Identify four main types of hazards to health and safety.

■ Explain the various routes of entry that chemical and biological hazards use to enter the human body.

■ Describe the short-term and long-term effects that hazards can have on an individual's health and safety.

■ Identify types of environmental hazards.

■ Describe the short-term and long-term effects of hazards on the environment.

Key Terms

Acute—short-term health effects.

Ceiling—a concentration that should not be exceeded by workers during an exposure period. Also called the Maximum Acceptable Ceiling.

Chronic—long-term health effects.

Dose-response relationship—the connection between the amount (dose) and the effect (response) that a substance can have on the body.

Permissible Exposure Limit (PEL)—an OSHA limit representing the maximum acceptable exposure of workers to a hazard over a specific period of time.

Routes of entry—the ways in which a hazardous substance can enter the body, such as inhalation through the nose, absorption through the skin or eyes, accidental ingestion, or through injection.

Short-Term Exposure Limit (STEL)—a concentration to which workers can be exposed for a short term (e.g., 15 minutes) before suffering any harm.

Threshold Limit Value (TLV)—a limit set by the American Conference of Government Industrial Hygienists, representing the maximum acceptable exposure of workers to a hazard over a specific period of time.

Time Weighted Average (TWA)—an average concentration of a chemical or a noise level to which an employee can be exposed over an eight-hour period, or 40 hours a week.

Introduction

In the process industries, there are hazardous substances and environmental factors that can cause injury, illness, or death if these substances are not handled properly. In addition, some of these hazardous substances can have short or long-term effects on the environment. Because of this, process technicians must be able to recognize and understand the hazardous substances associated with their operating facilities and how these substances can impact local communities. Technicians also share the responsibility to protect themselves, their co-workers, and the community from such hazards.

In order to reduce or prevent exposure to hazardous substances, the federal government has created a wide range of rules and regulations that the process industries must follow. In addition to these rules and regulations, the process industries also employ other preventive measures such as engineering and administrative controls, and the use of personal protective equipment (PPE) to protect or limit personnel exposure to hazardous substances.

Hazards Overview

This chapter provides an overview of various workplace hazards that process technicians might encounter. Each hazardous substance is briefly described, along with potential short-term and long-term effects. Other chapters in this text discuss these hazards in more detail.

Hazards Defined

A hazard is a substance (e.g., a chemical or a disease) or action (e.g., lifting a heavy object or hearing a loud noise) that can cause a harmful effect. There are many ways to classify hazards. However, following are the four most common classifications:

- Chemical hazards—any hazard that comes from a solid, liquid or gas element, compound, or mixture that could cause health problems or pollution.
- Physical hazards—any hazard that comes from environmental factors such as excessive levels of noise, temperature, pressure, vibration, radiation, electricity, or mechanical hazards (note: this is not the OSHA definition of a physical hazard).
- Ergonomic hazards—hazards that can create physical and psychological stresses because of forceful or repetitive work, improper work techniques, or poorly designed tools and workspaces.
- Biological hazards—a living or once-living organism, such as a virus, a mosquito, or a snake, that poses a threat to human health.

In addition to the four classifications mentioned above, physical and cybersecurity threats have emerged in recent times. Physical security includes security measures intended to protect specific physical assets such as pipelines, control centers, tank farms, and other vital areas. Cybersecurity involves security measures intended to protect information and information technology (e.g., computers) from unauthorized access or use. Both physical and cybersecurity are discussed in more detail later in this textbook.

CHEMICAL HAZARDS

A chemical is a substance with a distinct composition that is used in, or produced by, a chemical process. Some chemicals can be helpful (e.g., pharmaceuticals can help a sick person get well), while others can be harmful (e.g., benzene may cause cancer or other health effects). Chemicals take three basic forms: solid, liquid, or gas.

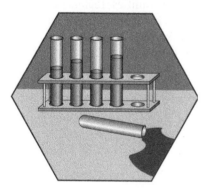

FIGURE 2-1 Chemicals can be Both Helpful and Hazardous

The physical form of a chemical (solid, liquid, or gas), in addition to its chemical composition, can determine how hazardous or non-hazardous a substance is. For example, a small quantity of water (a liquid) at room temperature would most likely have no impact on a process technician's clean, dry skin. However, if that water were heated into steam (a gas), serious burns or other injuries could occur.

Solids

Solids are substances with a definite volume and a fixed shape that are neither liquid nor gas, and that maintain their shape independent of the shape of the container. Solids can be broken down into smaller particles that can enter the body. We call this small, particulate matter dust, fibers, and fumes.

- Particulates—fine solid particles (e.g., dust, smoke, mist, fumes, or smog) that are suspended in air or liquid. Particulates can vary in size, shape, density, and electrical charge.

- Dust—small, airborne particles that are typically created when a material is reduced by mechanical methods such as grinding or pulverizing (e.g., grain dust).
- Fibers—strandlike particles with a length at least three times their width (e.g., asbestos fibers).
- Fumes—very fine airborne particles that are created when solid chemicals are heated, changed to vapor, and then condensed to solid form again (e.g., welding fumes).

Liquids

Liquids are substances that have a definite volume but no fixed shape, demonstrate a readiness to flow with little or no tendency to disperse, and are limited in the amount in which they can be compressed. Liquids can be broken down (atomized) into a mist.

- Mist—liquid droplets of various sizes that are created when liquids are agitated or sprayed (e.g., mist or spray produced as water falls through a cooling tower).

Gases

Gases are substances with a definite mass but no definite shape, whose molecules move freely in any direction and completely fill any container they occupy, and which can be compressed to fit into a smaller container. Gases can take the form of a vapor.

- Vapor—the gaseous form of a material that is normally solid or liquid at room temperature and pressure (e.g., water vapor or dry ice "fog").

Evaporation is the process by which a *liquid* (e.g., water) is changed into a vapor, and sublimation is the process by which a *solid* (e.g., dry ice) is changed directly into the vapor state.

CHEMICAL EXPOSURE

Process technicians can be exposed to chemical hazards in a variety of ways.

- Using chemicals to degrease equipment
- Working with a leaking pump, valve, or other piece of equipment
- Wearing clothing or personal protective equipment (PPE) that is inappropriate for the hazard (e.g., wearing cloth gloves instead of chemical resistant gloves to clean up a toxic liquid spill)
- Taking samples
- Breaking piping or hose connections

Exposure to chemicals can have harmful effects if appropriate personal protective equipment (PPE) is not worn and if proper procedures are not followed.

PHYSICAL HAZARDS

A physical hazard is any hazard that comes from environmental factors such as excessive levels of noise, temperature, pressure, vibration, radiation, electricity, or mechanical hazards. (Figure 2-2) (Note: This is not the OSHA definition of physical hazard.)

FIGURE 2-2 Physical Hazards Include Excessive Levels of Noise, Temperature, Pressure, Vibration, Radiation, Electricity, and Mechanical Hazards

In the process industries, workers face numerous physical hazards. For example, process technicians often work around equipment that is operating at extreme temperatures or pressures. This equipment often contains swiftly moving parts (e.g., gears or shafts), and is powered by electricity, high-pressure air (pneumatics), or high-pressure fluid (hydraulics). Additionally, some types of equipment can also produce varying levels of radiation.

Following are the most common physical hazards a process technician might encounter:

- Noise
- Temperature
- Radiation
- Pressure
- Energized equipment
- Hazardous atmospheres
- Heights

Physical hazards can occur when process technicians work near or around the following:

- Equipment that produces high levels of noise
- Exposed steam pipes with high temperatures, or cryogenic equipment with extremely low temperatures (both can cause burns)
- Pressurized tanks, vessels, or equipment
- Energized equipment (electric, pneumatic, hydraulic)
- Unshielded radiation or radioactive sources from processes, instrumentation, and testing

ERGONOMIC HAZARDS

Ergonomic hazards are hazards that can create physical and psychological stresses because of forceful or repetitive work, improper work techniques, or poorly designed tools and workspaces. Ergonomic hazards occur when equipment, tools, or workstations are not designed to match the needs of the workers or are used improperly. Ergonomic hazards can produce or aggravate musculoskeletal (muscle, joint, and bone) injuries or conditions.

FIGURE 2-3 Ergonomic Hazards can Create Physical and Psychological Stresses Due to Forceful or Repetitive Work, Improper Work Techniques, or Poorly Designed Tools and Workspaces

Most ergonomic hazards involve physical stresses placed on the body, either through a one-time task (e.g., lifting a heavy box), repetition of the same task (e.g., typing), or a combination of tasks.

Ergonomic hazards in the process industries fall into one of four main types:

- Manual material handling—moving heavy or bulky objects without using appropriate equipment or tools
- Poor posture—placing the body in awkward positions or moving incorrectly
- Improper use of tools—using the wrong tool for a particular task, or incorrectly using the right tool for a task

- Repetitive motions—using tools, adding manual quantities of chemical to a vessel, or frequently climbing stairs or ladders

Process technicians regularly face situations that can involve ergonomic hazards, such as the following:

- Working in tight spaces or in awkward positions
- Moving bulky or heavy equipment
- Working for prolonged periods on a computer
- Using tools repeatedly or incorrectly

BIOLOGICAL HAZARDS

Biological hazard is a living or once-living organism, such as a virus, a mosquito, or a snake, that poses a threat to human health. Workers in most of the process industries face less potential exposure to biological hazards than chemical, physical, or ergonomic hazards. However, the effects of biological hazards can be just as harmful and life threatening as those posed by other types of hazards.

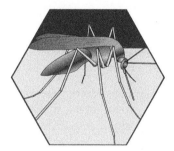

FIGURE 2-4 Biological Hazards are Hazards that Come from Living or Once Living Organisms

Following are some of the most common biological hazards that a process technician can be exposed to:

- Microorganisms—such as viruses, bacteria, and fungi
- Arthropods—such as crustaceans (e.g., crabs), arachnids (e.g., spiders) and insects (e.g., mosquitoes)
- Animals—such as alligators, bears, and poisonous snakes
- Plant allergens and toxins—such as pollen and poison ivy
- Protein allergens—such as food, urine, feces, blood, hair, or dander (dandruff or tiny particles of skin) from vertebrate animals

Unless working in a process industry that handles animals or plants, the most likely places a process technician would encounter biological hazards would be situations like the following examples:

- Coming into contact with bacteria (e.g., Legionella, which causes Legionnaires' disease) through improperly treated cooling towers
- Encountering a brown recluse or black widow spider in a warehouse
- Handling bodily fluids and used first-aid materials that could contain a blood-borne pathogen such as hepatitis or human immunodeficiency virus (HIV)
- Being bitten by a poisonous snake
- Being bitten by an infected mosquito (e.g., mosquitoes can transmit St. Louis encephalitis and West Nile virus to people)
- Encountering infected rodent droppings (e.g., Hanta virus) in a warehouse

Factors for Determining Chemical and Biological Hazards

Chemical and biological substances are considered hazardous based on the likelihood that the substance will cause harm. The following factors determine whether a substance is hazardous:

- Toxicity—how much of the substance is required to cause harm
- Route of entry—how the substance enters the body

- Dose—how much enters the body
- Duration—the length of time the body is exposed
- Interaction—other substances to which the body is exposed
- Response—how the body reacts to the substance
- Sensitivity—how one person reacts to the substance compared to the reactions of other people
- Frequency—how often a person is exposed to the substance

Toxicology is the study of substances and the harm they cause to living things. A substance's toxicity is the potential of that substance to cause harmful effects. Toxicity is only one factor in determining whether a hazard exists (e.g., some chemicals are designated as hazardous because of a risk of fire and/or explosion).

The effects of toxicity can strike a single cell in the body, or the entire body. Some toxic effects can cause visible damage or symptoms, while others strike silently, causing longer-term effects in the body's performance or function (e.g., cancer or cirrhosis of the liver).

Toxicity depends on three factors:

1. A substance's chemical structure
2. The extent of the body's absorption of the substance
3. The body's ability to change the substance into less-toxic substances (detoxification) so it can be eliminated from the body

The amount of the chemical required to cause harm determines the toxicity. When large amounts of a chemical are necessary to inflict harm, the chemical is considered to be relatively non-toxic. However, if only a small amount can be harmful, the chemical is considered toxic.

All substances, whether classified as hazardous or not, can harm a person when they enter the body in toxic amounts. For example, many substances that are beneficial to the body in small quantities (e.g., water) are toxic in larger quantities.

Substances become toxic to humans when an undesirable physiological condition develops in the body as a result of overexposure to the chemical or biological agent.

The following variables are imperative for determining toxicity:

1. What route of entry to the body did the substance take?
2. How large of a dose (concentration) of a substance entered the body?
3. What was the duration (how long) for which the body was exposed to this substance?
4. What is the chemical composition of the substance?
5. What is the general health of the exposed person?

Routes of Entry

Hazardous substances do not present a threat to a person until they come in contact with the body. There are four ways through which hazardous substances can enter the human body:

- Inhalation
- Absorption
- Injection
- Ingestion

Knowing what these routes are can help minimize the chances of exposure to hazardous substances.

INHALATION

Inhalation occurs when substances enter the body through the mouth or nose and are then transported into the lungs. Inhalation is the most common route of entry for hazardous substances. Hazardous chemical or biological agents, in forms such as dust, mist, or vapors, can be inhaled and cause problems in the nose, throat, or lungs (or be absorbed into the body through the lungs).

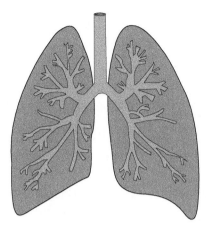

FIGURE 2-5 Inhalation is the Most Common Route of Entry for Hazardous Substances

Some hazardous substances, when inhaled, cause symptoms such as nose or throat irritation, coughing, chest pains, or general discomfort. Other hazards (e.g., asbestos and coal dust) can be inhaled without any immediate symptoms, but are still dangerous. In some cases, the presence of hazardous substances can be determined only by a physician.

The inhalation of hazardous substances can result in damage to the respiratory tract from direct exposure. Particles in the lungs might be coughed out, but enough could remain in the lu ngs to cause damage. Some hazardous substances are small enough to be absorbed into the bloodstream through the alveoli, tiny air sacs in the lungs. These substances could cause harmful effects in other parts of the body.

The lungs also have natural defense mechanisms such as nose hairs, mucous and cilia. The nose hairs serve as filters. Mucous is a sticky substance that can trap some particulates and vapors and then be expectorated (coughed up). Cilia sweep trapped particles from the windpipe upwards toward the throat so it can be cleared.

Process technicians must be familiar with the chemicals and hazardous materials they are working with and always use proper protection when handling these chemicals.

The proper use of respiratory protection, such as cartridge respirators or supplied air, will reduce the risk of exposure through inhalation, while engineering controls, such as ventilation, can also be used to minimize inhalation hazards.

ABSORPTION

Absorption is the process by which substances or particles are drawn into another structure. Absorption occurs when substances are brought into the body (absorbed), typically through the skin, although hazardous substances can also be absorbed through the eyes (Figure 2-6). When hazardous substances enter the body they can potentially get into the bloodstream.

The skin forms a protective barrier around the body, generally keeping hazardous substances out. However, some chemical and biological hazards can easily pass through the skin. Other hazards require a break in the skin to pass through (e.g., scratches, cracks, or cuts).

Temperature can also affect skin absorption. For example, higher environmental temperatures can cause blood vessels to widen (called vasodilation). This increases the surface area for absorption and allows more of the substance to enter the bloodstream.

Some hazardous substances will only cause redness or irritation of the skin and eyes, while caustic substances (e.g., strong acids and alkalis) can burn skin. Other substances (e.g., organic solvents) dissolve the skin's natural oils, making it dry, sensitive, or cracked.

Process technicians use many types of personal protective equipment (e.g., chemical protective gloves, protective clothing, and chemical goggles) to limit the risk of

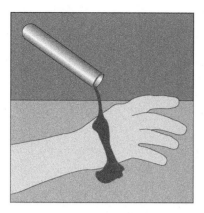

FIGURE 2-6 Absorption Occurs When Substances are Brought into the Body (absorbed), Typically Through the Skin

absorption exposure. Engineering controls such as enclosed sampling systems or automated chemical handling systems are also used to minimize exposure.

INJECTION

Injection occurs when substances are brought into the body through a puncture in the skin or an open wound. When the skin is punctured hazardous substances can enter the skin or the blood-stream.

Injection is fairly uncommon in the process industries. However, injection can occur if the skin is punctured by a syringe, a piece of broken glass, or some other sharp object in the production area.

Hands and feet are the most common body parts to sustain a puncture wound, so process technicians should always wear proper protective equipment (e.g., gloves and footwear).

INGESTION

Ingestion occurs when substances enter the body through the mouth and are transported to the stomach and gastric system. (Figure 2-7) Swallowing a hazardous substance can result in damage or irritation of the digestive system, or the absorption of the substance into the gastrointestinal tract.

While this is the least common route of entry for hazardous substances in the process industries, process technicians could be exposed if they eat, drink, or smoke in a contaminated area (e.g., an area filled with hazardous dust particles). They can also be exposed by a chemical splash to the face that can enter the mouth, or through the eyes if they handle their contact lenses with contaminated hands or in a contaminated environment. Clothing can also become contaminated with hazardous materials, which can then easily be transferred to the hands and mouth.

Process technicians can prevent accidental ingestion of hazardous substances by eating or smoking only in designated areas, by wearing gloves, and by washing their hands thoroughly before eating, drinking, chewing gum, smoking, or handling contact lenses.

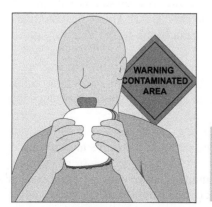

FIGURE 2-7 Ingestion Occurs When Substances Enter the Body Through the Mouth and are Transported to the Stomach and Gastric System

Doses and Routes of Entry

A dose refers to the amount of substance taken into or absorbed by the body. Generally speaking, the greater the dose, the greater a substance's effect on the body. The connection between amount and effect is called the **dose-response relationship**. The dose-response relationship is illustrated in Figure 2-8.

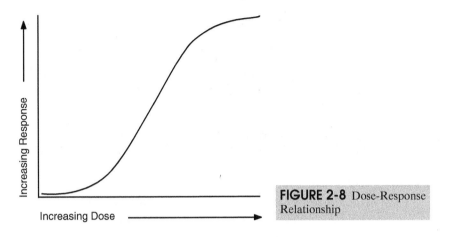

FIGURE 2-8 Dose-Response Relationship

Some chemicals are more toxic than others, making it necessary to limit the exposure. The **Threshold Limit Value (TLV)** of a substance is a limit set by the American Conference of Government Industrial Hygienists representing the maximum acceptable exposure of workers to a hazard over a specific period of time.

The threshold limit varies by hazardous substance, and even from person to person (e.g., some people might have a higher sensitivity to inhaling a substance such as the solvent toluene and may pass out quickly upon exposure).

A dose of a hazardous substance is expressed in terms of the route of entry it may take into the body:

- Inhalation—the dose is expressed in terms of the amount of the material inhaled per unit of volume breathed, such as parts per million (ppm)
- Absorption or injection—the dose is expressed as the amount of material absorbed or injected per unit of body surface area, such as milligrams per square centimeter of skin or tissue (mg/cm^2)
- Ingestion—the dose is expressed as the amount of material ingested per unit of body weight, such as milligrams per kilogram of body weight (mg/kg).

Dose-Response Relationship

Recall that all toxicological considerations of a hazardous substance are based on a predetermined dose-response relationship. Response refers to the physiological effect experienced by a person as a result of exposure to the given amount of substance.

The relationship between dose and response is expressed mathematically with the following formula:

$$(C) \times (T) = K$$

In this equation,
C represents the concentration (or dose)
T represents the duration of exposure
K represents the constant

In this equation, concentration (C) and duration (T) determine the severity of exposure. Calculating these variables using the dose-response equation $[(C) \times (T) = K]$ can help industrial hygiene and safety professionals predict safe limits for hazardous substances in the workplace (Figure 2-9).

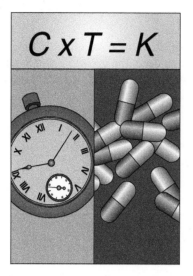

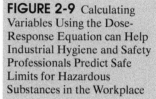

FIGURE 2-9 Calculating Variables Using the Dose-Response Equation can Help Industrial Hygiene and Safety Professionals Predict Safe Limits for Hazardous Substances in the Workplace

Although certain doses of hazardous substances can produce no harmful effects, at some point all substances can produce a lethal effect.

LETHAL DOSE AND LETHAL CONCENTRATION

A lethal dose (LD) is the amount of a hazardous substance that will cause death when ingested or absorbed. Lethal concentration (LC) is the amount of a hazardous substance that will cause death when inhaled. Both lethal dose and lethal concentration are expressed in terms of the percentage of organisms in an experimental population that die as a result of exposure. Typically, this percentage is expressed as either LD50 or LC50, depending on the route of entry in the test population. LD50 means that exposure to a particular amount of a hazardous substance through ingestion or absorption resulted in the death of 50% of the test population. LC50 means that exposure to a particular amount of a hazardous substance through inhalation results in the death of 50% of the test population.

Some substances require greater concentrations to cause a lethal dose, while others require far lower concentration (e.g., a substance with an LD50 of 500 grams is less hazardous than an LD50 of .02 grams). The dose threshold is the minimum dose required to produce a measurable response in the body.

Table 2-1 compares LD50 values for several substances process technicians might encounter.

TABLE 2-1   Comparison of LD50 Values in Order from Least Lethal to Most Lethal

Chemical	*LD50 (mg/kg)*
Ethyl alcohol (grain alcohol)	10,000
Sodium chloride (table salt)	4,000
Ferrous sulfate (iron used in vitamins, paints, and other products)	1,500
Morphine sulfate (medication used for pain)	900
Strychnine sulfate (used in rodent poisons and some medications)	150
Nicotine (alkaloid poison found in tobacco, medicines, and insecticides)	1
Black widow venom (the most venomous spider in North America)	0.55
Curare (alkaloid-containing plant product used as skeletal-muscle relaxant)	0.50
Rattlesnake venom (venomous snake found in North America)	0.24
Dioxin (extremely toxic chemical formed during the manufacturing of other chemicals and during incineration)	0.001
Botulinum toxin (bacterial neurotoxin which causes botulism)	0.0001

TABLE 2-2 Exposure Duration

Exposure Type	Exposure Duration	Exposure Amount
Acute	Less than 24 hours	Usually one exposure
Subacute	1 month	Repeated doses
Subchronic	1–3 months	Repeated doses
Chronic	More than 3 months	Repeated doses

DURATION OF EXPOSURE

The length of time that the body is exposed to a hazardous substance can determine the likelihood the body will be affected. Lower doses may not affect the body in the short term, but long-term exposure might cause harm, since some hazardous substances accumulate in the body, or the body does not have a chance to repair itself before the next exposure. The dose and duration factors combined are called the rate of exposure.

Some systems in the body, such as the lungs, liver, and kidneys, can detoxify chemicals and eliminate them from the body. However, if the rate of exposure exceeds the rate at which these systems can eliminate the hazardous substances, the concentration of that substance will begin to be stored in the body. In these cases hazardous substances become toxic chemicals (Table 2-2).

Chemical Composition of Hazardous Substances

The primary factor in the toxicity of a hazardous substance is its chemical structure. Substances that have similar structures often cause similar health effects. However, one slight change in a chemical structure can create striking differences in terms of health effects.

For example, amorphous silica can be present in the workplace at high levels because it has little effect on human health. However, once amorphous silica is heated, it changes into crystalline silica that can cause lung damage. Thus the permissible level of crystalline silica in the workplace is 200 times less than that for amorphous silica.

Workers in the process industries are routinely exposed to more than one type of chemical. Thus, process technicians must be aware of possible reactions and interactions between various chemicals.

A reaction occurs when chemicals combine to create a new substance or substances that can have properties very different from the original. For example, when household bleach and drain cleaner (lye) are mixed together, they create dangerous chlorine gas and hydrochloric acid.

The Material Safety Data Sheet (MSDS) for a chemical will often list its potential hazardous reactions and the incompatible substances that should not be mixed or stored with it. Companies are required by law to make an MSDS available for each potentially hazardous substance in the workplace.

Health and Sensitivity to Hazards

The impact a hazardous substance can have on an individual varies from person to person. The following are some of the main factors that affect sensitivity:

- Age
- Gender
- Genes (inherited traits)
- Diet
- Fitness
- Allergies
- Use of drugs, alcohol, or tobacco
- Pregnancy

Based on these factors and others, some people will experience the toxic effects of a hazardous substance at higher doses than others, while other individuals will have an allergic reaction to a certain hazardous substance at a lower-than-normal dose.

If a person has an illness (e.g., hepatitis) that affects the body's ability to detoxify substances, the effect of a hazardous substance may be amplified because of the body's inability to eliminate the substance.

EXPOSURE LIMITS

An exposure limit is a specific threshold that determines the concentration of a chemical to which a process technician can be safely exposed. Exposures exceeding the set threshold can cause significant harm to the worker.

Exposure limits vary, based on the substance and the potential concentration levels required for harm to occur. Some chemicals require extensive exposure, while others require little exposure. Exposure limits are set to protect the average worker. Some will exhibit signs of sensitivity to a hazardous substance at lower levels than others. Also, workers performing strenuous tasks or experiencing stress may breathe in more air and potentially inhale more of the hazardous substance.

Exposure limits are based on the best available data (such as laboratory testing and workplace exposure histories). This information may be incomplete for some substances, especially regarding long-term health effects. Long-term effects are often discovered only after workers have been exposed to a substance for many years.

The exposure limit usually represents the maximum concentration of a substance in the air before it is considered to be a health hazard. Exposure limits usually address inhalation of hazards. This may not limit the exposure through skin contact or ingestion (of dusts, mists, vapors, etc.). Workers exposed to such substances must be provided with, and must use, appropriate Personal Protective Equipment.

Process technicians must understand these exposure limits and how they impact their work when dealing with chemicals. The two main sources of exposure limit values are the American Conference of Governmental Industrial Hygienists (ACGIH) and the Occupational Safety and Health Administration (OSHA).

The ACGIH publishes a book listing Threshold Limit Values (TLVs) for a wide range of chemicals. TLVs are revised as new information is discovered about chemicals, and the TLV book is updated annually.

OSHA has also established its own set of exposure limits, called **Permissible Exposure Limits (PELs)**. The difference between a PEL and TLV is that the ACGIH can change TLVs every year, while PELs have remained fairly static. Being a government agency, OSHA does not have the same ability as the ACGIH to dynamically update the exposure limit values. Also, TLVs were the original basis for the OSHA PELs. Not every chemical will have a PEL or TLV, but there are more chemicals listed with TLVs than PELs.

Occupational Exposure Limit (OEL) is another term that process technicians might hear. This is a more generic term, used to refer to exposure limits set by other organizations or government agencies around the world. The term Occupational Exposure Standards is also used.

To further complicate the issue, individual states can either adopt the OSHA PELs or use lower published values (such as the TLVs). Manufacturers can also provide their own exposure limits for their products.

Efforts to develop one all-encompassing set of exposure limits have been attempted, but no single solution has been created. So, for exposure limits, there are three major concepts to understand:

- Time-Weighted Average (TWA)
- Short-Term Exposure Limit (STEL)
- Exposure Ceiling (C)

Exposure is most often measured using a **Time-Weighted Average (TWA)**. The TWA is the average airborne concentration of a chemical to which an employee can be exposed over an eight-hour period or a 40-hour work week. Exposure is determined by air monitoring, often conducted by an industrial hygienist technician using special monitoring devices. The measured level can exceed the TWA value at times, as long as the eight-hour average stays below the exposure limit value.

FIGURE 2-10 A Short-Term Exposure Limit (STEL) is the Concentration a Worker can be Exposed to Before Suffering Any Harm

For some substances, a **Short-Term Exposure Limit (STEL)** is set. The STEL is a concentration to which workers can be exposed for a short term (e.g., 15 minutes) before suffering any harm. STELs are typically set for chemicals that produce acute exposures. STELs can be difficult to measure, even with current technology.

Other substances require a maximum exposure limit that should never be exceeded. The **ceiling** (C) is a concentration that should never be exceeded by workers during an exposure period. This is sometimes also called the Maximum Acceptable Ceiling.

Process technicians must pay attention to the type of unit used with the exposure limit. Often values are listed in pairs, with different units of measure being used to specify the same exposure limit. For example, a chemical's exposure limit can list one unit that specifies the amount of the substance in gas form (e.g., ppm, or parts per million), while a second unit specifies the amount in solid or liquid form (e.g., mg/m^3, or milligrams per cubic meter).

One last important term to understand is Action Level (AL). Generally, an Action Level is a hazard threshold that, if exceeded, puts additional control measures into action. Action Levels can be set by a variety of government agencies, such as OSHA or the Environmental Protection Agency (EPA).

Action Levels often are set to prevent problems before an exposure limit level (e.g., PEL) is reached. In OSHA's case, action levels may be set at half as much as a PEL. The difference between an action level and a PEL is a margin of error that takes into account statistical variations in measurements. This margin ensures that a worker's exposure does not exceed the PEL by putting controls in place before the hazard level reaches the PEL.

CHEMICAL INTERACTIONS

In certain situations, workers are exposed to two or more chemicals at the same time. Sometimes, the effect of the exposure is as simple as taking into account the combined sum of the effects of the chemicals. Most often, however, the effects of two or more chemicals are greater than the sum of each individual chemical's effects. The chemicals act together to increase the risk.

There are even some cases where two substances can counteract each other's effects. This is how antidotes for poisons work. The antidote negates the poison's effects, reducing or eliminating the hazard entirely. In another example, some harmful caustics can combine with certain harmful acids to create salts, which are not harmful.

Another scenario is when a chemical does not cause harm on its own, but can make the effect of another chemical worse. For example, the solvent isopropanol poses no hazard to the liver by itself, but it can increase the potential for carbon tetrachloride to cause liver damage if a worker is exposed to a combination of the two chemicals.

Effects of Hazards on People

A response is the physiological effect experienced by a person after exposure to a hazardous substance. Physiological responses are determined based on when the response is experienced and the effect the substance has on the body.

WHEN SYMPTOMS APPEAR

The effects of exposure to hazardous substances develop either immediately (acute) or slowly (chronic) and may appear at the point of contact (local) or affect another area of the body (remote). The delay between the start of exposure and the appearance of a harmful effect from the exposure is called the latency period. The effects of hazardous substances may appear immediately or soon after exposure, or could take months or years to appear. Sometimes after repeated exposures to a substance, the body develops increased tolerance.

Acute effects are short-term health effects characterized by exposure to high concentrations of a hazardous substance over a short period of time, perhaps even a single exposure. For example, a person whose hand is exposed to hydrochloric acid will immediately experience a burn to the skin.

Chronic effects are long-term effects characterized by continued exposure to a hazardous substance over an extended period of time, anywhere from months to years. For example, a person exposed to low levels of carbon tetrachloride over a period of time without wearing proper respiratory protection might eventually develop liver cancer.

Some hazardous substances can cause both acute and chronic effects. For example, exposure to solvents on the job could cause acute effects such as headaches or dizziness, while long-term exposure could cause chronic effects such as liver or kidney damage.

Table 2-3 shows the differences between acute and chronic effects.

TABLE 2-3 Difference Between Acute and Chronic Effects

Acute Effects	*Chronic Effects*
Occur immediately or shortly after exposure	Occurs over a length of time
High dose exposure over a short period	Low dose exposure over a long period
Effects may be reversible	Difficult to reverse effects
Cause-and-effect relationship relatively easy to establish	Cause-and-effect relationship difficult to establish
Knowledge based on effect on humans	Knowledge based on animal testing

Some chronic effects such as cancer can have extremely long latency periods, possibly as long as 40 years. This can make the relationship between exposure and harmful effects difficult to determine. Thus, it is vital that process technicians understand what chronic effects could be caused by the substances they encounter on the job.

ACUTE (SHORT-TERM) EFFECTS

Acute or chronic refers to the time it takes an effect to appear and local or remote refers to the location where the effect occurs. When toxic substances cause damage to the body immediately or shortly after exposure, that damage is called an acute effect. The skin, eyes, nose, throat, and lungs are the most common areas for local effects to occur.

Short-term effects can include the following:

- Irritation—inflammation of body tissue
- Asphyxiation—interference in breathing or oxygenating the body
- Central nervous system depression—interference in the functioning of the central nervous system, such as dizziness or lack of coordination

CHRONIC (LONG-TERM) EFFECTS

Once in the body, toxic substances can travel through the bloodstream to internal organs and cause harmful effects (called systemic effects) to those organs.

Following are the organs most commonly impacted by systemic effects:

- Liver
- Kidneys
- Heart
- Nervous system/brain
- Reproductive system
- Blood

Cancer (carcinogenesis) results from the growth and spread of abnormal tissue in the body. Substances that cause cancer are called carcinogens. Cancer can have a long latency period (from 10 to 40 years). Determining if a substance is a carcinogen is difficult because of the latency period and the variety of substances that people are exposed to during that time.

Some chemicals can alter or damage the genes or chromosomes in the body. This alteration of genetic materials is called a mutation, and can lead to cancer or birth defects in future generations.

A mutagen is a substance that can cause a mutation. Some mutations can result in cancer. Most chemicals that are carcinogens are also mutagens. However, not all chemicals that are mutagens are carcinogenic.

Every cell in a body contains genetic material that instructs the cell how to function and reproduce, or make new cells. Hazardous substances can damage or mutate cells by altering the genetic material. If this occurs in the eggs or sperm these mutations can be passed on to the child as a birth defect.

Exposure to hazardous substances can have toxic reproductive effects, including the inability to have children (infertility or sterility), lowered sex drive, menstrual imbalances, spontaneous abortions (miscarriages), stillbirths, and birth defects in children. Substances that can affect the fetus in utero (within the womb) are called teratogens. Because of the risks teratogens pose, pregnant women need to be cautious, and often may be reassigned to a non-hazardous environment during their pregnancy.

Most chemicals have not been tested for reproductive effects in animals, so there is little information on the reproductive toxicity of chemicals. Currently, only a select few chemicals are known to cause birth defects or other reproductive effects in humans.

Local and Systemic Effects

Hazardous substances exert their influence on a body in one of two ways: as either a local effect or a systemic effect. Local effects occur on the area of the body exposed to the hazard. For example, acid can burn the skin it touches, while fumes can scar the lungs. Systemic effects occur when the hazardous substance enters the body through one of the routes of entry (e.g., ingestion or absorption) and is distributed to other parts of the body, typically organs. Some substances can produce both local and systemic effects (e.g., lead, mercury, manganese, and cadmium). Chlorinated hydrocarbons and carbon disulfide are also systemic poisons.

Toxic effects differ by organ, and some organs are more susceptible to toxic effects than others. The organ that is affected most by a hazardous substance is called the target organ. Target organs may include any of the following:

- Central nervous system—the most frequent target for hazardous substances; sensitive to organic solvents and metals such as lead, mercury, and manganese
- Lungs—the major route for hazardous substances to enter the body; affected by fumes, dust, and other airborne contaminants
- Skin—the body's largest organ; provides a protective layer to the body, but can absorb hazardous substances; can also be affected by irritants and corrosive substances

- Circulatory system (blood and heart)—affected by substances such as benzene, carbon monoxide, arsenic, and toluene
- Liver—the largest internal organ; responsible for purifying unwanted substances from blood; can be affected by solvents (e.g., carbon tetrachloride, toluene, chloroform, vinyl chloride)
- Kidneys—excrete waste; are susceptible to substances such as carbon tetrachloride, turpentine, lead, mercury, and cadmium
- Reproductive organs—vulnerable to many hazardous substances
- Muscles and bones—rarely the target of hazardous substances

Hazards to the Environment

Hazardous substances also impact the environment when they are released. Once these substances are released, they can negatively impact the environment by polluting the air, soil, water, or any combination of these.

AIR POLLUTION HAZARDS

Air pollution occurs when the concentration of natural and/or man-made substances in the atmosphere becomes excessive and the air turns toxic. Nature can produce pollutants in the form of forest fires, pollen, and volcanic eruptions. Man-made sources of air pollution include emissions from transportation, industry, and agriculture.

- Primary pollutants—gases, liquids, and particulates dispersed into the atmosphere through either man-made or natural processes. In the United States, the primary pollutants are the following:
  - Carbon monoxide (CO)
  - Sulfur dioxide (SO_2)
  - Nitrogen oxide (NO_X)
  - Volatile organic compounds (VOCs)
  - Particulate matter (PM, or miscellaneous matter)
- Secondary pollutants—primary pollutants that undergo a chemical reaction that transforms the pollutants into a different type of toxic material. In the United States, secondary pollutants are the following:
  - Ozone
  - Photochemical smog
  - Acid deposition (or acid rain)

Primary Air Pollutants in the United States

Carbon monoxide (CO), a colorless, odorless, and poisonous gas, accounts for most of the primary air pollutants in the United States (almost 40%). Carbon monoxide is produced by the incomplete combustion of carbon fuels. Because most modes of transportation

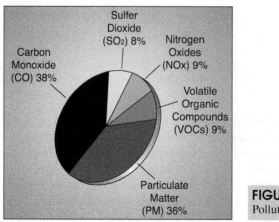

FIGURE 2-11 Primary Air Pollutants in the United States

(e.g., cars, buses, and boats) require the combustion of fuels, the transportation industry is the largest contributor to carbon monoxide in the atmosphere. (See 2-11.)

Particulate Matter (PM) consists of solid particles and liquid droplets emitted into the atmosphere. Particulate matter makes up about 35% of the primary air pollutants. Fine particulate matter (less than one micron in size) is created by condensation processes following combustion events (e.g., an internal combustion engine, a coal-fired power plant, or a forest fire). Mechanical processes (e.g., machining), create larger particulate matter that is greater than one micron in size. Construction, agriculture, and unpaved roads also produce much of the particulate matter in the environment.

Nitrogen Oxide (NO_X) gases form during the high-temperature combustion of fuel when nitrogen in the fuel or the air reacts with oxygen. Motor vehicles and power plants produce most of the NO_X that pollutes the air. NO_X contributes approximately 10% of the total primary air pollutants.

FIGURE 2-12 Motor Vehicles and Power Plants Produce Most of the Nitrogen Oxides that Pollute the Air

Volatile Organic Compounds (VOCs), also referred to as hydrocarbons, include a variety of solid, liquid, and gaseous compounds that are composed exclusively of hydrogen and carbon. The incomplete combustion of fossil fuels contributes to VOCs in the atmosphere (approximately 10% of the total primary air pollutants).

Sulfur Dioxide (SO_2) gases, which are colorless and corrosive, account for less than 10% of the primary air pollutants in the U.S. The combustion of fuels containing sulfur (e.g., coal and oil) creates SO_2. Power plants, pulp and paper mills, smelters and petroleum refineries generate most of the SO_2 emissions. Volcanic eruptions can also add a significant amount of sulfur to the atmosphere.

Secondary Air Pollutants in the U.S.

Ozone, an atmospheric layer at the very top of the stratosphere, shields Earth's inhabitants from lethal (harmful) levels of ultraviolet radiation. The ozone high up in the atmosphere can be destroyed by chlorofluorocarbons (CFCs), harmful substances which were banned in the United States and most other industrialized countries but are still used in many other countries, by creating the much-reported hole in this good ozone layer.

Ozone in high concentrations closer to the Earth can cause health effects (e.g., respiratory and eye irritation) in people and can harm plants, thereby reducing crop yields. This type of ozone occurs when sunlight breaks down nitrogen dioxide in the presence of VOCs and creates free atomic oxygen (O) that then combines with the abundant oxygen molecules (O_2) in the atmosphere to create ozone (O_3). Ozone can also speed up the degradation of synthetic materials such as paints and plastics.

NO_X, VOCs, O_3, and peroxyacetyl nitrates (PANs) mix to create photochemical smog. Peroxyacetyl nitrates are produced when NO_X and VOCs interact while influenced by solar radiation. PANs can cause eye and lung irritation in humans and can harm plants.

Acid deposition can result from several processes. Nitrogen oxides and sulfur dioxides can either fall to the ground through dry deposition, or these oxides can be converted to nitric acid (HNO_3) or sulfuric acid (H_2SO_4) through oxidation, changing to acid when it dissolves in water vapor—clouds—in the atmosphere. Nitric acid and sulfuric acid then combine with normal precipitation to create acid rain. Acid deposition, whether dry or wet, changes the pH balance in bodies of water and soil, which is harmful to plants and animals that are very sensitive to pH change. It also destroys buildings, monuments, and other structures through corrosion.

Before   After

FIGURE 2-13 A Statue Damaged by the Effects of Acid Rain

SOIL POLLUTION

Soil is the layer of earth that can support plant growth. Soil is composed of decaying organic materials and minerals (including microbes that help the organic materials decay). Soil pollution results when the fertile layer of soil used to produce the world's food is destroyed. For example, destruction can occur when the beneficial microbes in soil are damaged or destroyed by certain chemicals. Processes such as chemical contamination and salinization can also chance the composition of the soil or add unwanted and usually harmful materials to the soil.

Also, extensive irrigation can result in salinization, while overuse of pesticides and fertilizers creates chemical contamination. (Figure 2-14) These can all be toxic to plants or damage or inhibit plant growth.

Spills can also cause contamination. Improper handling and illegal disposal of industrial and household chemicals or waste can introduce a diverse range of hazardous materials into soil.

FIGURE 2-14 Extensive Irrigation and Overuse of Pesticides and Fertilizers can Damage or Inhibit Plant Growth

Acid rain, which begins as air pollution, can fall to the earth and deposit acid compounds in the soil. Leaking gas tanks and waste dumps can also leach hydrocarbons and other chemical pollutants into the soil. These contaminants can then infiltrate the groundwater and cause water pollution.

Other processes, such as erosion and exhaustion, "take away" from soil, removing or altering the fertile layer that is so crucial to life. Erosion occurs when large quantities of soil are moved or removed and the land is laid bare. Erosion happens naturally, but deforestation, strip-mining, and some farming methods accelerate to the process.

Mining can also bring extremely toxic heavy metals (e.g., arsenic) to the surface. Once on the surface, these metals can enter the food or water supply.

Exhaustion of the soil refers to the process of removing vital nutrients from the land, rendering it unfit for agricultural purposes. Agricultural activities, performed without proper planning, can degrade the soil (e.g., failure to rotate crops can deplete the soil of its natural nutrients).

WATER POLLUTION

Water pollution occurs when a body of water such as surface water, groundwater, or an aquifer is adversely impacted by the addition of large amounts of suspended or dissolved materials.

Sources of water pollution vary. Water quality can be degraded biologically, chemically, or physically.

Biological

The presence of pathogens in water sources can be attributed to sewage entering the water supply from city waste and septic systems, animal waste from farms and ranches, and storm-water runoff. These pathogens can cause disease if they enter drinking water or if people swim in contaminated water.

Animal and plant material left to decay in water create oxygen-demanding wastes that deplete the water of valuable oxygen. Fish and plants are then affected by this removal of oxygen from the water.

Chemical

Water-soluble inorganic and organic chemicals enter the water supply from landfills, underground storage tanks, acid rain, agricultural activities, illegal dumping, and accidental spills. People, animals, or plants that come in contact with these chemicals can suffer acute and/or toxic effects depending on the toxicity of the chemical.

Did You Know?

Arsenic is an element that is naturally found in geological formations, especially granites. As long as the arsenic remains bound to the granite, it causes no harm.

However, when the granite is disturbed, either through mining or natural erosion, the arsenic may be released into the soil, thereby contaminating groundwater. Thus, mining companies must monitor water quality during and after mining operations.

Physical

Suspended solids are usually generated through erosion and runoff; they can also result from large amounts of organic product decay. When solid particles are present in the water source, the penetration of light in the water is affected and photosynthesis is reduced. If these particles are organic based, these can also put a demand on the oxygen supply as microbes break them down.

Industrial facilities and power plants generate hot-water emissions, leading to thermal water pollution. Heat changes the nature of the aquatic system by altering its ability to retain dissolved oxygen and by adversely affecting both animal and plant life. Certain plants and animals are sensitive to temperature changes.

Effects of Hazards to the Environment

Air, soil, and water pollution can have significant short and long-term effects on the delicate balance of the environment. The interdependent nature of natural resources causes the entire ecosystem to become imbalanced when one variable, such as air quality, is negatively impacted by pollution.

SHORT-TERM EFFECTS

The following are just some of the short-term effects that pollution can have on the environment:

- Poor air quality (affecting respiration of living organisms)
- Contaminated water supply
- Sudden death of animal populations
- Affected plant growth and reduced crop yields
- Unusable, infertile soil
- Erosion
- Habitat destruction

LONG-TERM EFFECTS

The following are longer-term effects that pollution can have on the environment:

- Imbalanced ecosystems that cannot replenish resources
- Cancers in animal populations
- Mutations of living organisms
- Inability of living organisms to reproduce
- Large-scale death of animal populations
- Loss of habitat for living organisms
- Widespread destruction of Earth's ecosystems (e.g., rain forests, forests, wetlands, oceans, rivers and lakes, and coral reefs)

Summary

In the process industries, hazardous substances and environmental factors can cause injury, illness, or death if not handled properly. Some of these hazardous substances can have short or long-term effects on public health and the environment. In order to prevent injury or death, process technicians must be able to recognize and understand the hazardous substances and activities associated with their operating facilities and how these can affect individuals, the environment, and local communities.

In order to reduce or prevent exposure to hazardous substances, the federal government has created a wide range of rules and regulations that the process industries must follow. In addition to these rules and regulations, the process industries also employ other preventive measures such as engineering and administrative controls and the use of personal protective equipment (PPE) to protect or limit personnel exposure to hazardous substances. Technicians must use proper protective equipment and follow proper procedures in order to protect themselves, their co-workers, and the community.

Checking Your Knowledge

1. Define the following terms:
   a. Acute
   b. Ceiling
   c. Chronic
   d. Dose-response relationship
   e. Permissible Exposure Limit
   f. Short-Term Exposure Limit
   g. Threshold Limit Value
   h. Time Weighted Average
2. Explain the four hazard types: chemical, physical, ergonomic, biological.
3. *(True or False)* Chemical hazards can be in solid, liquid, or gas form.
4. List the five types of physical hazards discussed in the chapter.
5. Radio waves are an example of:
   a. Non-ionizing radiation
   b. Vibration
   c. Ionizing radiation
   d. Temperature
6. *(True or False)* Moving bulky or heavy objects without proper equipment or tools is an example of a physical hazard.
7. List five examples of biological hazards.
8. What is the most common route of entry by which a hazardous substance can get into a body?
   a. Absorption
   b. Inhalation
   c. Ingestion
   d. Injection
9. Which is more lethal, a substance with an LD50 of 100 grams, or a substance with an LD50 of 12 grams?
10. Compare and contrast acute effects and chronic effects.
11. A chemical burn on the skin is an example of a:
    a. Local effect
    b. Systemic effect
12. What is the primary air pollutant in the United States?
    a. Nitrogen oxide
    b. Carbon monoxide
    c. Sulfur dioxide
    d. Particulate matter

Student Activities

1. Write a one-page paper describing the differences between the terms "toxic" and "hazardous."
2. Choose one of the following instances in which hazardous substances were released from a controlled situation. Research the case, then write a two-page paper on the cause of the problem and the short and long-term effects on humans and the environment.
   - Love Canal
   - Chernobyl
   - Minamata, Japan
3. For each of the hazards listed below, fill in the type of hazard (chemical, physical, ergonomic, or biological) that the process technician faces (if more than one hazard type is present, list all relevant answers and explain why).

Hazard Description	Hazard Type(s)
a. Heat stress	
b. Prolonged work at a computer	
c. Soiled bandages left over from a first-aid incident	
d. Cleaning tools with a chemical degreaser	
e. Climbing to the top of a storage tank to check a relief valve	
f. Using X-rays for non-destructive equipment testing	
g. Vapor release	
h. Resin dust	
i. Lack of potable water at a work site	

Hazard Description	*Hazard Type(s)*
j. Adding bucket quantities of materials to a reactor	
k. Operating a grinder without goggles or a face shield	
l. Moving a 55-gallon drum without proper equipment	
m. A small acetone spill in the laboratory	
n. Working in an area where large mosquito populations are present	
o. Entering and remaining in a rail car for cleaning over a long period of time	
p. Welding arc	
q. Cavitating pump	
r. Leaking flange on a methanol pipeline	
s. Poor ventilation in a raw-material storage shed	
t. Hauling long lengths of hose	
u. Isolating a valve prior to maintenance	
v. Pouring hazardous materials without chemical splash goggles, a face shield, or respiratory protection	

Recognizing Chemical Hazards

Objectives

This chapter provides an overview of chemical hazards in the process industries, which can affect people and the environment.

After completing this chapter, you will be able to do the following:

- Identify the various chemical hazards (gases, liquids, and particulates) found in the process industries and discuss the potential effects such chemicals have on safety, health, and the environment.
- Identify specific categories of hazardous chemicals used in the process industries and describe the potential health and environmental hazards posed by each.
- Explain the purpose and function of labeling systems found in local process industries.
- Understand the purpose and components of an MSDS.
- Discuss the primary government regulations relating to chemical hazards.

Key Terms

Acid—a substance with a pH less than 7.0.

Base—a substance with a pH greater than 7.0; also referred to as alkaline.

Carcinogen—a cancer-causing substance.

Combustible liquid—any liquid that has a flashpoint at or above 100 degrees F, but below 200 degrees F.

Explosive—a substance that causes a sudden, almost instantaneous release of pressure, gas, and heat when subjected to sudden shock, pressure, or high temperature.

Flammable gas—any gas that, at ambient temperature and atmospheric pressure, forms a flammable mixture with air at a concentration of 10% by volume or less.

Flammable liquid—any liquid that has a flashpoint below 100 degrees F.

Flammable solid—any solid, other than a blasting agent or explosive, that is liable to cause fire through friction, absorption of moisture, spontaneous chemical change, or retained heat from manufacturing or processing, or which can be ignited readily and when ignited burns so vigorously and persistently that it creates a serious hazard.

Flashpoint—the minimum temperature at which a liquid gives off a vapor in sufficient concentration to ignite.

HAZCOM—OSHA 29 C.F.R. 1910.1200 Hazard Communication/Employee Right-to-Know, a standard to ensure that employees are aware of the chemicals they are exposed to in the workplace and the measures to take to protect themselves from such hazards.

Highly toxic—a substance that requires only a small amount of exposure to be lethal.

Inorganic compound—a chemical compound that does not contain carbon chains.

Mutagen—a chemical suspected to have properties that change or alter a living cell's genetic structure; mutagens can lead to cancer or birth defects if the egg or sperm is affected.

Neurotoxin—a poison that affects the nervous system.

Organic compound—a chemical compound that contains carbon chains.

Organic peroxide—an organic compound that contains the bivalent -O-O- structure and which can be considered to be a structural derivative of hydrogen peroxide, where one or both of the hydrogen atoms has been replaced by an organic radical.

Oxidizer—a chemical that can initiate or promote combustion in other materials.

pH—a measure of the amount of hydrogen ions in a solution that can react and indicate if the substance is an acid or a base.

Pyrophoric—a chemical that will ignite spontaneously in air at a temperature of 130 degrees F or below.

Sensitizer—an agent that can cause an allergic reaction.

Teratogen—a substance believed to have an adverse effect on human fetus development.

Toxic material—a substance determined to have an adverse health impact.

Water-reactive—any chemical that reacts with water to release a gas that is either flammable or that presents a health hazard.

Introduction

In the process industries, workers are routinely exposed to chemical hazards that can cause injury, illness, or death. Some hazards also pose a danger to the environment, causing potential short and long-term impacts. Process technicians must understand these hazards, including those posed by chemicals.

This chapter covers chemicals and the hazards they present, along with labeling/ identification, documentation (Material Safety Data Sheets), and government regulations.

Chemical Hazards Overview

Process technicians come into contact with chemicals on a regular basis, even if they do not directly work for a chemical manufacturer. OSHA estimates that at least 25% of all workers, regardless of the industry, come into contact with chemicals.

Chemical hazards may have the potential to produce adverse health effects on people, including illness, injury, and even death. Chemical hazards can also impact the environment. The effects of chemicals vary as widely as the chemicals themselves. The goal of this chapter is not to explain all potential hazards associated with every chemical. Rather, it is to provide enough information so process technicians can better understand the specific chemical hazard information they will be provided with on the job.

To understand chemical hazards, process technicians should understand chemical properties. (Figure 3-1) (Note: The fundamentals of chemistry are covered in other textbooks, including the Center for the Advancement of Process Technology's *Introduction to Process Technology*. This textbook presumes that the reader has some basic understanding of chemistry concepts.)

Chemicals can exist in different physical states: solid, liquid, or gas. Any chemical is most active in its gaseous state, and molecules readily move about in all directions. In the liquid state, the chemical's molecules still move around but not as actively as in the gaseous state. In a solid state, the chemical is in its most stable form, with its molecules barely moving.

Chemicals can enter the body in different ways depending on their physical state. For example, if a chemical is in a gaseous state, it is more likely to be inhaled into the respiratory system. If a chemical is in a liquid state, the likelihood is greater that it will be absorbed through the skin. Ingestion through the mouth can occur in either solid or liquid forms.

Chemical composition (the chemical structure of the substance) is one of the factors that determine the toxicity of a substance. In order to prevent injury, technicians must fully comprehend the concepts of chemical structure and toxicity.

Chemical compounds can be categorized as either organic or inorganic. **Organic compounds** are chemical compounds that contain carbon chains. Following are some examples of organic compounds:

- Aliphatics (e.g., methane)
- Aromatics (e.g., benzene)
- Halogenated hydrocarbons (e.g., methyl chloride)
- Ketones (e.g., methyl ethyl ketone—MEK)
- Esters (e.g., butyl acetate)
- Alcohols (e.g., methyl alcohol)
- Ethers (e.g., di-ethyl ether)

Hydrocarbons are one of the most common types of organic compounds. Following are the major classes of hydrocarbons:

- Aliphatic hydrocarbons
- Aromatic hydrocarbons
- Cyclic aliphatic hydrocarbons

Generally, organic compounds are flammable. Many are also toxic.

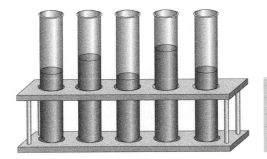

FIGURE 3-1 To Understand Chemical Hazards, Process Technicians Should Understand Chemical Properties

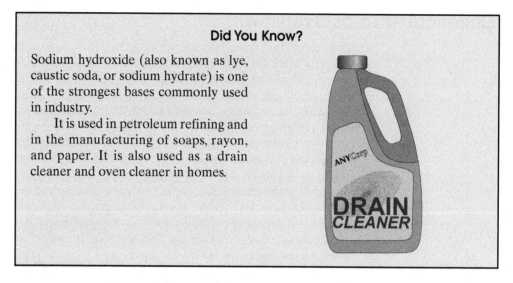

Did You Know?

Sodium hydroxide (also known as lye, caustic soda, or sodium hydrate) is one of the strongest bases commonly used in industry.

It is used in petroleum refining and in the manufacturing of soaps, rayon, and paper. It is also used as a drain cleaner and oven cleaner in homes.

Inorganic compounds are chemical compounds that do not contain carbon chains. Examples of inorganic compounds include the following:

- Acids (e.g., acetic acid)
- Bases (e.g., sodium hydroxide)
- Salts (e.g., sodium chloride)
- Metals (e.g., iron)

Acids and **bases** are ionic compounds that exhibit the toxic characteristic of corrosiveness. The strength of the compound's corrosiveness is based on its pH level. The **pH** level is a reference to the amount of hydrogen ions in a solution. It is measured with a scale ranging from 0 to 14. Acidic substances have a pH less than 7 on the pH scale, while basic substances have a pH greater than 7 on the scale. A neutral substance (e.g., pure water), which is neither acid nor base, has a pH of 7.

The closer to zero on the pH scale that a substance reaches, the more acidic it is (e.g., battery acid). The closer to 14 on the scale that a substance reaches, the more basic it is (e.g., drain cleaner).

Figure 3-2 lists some common items and their pH levels:

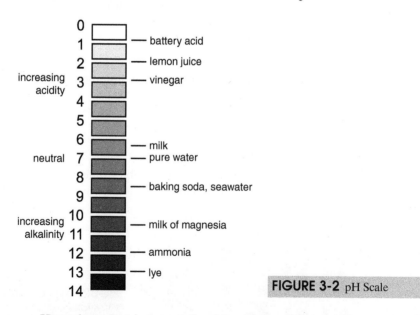

FIGURE 3-2 pH Scale

Hazardous chemicals are often identified and grouped into categories on the basis of their hazardous characteristics (e.g., flammable, corrosive, toxic, or unstable when exposed to outside forces such as pressure or temperature changes).

Some chemicals present physical hazards (e.g., flammability or combustibility) while others present health hazards (e.g., cancer). It is important for process technicians to understand this concept of categorization, since it provides a foundation of how materials should be labeled, what materials are incompatible with others, what materials can be safely stored alongside others, and how workers should protect themselves from the chemical hazards.

The three main government agencies that address chemical hazards are the Occupational Safety and Health Administration (OSHA), the Department of Transportation (DOT), and the Environmental Protection Agency (EPA). These three agencies use similar terminology when discussing chemical hazards, but the terms are not always interchangeable. For example, OSHA's terminology relates to employee health and safety, DOT's terminology relates to the safe transportation of materials or wastes, and the EPA's terminology relates to the protection of the environment. Thus, this chapter will first cover general chemical hazard categories, with later sections exploring OSHA, DOT, and EPA categories.

Flammable and Combustible Materials

Flammable and combustible materials (gases, liquids, and solids) pose physical hazards in the form of fire or explosion. These materials are classified as follows:

- **Flammable liquid**—any liquid that has a flashpoint below 100 degrees F. (When heated, flammable liquids can give off a vapor in sufficient concentrations to ignite. The minimum temperature at which this happens is called the **flashpoint.**)
- **Flammable gas**—any gas that, at ambient temperature and pressure, forms a flammable mixture with air at a concentration of 10% by volume or less
- **Flammable solid**—any solid, other than a blasting agent or explosive, that is liable to cause fire through friction, absorption of moisture, spontaneous chemical change, or retained heat from manufacturing or processing, or which can be ignited readily and when ignited burns so vigorously and persistently as to create a serious hazard
- **Combustible liquid**—any liquid that has a flashpoint at or above 100 degrees F, but below 200 degrees F

(Note: Chapter 6, *Fire and Explosion Hazards,* covers flammable and combustible hazards in more detail.)

Water-Reactive, Pyrophoric, and Explosive Materials

Water-reactive materials are chemicals that react with water to release a gas that is either flammable or presents a health hazard. (Figure 3-3) A **pyrophoric** material is a chemical that will ignite spontaneously in air at a temperature of 130 degrees F or below. An **explosive** is defined as a chemical that causes a sudden, almost instantaneous release of pressure, gas, and heat when subjected to sudden shock, pressure, or high temperature. These materials primarily pose physical hazards, in the form of fire or explosion.

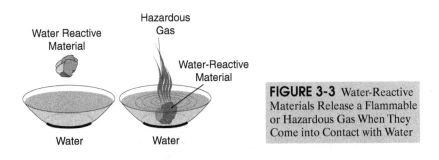

Water Reactive Material

Hazardous Gas

Water-Reactive Material

Water

Water

FIGURE 3-3 Water-Reactive Materials Release a Flammable or Hazardous Gas When They Come into Contact with Water

Organic Peroxides and Oxidizers

An **organic peroxide** is an organic compound that contains the bivalent -O-O- structure and which can be considered to be a structural derivative of hydrogen peroxide, where one or both of the hydrogen atoms has been replaced by an organic radical. Peroxides can explode if the temperature reaches a certain level.

An **oxidizer** is a chemical that can rapidly release oxygen, thereby initiating or promoting combustion in other materials. This can result in a fire or explosion. These materials primarily pose physical hazards in the form of fire or explosion.

Carcinogens, Mutagens, Teratogens, and Reproductive Toxins

Carcinogens are cancer-causing substances. **Mutagens** are chemicals suspected to have properties that change or alter a living cell's genetic structure. **Teratogens** are substances believed to have an adverse effect on the human fetus development. Reproductive toxins are chemicals that affect a person's ability to have children. These materials primarily pose health hazards.

Toxic, Highly Toxic, and Target Organ Effects

A **toxic material** is a substance that has been determined to have an adverse health impact. A **highly toxic** material requires only a small amount of exposure to be lethal.

Many toxic and highly toxic materials target specific organs in the body. If a toxic or highly toxic material contacts a remote location on the human body, is transferred through absorption and is then carried through the bloodstream to another area of the body where it adversely impacts an organ, this material is said to cause target organ effects. For example, **neurotoxins** are poisons that affect the nervous system. These materials primarily pose health hazards.

Other Chemical Hazards

The following are other types of chemical hazards:

- Asphyxiants—substances that disrupt breathing and can result in suffocation
- Irritants—substances that cause irritations to skin, eyes, upper respiratory tract, nose, mouth, and throat
- Depressants, anesthetics, or narcotics—substances that can affect the body's nervous system
- Allergens—substances that cause an allergic reaction or unhealthy response by the immune system
- **Sensitizers**—agents that can cause an allergic reaction

(Note: These hazards are discussed in more detail in other chapters of this textbook.)

Detection

Process technicians must be constantly aware of their environment, using their senses to detect potential chemical hazards. For example, you can see some vapors as they are being released. You can smell strange odors. You can hear out-of-the-ordinary sounds. You can feel vibrations. You can sense odd tastes with your tongue and sensitive body parts (e.g., nose, mouth, eyes, and skin) can feel irritants.

However, process technicians should not rely on their senses alone. Sometimes, by the time process technicians can sense a chemical hazard (e.g., a nitrogen-deficient atmosphere), it is too late. This is where monitoring equipment proves valuable. Portable or permanent, continuously running monitoring devices can sample environments and detect specific chemical hazards. These monitors can then alert process technicians to potential chemical hazards in time to take action.

Hazardous Chemical Labeling and Documentation

Hazardous chemicals must be properly labeled and documented so that workers can identify chemicals they encounter, understand the hazards they present, and know how to protect themselves.

OSHA has set forth some very specific guidelines that apply to most chemical products. However, some substances are governed by other regulatory agency rules, so they are excluded from the labeling requirements (e.g., pesticides, foods, alcohol, drugs, cosmetics, tobacco products, and wood products). Appropriate documentation, called a Material Safety Data Sheet (MSDS) must be provided for hazardous chemicals as a ready reference for individuals working with a particular chemical, or who might be exposed to a particular chemical during an emergency situation.

OSHA HAZARD COMMUNICATION STANDARD (HAZCOM)

OSHA established the Hazard Communication standard (**HAZCOM** – also known as the "Worker Right to Know" law) to make sure that employees are aware of the chemicals they are exposed to in the workplace. OSHA states that this standard "covers both physical hazards (such as flammability) and health hazards (such as irritation, lung damage, and cancer)." HAZCOM can help process technicians understand the chemicals they encounter, their hazards, and why certain precautions must be taken on the job. The intent is that informed process technicians are more likely to remain safe than those who are not informed about hazards.

All employers are responsible for notifying their employees about chemicals they could be exposed to during normal work situations or emergency situations. Employers are classified as chemical users, chemical manufacturers, distributors, or importers. This classification determines the specific requirements the employer must follow when communicating chemical information to employees.

OSHA defines chemical users as "employers that 'use' hazardous chemicals." "Use" in this case means to package, handle, react, or transfer the chemicals. The employer must identify "any situation where a chemical is present in such a way that employees may be exposed under normal conditions or used in a foreseeable emergency."

Work operations such as laboratories and warehouses in which chemicals are handled only in sealed containers, are required to keep labels on containers as they are received and maintain MSDSs. In addition, employees must be given access to the MSDSs and be provided with information and training.

Chemical manufacturers and importers are required to notify employees of all chemicals to which they might be exposed and their subsequent hazards. Also, these employers must evaluate the hazards associated with their chemicals, and then notify customers of these hazards using labels and MSDSs.

There are four requirements identified by the HAZCOM standard regarding the communication of chemical hazards:

1. Written hazard communication programs
2. Labels and other warning forms
3. Material Safety Data Sheets (MSDSs)
4. Employee information and training

Written hazard communication programs are plans that identify how the employer will ensure that employees are notified of hazardous chemicals to which they could be exposed. The plan outlines how the OSHA standard will be implemented within that particular facility or company. The plan should describe how every level of an organization will inform employees about chemical hazards. Employees can be involved in writing the plan. OSHA can ask for the written plan if it conducts an audit of the organization.

The American National Standards Institute (ANSI) has developed guidance for labeling chemicals. Manufacturers follow the ANSI standards to ensure compliance with the OSHA labeling standard. All hazardous chemicals must be labeled, tagged, or marked with the identity of the chemical and all appropriate hazard warnings.

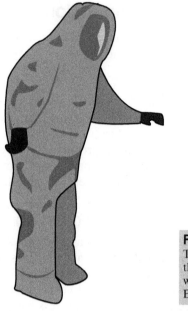

FIGURE 3-4 Process Technicians Must Understand the Procedures for Working with Hazardous Chemicals and Emergency Operations

Chemical manufacturers are responsible for affixing a label or tag to the container. This label or tag identifies the chemical, hazard warnings (e.g., eye irritant), and the manufacturer's name and address. Employers that use a chemical are responsible for making sure that labels on hazardous chemical containers are not removed or defaced.

An MSDS must be maintained for each hazardous chemical in a facility. Employees are responsible for familiarizing themselves with the MSDS for each chemical they might encounter.

Employee training is essential to the success of the HAZCOM standard. The written plan, labels, and MSDS are all components that employees must be trained to use. New employees typically are trained on initial assignment. Employees moving to new areas can be updated on the different chemical hazards posed by the new assignment.

During training, employees are provided with the information to help them do the following:

- Identify the chemical hazard.
- Locate resources, such as the MSDS.
- Determine how to protect themselves against the hazard.
- Determine how to detect the presence of a release.
- Respond to accidental exposure or emergencies.

Process technicians must understand the procedures for working with hazardous chemicals and emergency operations (e.g., spills and releases), as outlined in the company HAZCOM program and related plans. (Figure 3-4) They must understand how to handle, use, store, and transport chemicals, as necessary. Also, Personal Protective Equipment (PPE) selection, use, and maintenance are key issues relating to chemical hazard protection. (Note: Personal Protective Equipment is covered in more detail later in this textbook.)

EPA TOXIC SUBSTANCES CONTROL ACT (TSCA)

The Environmental Protection Agency's Toxic Substances Control Act (TSCA) permits the agency to review existing and new chemicals to identify potentially dangerous products or uses which should be subject to federal control. The Environmental Protection Agency (EPA) has the authority to require the manufacturers and processors of chemicals to conduct research to determine the effects of potentially dangerous chemicals on living creatures and report the results. If the EPA determines that using these dangerous chemicals does present an unreasonable risk to human health or the environment, the agency can regulate the chemical usage. EPA regulations can range from requiring proper product labeling to the potential ban of the substance.

The TSCA requires manufacturers and processors to conduct tests for existing chemicals under the following conditions:

- Their manufacture, distribution, processing, use, or disposal may present an unreasonable risk of injury to health or the environment; or they are being produced in substantial quantities and the potential for environmental release or human exposure is substantial or significant
- No data exists or existing data are insufficient to predict the effects of human exposure and environmental releases

Manufacturers must prevent future risks through pre-market screening and regulatory tracking of new chemical products. They must also control unreasonable risks (already known, or as they are discovered) for existing chemicals. Additionally, manufacturers must gather and disseminate information about chemical product use, and possible adverse effects to human health and the environment.

EPA Emergency Planning and Community Right-To-Know Act (EPCRA)

The EPA's Emergency Planning and Community Right-To-Know Act (EPCRA) authorizes state and local entities to prepare communities to respond in the event of a hazardous chemical release. Each state has a State Emergency Response Commission (SERC), which oversees Emergency Planning Districts. Within these districts are Local Emergency Planning Committees (LEPCs). Local first responders, health officials, government representatives, and industry members serve on these local committees.

Facilities must notify local authorities about any accidents at their location that could pose a threat to community and industrial neighbors. Facilities that use or manufacture hazardous substances must report releases and spills every year to the government.

LEPCs must plan in advance for possible chemical releases caused by a spill, explosion, or fire. Plants must use the EPA's list of "extremely hazardous substances" to determine which chemicals on the list are in their facility. The EPA list also establishes Threshold Planning Quantities (TPQs) for these chemicals. Facilities must share information with LEPCs about all chemicals that exceed the published TPQ. LEPCs and facilities that handle extremely hazardous substances must develop emergency response plans and employee training programs for everyone who will be called on to respond in a related emergency.

EPCRA mandates that any sudden releases of hazardous or extremely hazardous substances, which exceed the established Reportable Quantity (RQ), must be immediately reported to local, state, and federal officials. Releases of this type must also be reported to a National Response Center, as specified by CERCLA. A company can be subject to heavy fines if such releases are not immediately reported to authorities.

EPCRA also requires facilities to submit MSDSs for each hazardous chemical in their inventory to the local fire department, LEPC, and SERC. A hazardous chemical inventory must also be submitted to these same agencies. The inventories must provide the following information:

- Estimates of the maximum amount of each inventoried chemical present at any time throughout the year
- Estimates of the average daily amount of each inventoried chemical
- The location of each inventoried chemical

The EPA's Resource Conservation and Recovery Act (RCRA) describes solid and hazardous waste management. The Comprehensive Environmental Response, Compensation and Liability Act (CERCLA) created a superfund hazardous substances cleanup program. CERCLA was enlarged and re-authorized by the Superfund Amendments and Reauthorization Act (SARA). RCRA, CERCLA, SARA, and EPCRA are described in more detail in Chapter 16, *Recognizing Environmental Hazards*.

DOT MATERIAL CLASSIFICATION

The United States and many other countries throughout the world have developed a Globally Harmonized System for the Classification and Labeling of Chemicals (GHS). For example, the Department of Transportation regulates the shipment of hazardous materials within the United States. The United Nations uses a globally recognized system of classifying and labeling dangerous goods, which the DOT also recognizes.

The GHS system describes the following procedures for hazardous materials:

- Identification and classification
- Labeling, marking, and packing
- Documentation
- Emergency response

The system does not indicate the severity of the hazard, only its classification. The following are the recognized classifications:

Class 1: Explosives

Class 1 contains explosives that fall into six sub-classes.

- 1.1 Explosives with a mass explosion hazard (e.g., trinitrotoluene and nitroglycerine)
- 1.2 Explosives with a severe projection hazard
- 1.3 Explosives with a fire, blast, or projection hazard but not a mass explosion hazard
- 1.4 Explosives with a minor fire or projection hazard (e.g., ammunition and most consumer fireworks)
- 1.5 Blasting agents
- 1.6 Extremely insensitive articles

Figure 3-5 is an example of a Class 1 label.

Class 2: Gases

Class 2 contains gases that are compressed, liquefied, or dissolved under pressure, including gases with subsidiary risks (e.g., poisonous, corrosive).

- 2.1 Flammable gas (e.g., acetylene, hydrogen)
- 2.2 Non-flammable gases that are not poisonous (e.g., oxygen, nitrogen)
- 2.3 Poisonous gases (e.g., chlorine, hydrogen cyanide)

Figure 3-6 through Figure 3-8 are examples of Class 2 labels.

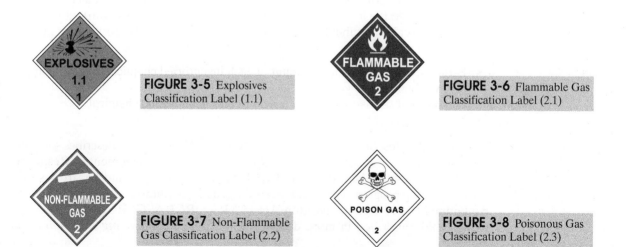

FIGURE 3-5 Explosives Classification Label (1.1)

FIGURE 3-6 Flammable Gas Classification Label (2.1)

FIGURE 3-7 Non-Flammable Gas Classification Label (2.2)

FIGURE 3-8 Poisonous Gas Classification Label (2.3)

Class 3: Flammable Liquids

Class 3 contains flammable liquids that fall into three sub-classes:

- 3.1 Highly flammable liquids with a boiling point below 35 degrees C (e.g., carbon disulfide, diethyl ether)
- 3.2 Flammable liquids with a flashpoint of less than 23 degrees C and a boiling point above 35 degrees C (e.g., acetone)
- 3.3 Liquids with a flashpoint above 23 degrees C but not exceeding 61 degrees C and a boiling point greater than 35 degrees C (e.g., kerosene)

Figure 3-9 is an example of a Class 3 label.

Class 4: Flammable Solids

Class 4 contains flammable solids.

- 4.1 Flammable solids that can be easily ignited and are readily combustible (e.g., magnesium, strike-anywhere matches)
- 4.2 Spontaneously combustible substances (e.g., aluminum alkyls, white phosphorous)
- 4.3 Substances that emit a flammable gas when wet, or react violently with water (e.g., sodium, calcium, potassium)

Figure 3-10 through Figure 3-12 are examples of Class 4 labels.

Class 5: Oxidizing Agents and Organic Peroxides

Class 5 contains oxidizing agents.

- 5.1 Oxidizing agents other than organic peroxides (e.g., ammonium nitrate, hydrogen peroxide)
- 5.2 Organic peroxides, in either liquid or solid form (e.g., benzoyl peroxides, cumene hydroperoxide)

Figure 3-13 and Figure 3-14 are examples of Class 5 labels.

FIGURE 3-9 Flammable Liquids Classification Label (3.1)

FIGURE 3-10 Flammable Solids Classification Label (4.1)

FIGURE 3-11 Spontaneously Combustible Solids Classification Label (4.2)

FIGURE 3-12 Dangerous or Reactive Flammable Solids Classification Label (4.3)

FIGURE 3-13 Oxidizing Agent Classification Label (5.1)

FIGURE 3-14 Organic Peroxide Classification Label (5.2)

FIGURE 3-15 Class 8 Classification Label (Corrosive Substances)

FIGURE 3-16 Class 9 Classification Label (Miscellaneous Dangerous Substances)

Class 8: Corrosive Substances

Class 8 contains corrosive substances.

- 8.1 Acids (e.g., sulfuric acid, hydrochloric acid)
- 8.2 Alkalies (e.g., potassium hydroxide, sodium hydroxide)

Figure 3-15 is an example of a Class 8 label.

Class 9: Miscellaneous Dangerous Substances

Class 9 substances are hazardous substances that do not fall into other categories (e.g., asbestos and dry ice). Figure 3-16 is an example of a Class 9 label.

A variety of regulations apply to shipping dangerous goods. For example, some materials cannot be packaged together, such as flammable materials and oxidizers. Radioactive materials require special handling—as well as materials being packaged as air cargo. Most companies will train workers on these regulations, as required.

The DOT classification diamonds are used on labels, placards (for trucks or rail cars), and other materials relating to the hazard. Guidelines must be followed for use and placement of labels, placards, and documentation. Labels must be made of a durable material and be prominently placed in plain sight, along with package orientation markings. Hazardous materials must be labeled with primary and subsidiary packing labels.

Accompanying documentation and shipping papers, including MSDSs, must be placed in locations for easy access when the material is being transported. Various information is included in the documentation and shipping papers, such as material description; name and address of the shipper; hazard classification; UN classification number; quantity; volume; precautions; directions for how to handle fires, spills, and leaks; and first-aid procedures.

Incorrectly classified materials and improper documentation or shipping papers can cause shipping delays. Criminal penalties may be imposed for intentional misclassification.

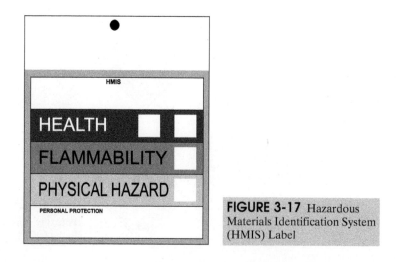

FIGURE 3-17 Hazardous Materials Identification System (HMIS) Label

Other Hazard Labeling Systems

Two other materials labeling systems are the Hazardous Materials Identification System (HMIS) and the National Fire Protection Association (NFPA) National Fire Code Section 704.

HAZARDOUS MATERIALS IDENTIFICATION SYSTEM (HMIS)

HMIS is a labeling system developed by the National Paint and Coating Association that lists five essential sections: chemical name, health hazard, flammability hazard, reactivity hazard, and Personal Protective Equipment (PPE) (Figure 3-17 and Table 3-1). Each section is color coded: blue (health), red (flammability), orange (physical hazard), and white (PPE). Each section except for the PPE section has an associated severity ranking from 0 to 4 (0 is the lowest rank and 4 is the highest, or most hazardous). The health section can have an asterisk (*) next to materials known to be carcinogens or to produce adverse effects after long-term exposure. The PPE block displays a letter, which corresponds to an index listing the appropriate PPE required. Icons are also used to provide a quick visual identification of specific PPE required, target organs, and physical hazards.

TABLE 3-1   Hazardous Materials Identification System (HMIS) Notations for Personal Protective Equipment (PPE)

Index Letter	PPE Required
A	Safety glasses
B	Safety glasses and gloves
C	Safety glasses, gloves, and an apron
D	Face shield, gloves, and an apron
E	Safety glasses, gloves, and a dust respirator
F	Safety glasses, gloves, apron, and a dust respirator
G	Safety glasses and a vapor respirator
H	Splash goggles, gloves, apron, and a vapor respirator
I	Safety glasses, gloves, and a dust/vapor respirator
J	Splash goggles, gloves, apron, and a dust/vapor respirator
K	Airline hood or mask, gloves, full suit, and boots
L-Z	Custom PPE, as specified

NATIONAL FIRE PROTECTION ASSOCIATION (NFPA)

The National Fire Protection Association (NFPA) created a National Fire Code. Section 704 of the code provides a system for identifying the hazards of materials. Section 704 allows people to easily visually identify the hazards associated with a substance. NFPA 704 uses a color-coded diamond along with a numbering system. The colors are red (flammability), blue (health), yellow (reactivity), and white (special information). Each color has an associated severity ranking from 0 to 4 (0 is the lowest rank and 4 is the highest, or most hazardous). Products and packaging must be clearly marked with labels using this diamond. Figure 3-18 shows an example of an NFPA diamond.

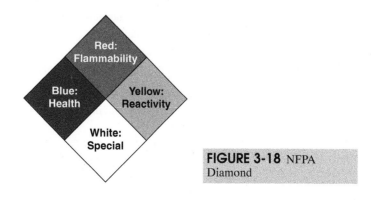

FIGURE 3-18 NFPA Diamond

Understanding How to Read an MSDS

At a facility, workers are notified about chemical hazards in a variety of ways: safety signs and warnings in areas where the chemicals are stored or used (e.g., warehouse or vessel), labels on containers, placards on trucks and rail cars, and documentation. Two important forms of documentation are a hazardous chemical inventory and Material Safety Data Sheets.

Recall that an MSDS is a vital document that provides key safety, health, and environmental information about a chemical. (Figure 3-19) Chemical manufacturers are responsible for creating MSDSs for their products and providing them with all chemicals they ship to users.

FIGURE 3-19 MSDS Sheets Provide Key Safety, Health, and Environmental Information About a Chemical

When a process technician starts work at a new facility or work area, one of the first and most important tasks is to review the hazardous chemicals inventory and the MSDSs. The process technician must also know where to locate an MSDS later. All MSDSs must be easily accessible at all times, and must be kept up-to-date since outdated MSDSs can cause great harm during an emergency situation.

It is critical that process technicians understand how to read an MSDS. Although the look and format of MSDSs can vary, the OSHA HAZCOM standard lists standard sections of information that must be documented. (Table 3-2)

The following are the sections contained within an MSDS:

TABLE 3-2   Material Safety Data Sheet (MSDS) Sections

Section Number	Contents
1	**Manufacturer's information**
	Name and contact information
	Address
	24-hour emergency telephone number
	Telephone number
	Date prepared
	Signature of preparer (optional)
2	**Hazardous ingredients and identity information**
	Components, including chemical identity and common names
	OSHA PEL
	ACGIH TLV
	Other recommended limits
	Percent of chemical in the composition (optional)

3   **Physical and chemical characteristics**
Boiling point
Vapor pressure
Vapor density
Solubility in water
Appearance and odor
Specific gravity
Melting point
Evaporation rate

4   **Fire and explosion data**
Flashpoint and method used
Flammable limits
Lower Explosive Limit and Upper Explosive Limit (LEL/UEL)
Extinguishing media
Special firefighting procedures
Unusual fire and explosion hazards

5   **Reactivity data**
Stability (stable or unstable) and conditions to avoid
Incompatibility (materials to avoid)
Hazardous decomposition or byproducts
Hazardous polymerization (may occur or will not occur) and conditions
to avoid

6   **Health and hazard data**
Routes of entry (inhalation, absorption, or ingestion)
Health hazards (acute and chronic)
Carcinogenicity (i.e., likelihood to cause cancer)
Signs and symptoms of exposure
Medical conditions generally aggravated (made worse) by exposure
Emergency and first-aid procedures

7   **Precautions for safe handling and use**
Steps to be taken in case material is released or spilled
Waste disposal method
Precautions to be taken when handling or storing
Other precautions

8   **Control measures**
Respiratory protection (specify type)
Ventilation
Protective gloves
Eye protection
Other protective clothing or equipment
Work/hygienic practices

If the process technician ever has any questions or does not understand the information on an MSDS, it is crucial to contact a supervisor and resolve the issue immediately.

One potential issue with MSDSs is trade secrets. Companies can have proprietary chemical identities which, if listed on an MSDS, would reveal their trade secret for all to view. OSHA handles this issue by allowing companies to withhold these chemical identities, but only if they meet certain criteria established in the HAZCOM standard. Even then, they must disclose chemical identities and hazards to health care professionals, as governed by rules in the standard.

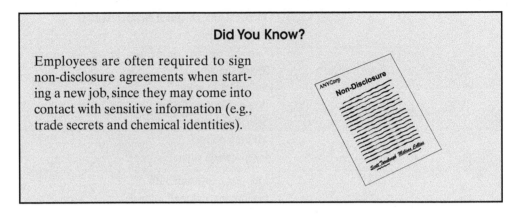

Did You Know?

Employees are often required to sign non-disclosure agreements when starting a new job, since they may come into contact with sensitive information (e.g., trade secrets and chemical identities).

Government Regulations Relating to Chemical Hazards

There are a variety of government regulations and private organization guidelines relating either directly or indirectly to chemical hazards:

- OSHA 29 CFR 1910.1200 Hazard Communication/Employee Right-to-Know (HAZCOM)
- OSHA 29 CFR 1910.119 Process Safety Management of Highly Hazardous Materials (PSM)
- OSHA 29 CFR 1910.120 Hazardous Waste Operations and Emergency Response (HAZWOPER)
- OSHA General Industry Standards
  - 1910 Subpart G: Health and Environmental Controls
  - 1910 Subpart H: Hazardous Materials
  - 1910 Subpart I: Personal Protective Equipment
  - 1910 Subpart J: General Environment Controls
  - 1910 Subpart L: Fire Protection
  - 1910 Subpart M: Compressed Gas/Air
  - 1910 Subpart N: Materials Handling and Storage
  - 1910 Subpart Z: Toxic and Hazardous Substances
- EPA Toxic Substance Control Act (TSCA)
- EPA Emergency Planning and Community Right-To-Know Act (EPCRA)
- EPA Resource Conservation and Recovery Act (RCRA)
- EPA Comprehensive Environmental Response, Compensation and Liability Act (CERCLA)
- EPA Superfund Amendments and Reauthorization Act (SARA)
- DOT Hazardous Materials Classification System
- DOT Hazardous Materials Handling: Loading and Unloading
- DOT Hazardous Materials Packaging and Marking
- National Fire Protection Association (NFPA) Hazardous Materials Storage Requirements
- American Conference of Governmental Industrial Hygienists (ACGIH) Threshold Limit Value (TLV) booklet

Many of these regulations and guidelines are discussed in more detail later in this textbook.

Summary

In the process industries, workers are routinely exposed to chemical hazards that can cause injury, illness, or death. Some hazards also pose a danger to the environment, causing potential short and long-term impacts.

Hazardous substances are often identified and grouped into categories based on their hazardous characteristics. The most common categories include flammable and combustible materials; water reactive, pyrophoric, and explosive materials; organic peroxides and oxidizers; and carcinogens, mutagens, teratogens, and reproductive toxins.

Flammable, combustible, water reactive, pyrophoric, explosive, and oxidizing materials primarily pose physical hazards in the form of fire or explosion. Carcinogens, mutagens, teratogens, and reproductive toxins primarily pose health hazards in the form of cell alteration or destruction.

Process technicians must be familiar with the different types of chemical hazards, know how to identify and detect them, and know how to protect themselves from exposure. Technicians must also be familiar with governmental regulations that govern hazardous chemicals.

Checking Your Knowledge

1. Define the following terms:
   a. Chemical hazard
   b. Combustible liquid
   c. Explosive
   d. Flammable solid
   e. Flashpoint
   f. Inorganic compound
   g. Neurotoxin
   h. Oxidizer
   i. Sensitizer

2. *(True or False)* There is a greater likelihood that a chemical in a liquid state will be accidentally inhaled rather than absorbed.

3. A substance with a pH of 9 is considered:
   a. An acid
   b. A base
   c. Neutral

4. *(True or False)* Pyrophoric materials are chemicals that react with water to release a gas that is either flammable or presents a health hazard.

5. _____ is a chemical that can initiate or promote combustion in other materials.

6. Which are cancer causing substances?
   a. Carcinogens
   b. Mutagens
   c. Teratogens
   d. Reproductive toxins

7. *(True or False)* Highly toxic materials require only a small amount of exposure to be considered lethal.

8. According to OSHA's HAZCOM standard, which of the following is/are responsible for researching chemical hazards and providing that information to customers?
   a. Chemical users
   b. Laboratories and warehouses
   c. Chemical manufacturers and importers
   d. All of the above

9. Chemical manufacturers are responsible for affixing a label or a tag to a container that identifies the:
   a. Chemical
   b. Hazard warnings
   c. Manufacturer's name
   d. All of the above

10. *(True or False)* Employees are responsible for familiarizing themselves with the MSDS for each chemical they might encounter.

11. Which of the following information is NOT included on an MSDS?
    a. Manufacturer information
    b. Health and hazard data
    c. Chemical inventory at facility
    d. Fire and explosion data

12. *(True or False)* All MSDSs must be locked away to protect trade secrets.

Student Activities

1. Look around your house for products that contain chemicals, and review the product warning labels. Make a list of the products and the potential hazards, and specify if you took any

action based on what you learned (e.g., changing how or where chemicals in your house were stored)?

2. Locate an MSDS for a chemical, and then review it with your fellow students. Discuss topics such as "Does the information sufficiently inform readers of the hazards?" and "What should be done in case of an emergency, such as a spill?"

3. Use the Internet to research OSHA's HAZCOM standard. Write a two-page summary of your findings including what the standard covers, how companies can implement it, and what potential penalties exist.

4. Use pH test strips to determine the pH of common items such as those listed below.
   a. Lemon juice
   b. Apple juice
   c. Tomato juice
   d. Black coffee
   e. Corn
   f. Pure or distilled water
   g. Seawater
   h. Baking soda
   i. Milk of magnesia
   j. Ammonia
   k. Soapy water
   l. Bleach
   m. Carbonated beverage
   n. Drain cleaner

5. Locate NFPA diamonds in places like gas stations, propane filling stations, and areas with air conditioning systems. Use what you know about the NFPA diamond to determine the hazards associated with each. Report your findings.

Recognizing Biological Hazards

Objectives

This chapter provides an overview of biological hazards in the process industries, which can affect people and the environment.

After completing this chapter, you will be able to do the following:

- Identify the following potential biological hazards found in the process industries and discuss their potential effects on safety, health, and the environment:

  Microorganisms (such as viruses and bacteria)

  Arthropods (arachnids and insects)

  Poisonous snakes

  Plant allergens and toxins

  Protein allergens from vertebrate animals

- Describe how bloodborne pathogens can affect the human body.
- Describe government regulations and industry guidelines that address biological hazards.

Key Terms

Algae—simple, plantlike organisms which grow in water, contain chlorophyll, and obtain their energy from the sun and their carbon from carbon dioxide (through photosynthesis).

Allergen—a substance that causes an allergic reaction or unhealthy response by the body's immune system.

Anaphylaxis—a rare, life-threatening, allergic reaction that can result in shock, respiratory failure, cardiac failure, or death if left untreated.

Arachnid—a class of arthropod that has four pairs of segmented legs (i.e., the legs are made up of sections) and includes scorpions, spiders, and ticks.

Arthropod—a type of animal that has jointed limbs and a body made up of segments, such as crustaceans (crabs), arachnids (spiders), and insects (mosquitoes).

Bacteria—single-celled, microscopic organisms that lack chlorophyll and are the most diverse group of all living organisms.

Bloodborne pathogens—pathogenic microorganisms that are present in human blood and can cause disease in humans. These pathogens include, but are not limited to, hepatitis B virus (HBV) and human immunodeficiency virus (HIV).

Exposure incident—an incident involving the contact of blood (or other potentially infectious materials) with an eye, mouth, other mucous membrane, or non-intact skin, which results from the performance of an employee's duties.

Fungi—plantlike organisms that obtain nutrients by breaking down decaying matter and absorbing the substances into their cells. They are similar to algae but do not contain chlorophyll.

Host—an organism whose body provides nourishment and shelter for another, smaller organism.

Infectious—capable of infecting or spreading disease.

Microorganism—a very small form of life, often viewable only through a microscope, that includes viruses, bacteria, algae, and fungi.

Pathogen—a specific cause of a disease, such as bacteria or a virus.

Protein allergen—an allergen caused by substances produced by vertebrate animals, including blood, feces, hair, and dead skin.

Toxin—a poisonous substance that can harm living organisms.

Universal precautions—an approach to infection control. According to the concept of Universal Precautions, all human blood and certain human body fluids are treated as if known to be infectious for HIV, HBV, and other bloodborne pathogens.

Venom—a poisonous substance created by some animals (such as snakes and spiders) and transmitted to prey or an enemy by biting or stinging.

Introduction

The practice of biological safety began as a result of biological warfare research conducted during the 1950s and 1960s at Fort Detrick in Maryland. Concern for worker and community safety led to the implementation of exhaustive precautions to prevent the exposure and release of biological hazards. These precautions provided the foundation of today's biological safety programs.

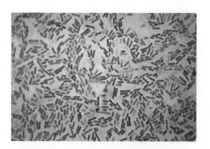

FIGURE 4-1 Microorganisms are One of the Biological Hazards a Process Technician may Encounter

Biological hazards are hazards caused by living organisms or parts or products of organisms that can generate an adverse response when they come into contact with the body.

Following are the most common biological hazards a process technician might face:

- **Microorganisms** (e.g., viruses, bacteria, and fungi)
- **Arthropods** (e.g., spiders and insects)
- Poisonous snakes (e.g., rattlesnakes, cottonmouths, copperheads, and coral snakes)
- Plant allergens and toxins (e.g., pollen, poison ivy, and poison oak)
- Protein allergens (e.g., urine, feces, blood, hair, or dander from vertebrate animals)

Biological Hazards Overview

The potential for most process technicians to become exposed to biological hazards is minor when compared to the potential for exposure to physical or chemical hazards in the process industries (with the exception of technicians working in the food processing industry, who could face biological hazards on a regular basis). However, many biological agents (e.g., anthrax, smallpox) can be used as bio-weapons, delivered through various methods such as the mail or air dispersal as part of a terrorist attack or other similar acts (discussed in more detail in Chapter 14, *Physical Security and Cybersecurity*).

The effects of exposure to biological agents can be just as toxic and life-threatening as the effects of exposure to many physical and chemical agents. This chapter addresses the more common biological hazards to which a process technician may be exposed, and some of the precautions a worker can take to prevent exposure.

The goal of biological safety is to minimize worker exposure to biological hazards through anticipation, recognition, evaluation, and control of these hazards.

Process technicians may encounter biological hazards through many routine tasks or emergency situations. Some examples include the following:

- Encountering a black widow spider while inspecting equipment
- Being bitten by mosquitoes while working outside during the night shift
- Stumbling across a snake or other animals while servicing equipment in a remote area of the plant
- Being exposed to infectious body fluids while attending to an injured co-worker
- Coming into contact with harmful bacteria in the cooling water while performing maintenance on a cooling tower
- Coming into contact with mouse droppings while cleaning a warehouse

MICROORGANISM HAZARDS

Microorganisms (also called microbes) are very small forms of life that can be seen only with the aid of a microscope. Examples of microorganisms include viruses, bacteria, algae, fungi, and protozoans.

Process technicians are typically not exposed to microorganisms any more than workers in most other occupations, but they should still be aware of what microorganisms are and how they can be hazardous. (Figure 4-1) Because it is beyond the scope of

---

Did You Know?

Legionnaires' disease is a form of pneumonia. It is called Legionnaires' disease because the first known outbreak occurred in a hotel that was hosting an American Legion convention.

In that outbreak, approximately 221 people contracted this previously unknown type of bacterial pneumonia, and 34 people died. The source of the bacterium was found to be contaminated water used to cool the air in the hotel's air-conditioning system.

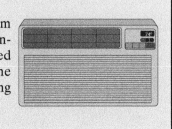

Did You Know?

Viruses are between living and non-living. If they're floating around in the air or sitting on a doorknob, they are inert and are as alive as a rock.

But if they come into contact with a suitable plant, animal or bacterial cell (called a host), they spring into action and infect the cell, much like a pirate hijacking a ship.

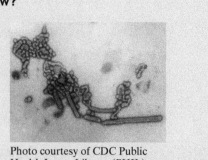

Photo courtesy of CDC Public Health Image Library (PHIL)

---

this text to discuss every microorganism and its hazards, this chapter is intended to provide a general overview. However, workers at a facility are urged to observe contractors, visitors, and other workers for signs of symptoms, infections, contaminations, or skin irritations and report these to supervisors or medical personnel.

VIRUSES

Viruses, the simplest of all organisms, cannot live and reproduce outside of a **host** (an organism whose body provides nourishment and shelter for another).

Viruses are parasites because they are organisms that live in or on the living tissue of a host organism at the expense of that host. They are also **infectious,** meaning they are capable of infecting or spreading disease.

Viruses can be found anywhere there are host cells to infect. Viruses live to reproduce.

Most viruses harm specific types of organisms, but cause no reaction in others. For example, some viruses target plants but do not affect humans. However, some diseases such as bird flu, Hanta virus, and Ebola have crossed over from one species to another.

Viruses are unaffected by antibiotics, the drugs used to kill bacteria. Instead these infections are best controlled by preventing transmission, administering vaccines (substances that help the immune system respond to and resist disease), and in some cases administering antiviral drugs.

In the workplace, government regulations address two types of viruses: the Human Immunodeficiency Virus (HIV), which can lead to Acquired Immune Deficiency Syndrome (AIDS), and the hepatitis B virus (HBV), which can cause liver problems. HIV and HBV are categorized as bloodborne pathogens under OSHA regulations. Bloodborne pathogens are described in a separate section later in this chapter.

BACTERIA

Bacteria (single-celled, microscopic organisms that lack chlorophyll) are the most diverse group of living organisms, and are more complex than viruses. Bacteria are found all over the world, from the dirt to water to air. They can live in temperatures above the boiling point or below freezing. Bacteria can multiply rapidly. A majority of

---

Did You Know?

We are all exposed to mold on a daily basis without evident harm. However, when present in large quantities, mold can cause allergic reactions and respiratory ailments.

Proper protective equipment (e.g., gloves, a respirator, and eye protection) should always be used when working in areas where large quantities of mold are present.

them have no effect or are even beneficial to humans, animals, and plants. However, some can cause harm.

Although it is unlikely, a process technician could come into contact with harmful bacteria (like Legionella, which causes a pneumonia-like disorder called Legionnaire's disease) during tasks involving equipment that uses hot or cold water systems, such as heat exchangers or cooling towers. As long as equipment is properly designed, operated, and maintained, it is not likely that a process technician will be exposed to harmful bacteria.

ALGAE

Algae (simple plantlike organisms that grow in water) contain chlorophyll and obtain their energy from the sun and their carbon from carbon dioxide (through the process of photosynthesis). Algae produce a considerable amount of oxygen during photosynthesis. Algae are found in fresh and salt water around the world. Some forms of algae, (e.g., blue-green algae) can be harmful to people, causing symptoms such as rashes, headaches, nausea, and vomiting. The likelihood of encountering harmful algae while working as a process technician, however, is very small.

FUNGI

Fungi (plantlike organisms that obtain nutrients by breaking down decaying matter and absorbing the substances into their cells) are similar to algae, although they do not contain chlorophyll. Fungi grow best in moist, dark areas. Examples of fungi include yeasts and molds.

Fungi can be helpful or harmful, depending on the type. For example, the antibiotic penicillin comes from a fungus (*Penicillium notatum*). The yeast we use to make bread and brew beer is also a fungus. Both of these are helpful fungi.

A few types of fungi, however, can harm humans by causing respiratory problems and other ailments (e.g., sick building syndrome and athlete's foot). However, most process technicians will not encounter these harmful types of fungi during normal operations.

PROTOZOA

Protozoa (one-celled animals that are larger and more complex than bacteria) hunt other by microorganisms, such as bacteria. Protozoa produce nitrogen as they eat. Amoebas are a type of protozoa. The vast majority of protozoa are not harmful, but a few can cause disease (e.g., malaria, which is transmitted by mosquitoes). Protozoa can cause a wide range of problems, from ulcers to diarrhea and more serious symptoms. The likelihood that a process technician will encounter harmful protozoans during normal operations is remote.

ARTHROPODS

Arthropods are a group of animals that have jointed limbs, a segmented body, and a hard exoskeleton. Examples of arthropods include crustaceans (e.g., crabs), **arachnids** (e.g., spiders and scorpions), and insects (e.g., mosquitoes and ants).

Process facilities provide good habitats for arachnids and insects because the various warehouses, tanks, and equipment provide an ideal environment in which to thrive. Process technicians may come across arachnids in dark recesses or insects when performing routine tasks around the facility. Because of this, individuals with known allergies to stings may wish to carry an epinephrine injecting device called an EpiPen® in the event they are bitten or stung. Individuals who carry an EpiPen® should notify coworkers, so they can assist you if you are bitten and unable to get to your epinephrine injecting device.

SPIDERS

All spiders are venomous, but most species cause only a mild reaction (slight pain and swelling) if they bite a person. In North America, the black widow spider and brown recluse spider are the only two highly venomous spiders. Spiders are rarely aggressive, biting people only when threatened or hurt.

Characteristics of black widow spiders include a shiny black appearance with long legs and a distinctive reddish-orange hourglass marking underneath the body. Typically, black widows grow to about one inch in length. Black widows prefer to nest in dark areas, especially in corners. (Figure 4-2)

Black widow bites cause immediate swelling and pain. The bite victim can also experience muscle cramps for six to 24 hours. Victims rarely die from a bite, but immediate medical attention must be sought.

FIGURE 4-2 Black Widow Spider

Courtesy of CDC Public Health Image Library (PHIL)

The brown recluse spider features a golden brown appearance, long legs, and a dark, fiddle-shaped mark on its head. Typically, brown recluse spiders grow to about $\frac{1}{2}$ inch in length. Brown recluses, which are primarily found in the midwestern United States, prefer living in dark areas and underneath objects. (Figure 4-3)

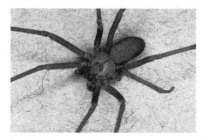

FIGURE 4-3 Brown Recluse Spider

Courtesy of CDC Public Health Image Library (PHIL)

Brown recluse spider bites often do not hurt immediately. Pain can occur later and a blister can form anywhere from four to eight hours after the bite. Victims rarely die from a bite, but immediate medical attention must be sought. If possible, the spider should be killed and brought with the victim for identification, since the tissue around a brown recluse bite can be severely damaged by the toxin and could require a skin graft.

To reduce the risk of a spider bite, visually inspect any dark, confined areas with a flashlight before reaching your hand into the area. Make sure to wear tough work gloves and long sleeves. If you spot a black widow or brown recluse spider, spray it with insecticide if the work environment permits. If not, use a long-handled tool to remove or kill the spider.

If you are bitten by a spider, place ice on the bite to slow the spread of the **venom,** and then report the bite to your supervisor and seek medical attention.

TICKS

Ticks, which are relatives of spiders and scorpions, have oval bodies with eight legs and live throughout North America. They feed on the blood of animals. Ticks can carry diseases (e.g., Lyme disease) that can be transferred to humans as the ticks feed. (Figure 4-4)

Ticks cannot move quickly, jump, or fly, so they cling to tall grass, brush and shrubs, or low-hanging trees, and wait for an animal to come by and come into contact with them. Individuals working in areas with this type of vegetation should avoid brushing

FIGURE 4-4 Deer Ticks can Spread Lyme Disease
Courtesy of CDC Public Health Image Library (PHIL)

up against it if possible and should check clothing and skin for ticks on a regular basis. You may also choose to wear insect repellant with DEET as an ingredient.

If you locate a tick on your body, you should have an onsite medical staff member remove it for you. If you do not have access to medical staff, remove the tick using small-tipped tweezers. Grasp the tick with the tweezers as close as possible to the skin, and then slowly but firmly pull the tick off the skin. If you do not have tweezers, you can cover the tick with petroleum jelly or some other similar substance (just make sure it will not harm your skin) to suffocate it. However, the tick will take awhile to die this way. It is best to use tweezers. Do not try to burn the tick off or kill it while it is still attached to the skin.

SCORPIONS

Scorpions tend to live in desertlike climates and other areas around the southwestern United States, but they can also be found in forests, prairies, grasslands, mountains, or along seashores. There are 20 different types of scorpions in the U.S. (Figure 4-5)

Scorpions range in color from tan to dark brown. They have eight legs and a flat body with two crab claw-like pinchers. The curved tail has a stinger on the end that the scorpion uses to kill prey and defend itself. Scorpions are typically nocturnal, meaning they are active mostly at night. During the day, they hide underneath rocks or in dark crevices.

Scorpion stings cause symptoms similar to black widow bites (e.g., pain and swelling). Victims rarely die from a sting, but immediate medical attention must be sought.

To reduce the risk of a scorpion sting, visually inspect any dark, confined areas with a flashlight before reaching your hand into the area. Make sure you wear tough work gloves and long sleeves. If you spot a scorpion, spray it with insecticide if the work environment permits it. If not, use a long-handled tool to remove or kill the scorpion. When changing clothes or putting on shoes or boots, inspect for scorpions prior to putting them on.

FIGURE 4-5 Scorpion
Courtesy of CDC Public Health Image Library (PHIL)

Insects

Process technicians may come into contact with a variety of insects during their normal job duties. Some of the most common insects include mosquitoes, fire ants, and bees.

MOSQUITOES

Mosquitoes can carry many diseases including St. Louis encephalitis and West Nile virus (WNV). (Figure 4-6) Infected mosquitoes can transmit these diseases to humans through a bite. St. Louis Encephalitis is an illness that causes the brain to become swollen and inflamed. West Nile virus is an illness that causes a range of symptoms, from mild ones similar to the flu to life-threatening ones that can cause neurological effects.

In temperate regions of North America, St. Louis encephalitis and WNV are believed to be seasonal, occurring most often in summer and continuing until fall when cold weather comes and mosquitoes die off. In southern states, mosquito populations can remain high year-round, so the threat of infected mosquitoes remains high.

With St. Louis encephalitis, symptoms can occur anywhere from five to 15 days after the bite. Mild cases can result in symptoms such as fever and headache. Severe cases can result in high fever, neck stiffness, disorientation, coma, tremors, convulsions, and spastic paralysis. Some cases can lead to death, especially in older people.

With WNV, approximately four out of five people infected with it will not show any symptoms. Only 1 in 1,000 dies from the disease. If you are bitten by a mosquito carrying WNV, any symptoms that develop will occur from three days to two weeks after the bite. Symptoms can include flu-like symptoms such as fever, aches, nausea, headaches, and possibly a rash or swollen lymph glands. These symptoms can last between a few days to several weeks.

In extreme cases of WNV, symptoms can include high fevers, neck stiffness, severe headaches, confusion, disorientation, vision loss, tremors, paralysis, and coma. These symptoms can last several weeks, while neurological effects may be permanent.

If you suspect you have St. Louis encephalitis or WNV, seek medical attention immediately. There is no specific treatment for St. Louis encephalitis, so treatments will vary based on the symptoms and complications. There is also no specific medical treatment for WNV. In mild cases, the symptoms will pass without treatment. Severe cases require medical attention (which can include hospitalization and supportive treatments, such as IVs, help with breathing, and nursing attention).

Risk of exposure to St. Louis Encephalitis and WNV is low, but technicians should still take steps to protect themselves. For example, mosquitoes feed most often during daybreak or dusk, so technicians should stay inside during those times, if possible. Technicians who must go outside during these times should wear thick, long-sleeve clothing. For extra protection, a technician can use an insect repellant, particularly one containing at least 20% of the ingredient DEET. However, it is important to check with a supervisor prior to application since insect repellant can be detrimental to certain types of clothing and Personal Protective Equipment.

The application of insect repellent should be limited to exposed skin to avoid overuse of the product. Product directions and application instructions should always be followed. Individuals who are sweating may need to re-apply repellant periodically. Repellant should not be applied to cuts, wounds, or irritated skin. Repellant should also not be sprayed on the face. Instead, the repellent should be sprayed onto the hands and then rubbed on the face, avoiding the mouth and eyes.

Stagnant water, which can collect at various areas around the plant, is the perfect breeding ground for mosquitoes. Technicians who must work in areas with standing water should protect themselves as much as possible.

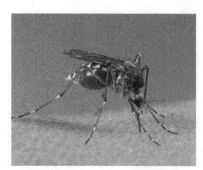

FIGURE 4-6 Mosquito

Courtesy of USDA Agricultural
Research Service

FIRE ANTS

Fire ants are approximately 1.5 mm long, red and black in color, extremely aggressive, and inflict a painful sting. Fire ant stings cause small, itchy blisters to form. Nests are shaped like a dome, and if disturbed, the ants send signals to swarm and attack. (Figure 4-7)

If you encounter fire ants, do not disturb the mound. If ants come in contact with the skin, do not move rapidly; this causes more ants to attack. Calmly brush or rinse them off. Apply a cold compress to the affected areas. Wash the area gently with soap and water, leaving the blisters intact. Avoid scratching. If you are allergic, seek medical attention.

FIGURE 4-7 Fire Ants

Courtesy of USDA Agricultural Research Service

BEES

Much media coverage has been given to Africanized bees (also called "killer bees"), which are much more aggressive than other honeybees and have killed people and animals. Most deaths occur when the person is unable to get away from the bees quickly.

Bees swarm in the spring and fall, looking for a new location to make a hive. They like to build hives in crevices and holes. If you encounter bees, do not swat at them. If they start stinging you, run away and get indoors as quickly as possible. Honeybees might chase for up to 50 yards (Figure 4-8), while Africanized bees might continue to chase for 150-200 yards or more.

If you are stung by a bee, gently brush off any stingers with your fingernail or the edge of a driver's license or credit card. Do not try to pinch or pull the stinger out; this can squeeze more venom into the wound. Apply a cold compress to the affected area. If you are allergic or are stung repeatedly, seek medical attention.

FIGURE 4-8 Honey Bees

Courtesy of USDA Agricultural Research Service

Poisonous Snakes

Four types of poisonous snakes reside in North America: rattlesnakes, cottonmouths (also referred to as water moccasins), copperheads, and coral snakes.

Rattlesnakes range across North America and reach lengths anywhere from 15 inches to 100 inches. Rattlesnakes vary in color and markings, but are generally gray, green, or tan with dark markings and a distinctive rattle on the tail. (Figure 4-9)

FIGURE 4-9 Rattlesnake

Courtesy of CDA Public Health
Image Library (PHIL)

Cottonmouth snakes (also known as water moccasins) are considered very aggressive and dangerous, and reside in the southern/central United States in lowland areas near water. Cottonmouths can reach lengths up to 75 inches, and are mostly dark in color. (Figure 4-10)

FIGURE 4-10 Cottonmouth
(i.e., Water Moccasin)

Courtesy of CDA Public Health
Image Library (PHIL)

Copperhead snakes reside in many southern, central, and lower eastern states. Their copper coloring features reddish brown cross bands on their backs. Copperheads can reach lengths of 20 to 54 inches. (Figure 4-11)

FIGURE 4-11 Copperhead

Courtesy of CDC Public Health
Image Library (PHIL)

Coral snakes live in many southern and central states. They grow from 13 to 48 inches long, and have distinctive red, yellow, and black bands (the red and yellow bands touch, prompting the familiar saying "red and yellow kill a fellow"). They are extremely venomous. (Figure 4-12)

FIGURE 4-12 Coral Snake

Courtesy of CDC Public Health
Image Library (PHIL)

Snakes like to make their home under cover or in cramped, covered spaces. Most poisonous snakes are not aggressive and will not strike if left alone. If you encounter a snake while working, freeze in place immediately, then slowly back away. A general rule is to keep at least six feet between you and a snake, since it can typically strike at a distance equal to only half of its body length.

According to the American Red Cross, you should take the following steps if bitten:

- Wash the bite with soap and water.
- Immobilize the area with the bite (typically an arm or leg) and keep it lower than the heart.
- Get immediate medical attention. [Note: It will help the medical staff if the type of snake can be identified. Phone ahead so the proper antivenin (antitoxin) can be obtained before the snakebite victim arrives].

If immediate medical care cannot be reached within about 30 minutes, two other options may be tried. A bandage can be lightly wrapped about two to four inches above the bite, if feasible. This can help slow the spread of venom. However, the person administering first aid should make sure the bandage is not wrapped so tightly that blood flow is cut off. The bandage should be loose enough to slip a finger under it. Also, some snakebite kits come with a suction device to help draw venom from the wound. Follow the instructions with the kit, but do not make any incisions. Tourniquets, ice, incisions, or any other treatments are discouraged.

To avoid snakebites, visually inspect any dark, confined areas with a flashlight before reaching your hand into the area. Make sure you wear tough work gloves and long sleeves. If you are walking through tall grass, watch where you step and wear thick boots. In general, stay alert.

Animals

Animals you might encounter on the job can also present a hazard. For example, bites from animals such as mice, raccoons, dogs, and cats can be painful and dangerous. Animals can carry rabies, and they can infect you through a bite.

Alligators are also considered dangerous. They can move very quickly over a short distance and on rare occasions have been known to attack people.

In remote work areas, you could encounter predators such as mountain lions and bears. Even animals like elk and caribou can be dangerous and attack if you come too close or provoke them.

The best recommendation is to leave plenty of distance between yourself and any animals, whether wild or domesticated. In most cases, if you leave them alone they will leave you alone. If you have to reach into dark spaces, make sure you wear thick gloves and use a flashlight to illuminate the area first.

Did You Know?

Process technicians working in remote regions of Alaska must watch out for polar bears and caribou as they make their rounds!

Photo courtesy of U.S. Fish and Wildlife Service

Allergens and Toxins

The human immune system protects the body against harmful substances like viruses and bacteria. However, some people have overly sensitive immune systems that react to generally harmless substances, such as pollen and dust. This is called an allergic reaction and results in a misdirected response by the immune system to the **allergen** (the substance causing the reaction, such as pollen). Types of immune reactions can vary from rash or hives to life-threatening **anaphylaxis** (a life-threatening allergic reaction that can result in shock, respiratory failure, cardiac failure, or death if left untreated).

An allergic reaction triggers the immune system to produce antibodies to fight the allergen. Antibodies initiate the release of histamine and other chemicals that cause the symptoms of the allergic response. Allergens can include pollen, mold, dust, animal dander (tiny cells of dead skin), certain foods (e.g., peanuts and strawberries), cockroach droppings, and more. The types of substances that cause allergic reactions, and the amount of immune system response, vary from person to person. It is important to know this because some materials and additives used in processes can potentially cause reactions (e.g., itching, rash, shortness of breath, or anaphylaxis) in some individuals.

PLANT ALLERGENS

In this section, we will briefly address plant allergens. (Figure 4-13) Later sections will address protein allergens (reactions to animal-based substances such as skin and hair) and plant toxins (e.g., poison ivy).

Allergic reactions to plants can produce the following symptoms:

- Itching (e.g., eyes, nose, mouth, skin)
- Coughing
- Sneezing
- Difficulty breathing
- Nasal congestion or runny nose
- Watering eyes
- Headache
- Impaired smell

Symptoms can appear anywhere from a few minutes to an hour after exposure. The severity of symptoms varies by person. Some highly sensitive individuals may experience hives, rashes, rapid heart rate, a drop in blood pressure, and even anaphylaxis.

Allergies are common, and can be influenced by heredity and environmental conditions. Pollens from trees, grass, and flowers are the most common plant allergens. Conditions that promote the distribution of allergens are hot, dry, and windy weather. Wet weather washes the pollens to the ground. Trees tend to produce pollen in spring,

FIGURE 4-13 Plant Allergens

Courtesy of CDC Public Health Image Library (PHIL)

grasses and flowers produce pollen during summer, and ragweed and other late-blooming plants produce pollen during late summer to early fall.

Treatment for allergic symptoms varies, but the best approach is to avoid exposure as much as possible. Check with your doctor to determine your specific allergens.

During days with high allergen counts (usually reported in the local news), stay in an air-conditioned environment as much as possible. If you must work outside, consider wearing protective goggles and a mask or filter over your mouth and nose.

Make sure your employer is aware of any allergies you have. Also, know what the potential side effects of your allergy medication are, such as drowsiness and slower reaction times. Such effects can impact safety.

PLANT TOXINS

Certain plants, such as poison ivy, poison oak, and poison sumac can cause allergic reactions, with symptoms such as redness, swelling, and an itchy, blistery rash.

These fragile plants create a natural defense: an oily substance that causes extreme irritation to people and animals that come into contact with it. Almost 75% of people who come into contact with the oil are allergic to it.

Poison ivy grows as a low vine or shrub. Its three green, shining leaflets with notched edges make it easy to identify. In fall, the leaflets turn bright red, then yellow.

Poison ivy grows throughout the United States. (Figure 4-14)

FIGURE 4-14 Poison Ivy
Courtesy of CDC Public Health
Image Library (PHIL)

Poison oak is a shrub with three to seven lobed leaflets, with a hairy appearance underneath each leaflet. It grows mainly in the northwestern and western states. (Figure 4-15)

Poison sumac is a shrub with smooth leaflets and white, drooping berries. It grows primarily in the eastern and central United States in damp environments. (Figure 4-16)

The following are some common symptoms of plant toxins:

- Redness
- Swelling
- Rash
- Blisters
- Itchiness

FIGURE 4-15 Poison Oak

FIGURE 4-16 Poison Sumac

Rashes caused by plant toxins typically disappear in one to two weeks. If you are exposed, avoid scratching the rash and keep your hands away from your eyes, mouth, or other sensitive areas. A variety of treatments are available from drugstores or a doctor if a reaction occurs. If you experience severe redness, rash on your mouth, eyes, genitals, or over a large area of your body, seek medical attention.

If you experience any of the following symptoms after exposure, seek immediate medical attention:

- Swollen lips, throat, or tongue
- Bluish lips or mouth
- Difficulty breathing
- Difficulty swallowing
- Weakness or dizziness

Toxic plants are often hard to control. Even if they are sprayed or uprooted, new plants can grow nearby. Toxic plants should never be burned, since the oil can be carried by the smoke and can cause a serious reaction to anyone inhaling it.

It is best to avoid contact with these plants, if possible. However, the oil can get on shoes or clothing that brushes against the plant. Then, when you touch or remove the clothing, your skin comes into contact with the oil. Also, if you have the oil on your skin and touch another person within 15 minutes of exposure, you can transfer the oil.

If you suspect that you have come into contact with a poisonous plant, you should take the following steps.

- Remove all affecting clothing and shoes (using gloves if possible) and wash them in hot water and a strong detergent.
- Thoroughly wash affected skin with soap and water immediately, at least several times. It takes approximately 15 minutes for the oil to penetrate the skin and cause damage.

Protein Allergens

Protein allergens cause similar reactions to those listed in the Plant Allergen section. Animal urine, feces, blood, hair, and dander (dead skin) can all cause allergic reactions in people. Foods can also cause allergic reactions. For example, peanuts, milk, shellfish, and strawberries are just some of the foods that can cause an allergic reaction.

If you work around food products or come into contact with other potential protein allergens (such as mouse droppings in a warehouse) watch carefully for signs of an allergic reaction in some people. Latex gloves can also cause an allergic reaction. If you have a latex allergy and must wear such gloves for your job, ask your employer about latex-free gloves. There are also certain chemicals (such as those in the amine family) to which a person can become allergic; these are referred to as sensitizers.

Did You Know?

Hantavirus pulmonary syndrome (HPS) is a deadly disease transmitted by infected rodents (specifically deer mice) through urine, droppings, or saliva. Humans can contract the disease when they breathe in the aerosolized virus.

Rodent control in and around the home and work environments remains the primary strategy for preventing hantavirus infection.

Photo courtesy of CDC Public Health Image Library (PHIL)

Bloodborne Pathogens and Government Regulations

Bloodborne pathogens refer to microorganisms (such as viruses or bacteria) present in human blood that can cause disease in people. Examples of such **pathogens** include Hepatitis B virus (HBV), Human Immunodeficiency Virus (HIV), malaria, and syphilis.

HIV and Hepatitis B, along with other bloodborne pathogens, can be transmitted through contact with blood and other potentially infected body fluids and materials (e.g., semen, vaginal secretions, saliva, mucus or any body fluid visibly contaminated with blood).

The most common way a process technician can be exposed to bloodborne pathogens on the job is through first aid. Some plants require process technicians to become certified in first aid and to agree to provide first aid to any co-worker who may be in need. Although no hard evidence exists that HIV can be transmitted through saliva, giving CPR to a victim is considered another potential way to be exposed.

OSHA created a standard covering occupational exposure to blood and other infectious materials, 29 CFR 1910.1030 (the Bloodborne Pathogens Standard). The standard's purpose is to limit exposure to blood and other infectious materials, as a precaution against the transmission of bloodborne pathogens from one person to another.

The standard applies in situations where it can be reasonably anticipated that an employee's job duties might cause exposure to blood or other infectious materials. OSHA created three categories of work-related tasks to clarify this:

- Category I—job tasks routinely expose the employee to blood and other infectious materials.
- Category II—job tasks do not involve routine exposure of the employee to blood and other infectious materials, but some parts of the job may involve performing Category I tasks.
- Category III—job tasks do not normally involve exposing the employee to blood or other infectious materials.

Many process industry jobs fall under Category III. However, depending on the plant, some job tasks might fall into Category II.

No matter the job task, it is still possible to be exposed to blood or other infectious materials if a co-worker is injured. Always assume that all body fluids are contaminated and act accordingly. The following sections describe HIV and HBV, then discusses universal precautions and Personal Protective Equipment that can be used to limit exposure.

Other aspects of bloodborne pathogens, such as legal implications, company policies and procedures, AIDS education, employee rights, and related topics are outside the scope of this textbook.

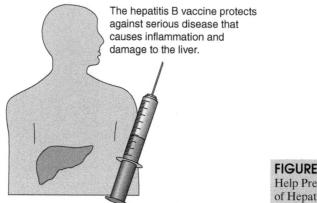

The hepatitis B vaccine protects against serious disease that causes inflammation and damage to the liver.

FIGURE 4-17 Vaccines can Help Prevent the Transmission of Hepatitis B

HEPATITIS B

There are several different types of hepatitis. The OSHA standard addresses the hepatitis B virus (HBV). Hepatitis B, transmitted most often through "blood to blood" contact, infects the liver and causes inflammation or more serious conditions such as cirrhosis or liver cancer. There is a vaccine for HBV. (Figure 4-17) Many health workers and first responders are required to get this vaccine.

Each year, around 300,000 Americans become infected with HBV. Only a small percentage of those cases result in a fatality. Hepatitis B has no specific treatment or cure. A person who is infected can develop antibodies that fight the infection and protect the person from getting HBV again.

Since there are different types of hepatitis (e.g., hepatitis A and C), having contracted HBV does not protect someone from those other types. The hepatitis B virus can survive for up to seven days in dried blood. This puts anyone who comes into contact with fresh or dried blood at risk. HBV symptoms can take anywhere from one to nine months to occur.

Early symptoms of HBV resemble a mild flu and can include the following:

- Fatigue
- Stomach pain
- Appetite loss
- Nausea

As the HBV infection progresses, symptoms can include the following:

- Jaundice (yellow appearance of eyes and skin)
- Darkened urine

If you think you have been exposed to hepatitis B, check with a doctor immediately, especially if you are experiencing any symptoms.

HUMAN IMMUNODEFICIENCY VIRUS

The Human Immunodeficiency Virus (HIV) is a disease that attacks and weakens a person's immune system, preventing it from fighting other deadly diseases. Currently, there is no vaccine to prevent HIV, and no cure.

HIV causes AIDS (Acquired Immune Deficiency Syndrome). AIDS has no known cure and is fatal. From the time a person contracts HIV, it can be many years before AIDS develops. During this latency period, the person may be unaware of having the infection and not show any symptoms. However, the person's body fluids still contain the virus and pose an infection hazard for those that may come into contact with those fluids. Although there is currently no cure for AIDS and it is fatal, treatments to lessen its effects have been developed in recent years.

Although the chances of contracting HIV in a work environment are slim (around 0.4%), proper precautions should be taken to avoid exposure. HIV does not survive long outside a human body, so fresh blood or other potentially infectious materials

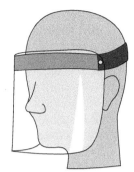

FIGURE 4-18 Face Shields Help Prevent the Transmission of Bloodborne Pathogens

pose the greatest HIV risk to first responders (fire fighters, EMTs, police) and medical personnel. However, anyone providing immediate first aid to a victim is also at risk.

Symptoms of HIV vary, but can include any of the following:

- Fever
- Weakness
- Headaches
- Diarrhea
- Nausea
- Sore throat
- Weight loss
- Coating on tongue
- Swollen lymph glands

Often, individuals infected with HIV do not display any symptoms shortly after exposure. These individuals are said to be asymptomatic. Those that do exhibit symptoms often report flu-like conditions.

If you think you have been exposed to HIV or are experiencing any symptoms, check with a doctor immediately.

TRANSMISSION OF HIV AND HBV

HIV and hepatitis B are most commonly transmitted in the following ways:

- Sexual contact
- Sharing of hypodermic needles
- From mothers to babies before or during birth
- A puncture from a contaminated needle, broken glass, or other sharp object

FIGURE 4-19 Sharps and Biohazardous Materials should Always be Placed in Appropriate Containers

Courtesy of CDC Public Health Image Library (PHIL)

- Contact between damaged or broken skin and infected bodily fluids and materials
- Contact between mucous membranes and infected bodily fluids and materials

Anytime there is blood-to-blood contact or contact with other bodily fluids, there is a chance for infection. That is why it is critical to protect yourself against punctures from contaminated needles, broken glass, and other sharp objects.

Skin can provide an impervious barrier to bloodborne pathogens. However, skin can be damaged in various ways, (e.g., cuts, wounds, open sores, acne, sunburn, blisters, and abrasions). If bloodborne pathogens come into contact with broken or damaged skin, there is a chance of infection.

Bloodborne pathogens can also be transmitted by contact with mucous membranes, the internal, mucous-secreting lining of the mouth, nasal sinuses, eyes, stomach, intestines, and many other parts of the body.

For example, if blood or other bodily fluids were to splash your face, bloodborne pathogens could enter through your eyes, nose, or mouth. Eyes are considered a vulnerable body part for the transmission of bloodborne pathogens. Thus, many medical personnel wear face shields (Figure 4-18) or protective glasses when working in situations where blood or body fluids could be splashed or atomized.

UNIVERSAL PRECAUTIONS

Universal precautions are a strategy to prevent exposure to bloodborne pathogens, such as HIV and hepatitis B. The basic premise of universal precaution is as follows:

> Always treat blood and other bodily fluids or materials as if they are infectious, and avoid any direct contact with them.

Use this approach with every individual and for every source of blood or other potentially infectious materials, and in every situation use proper procedures for minimizing exposure.

Common universal precaution methods include using Personal Protective Equipment (e.g., disposable clothing, gloves, face shield), disinfecting contaminated surfaces, disposing of contaminated materials properly in a biohazard container (Figure 4-19), and removing any contaminated clothing and isolating it so it can be disposed of or properly laundered.

Along with the universal precaution strategy, there are some basic procedures to follow such as hand washing, not eating or drinking in work areas where bloodborne pathogens might be present, and practicing proper decontamination or sterilization procedures.

Hand washing is a critical practice to prevent transmission of bloodborne pathogens. Always wash your hands immediately after removing any gloves or other Personal Protective Equipment (such as face shields or aprons).

If you are ever exposed to a bloodborne pathogen, thoroughly wash your hands and other exposed skin immediately, using a soft, antibacterial soap. (Figure 4-20)

FIGURE 4-20 Wash with a Non-Abrasive Antibacterial Soap to Help Prevent the Transmission of Bloodborne Pathogens

Avoid using a harsh, abrasive soap since it could damage your skin or open existing sores or wounds, thereby increasing the risk of infection.

Your facility should provide hand washing-equipment and supplies. Familiarize yourself with hand washing stations close to your work area. If you do not have access to such a station, you can use an antiseptic cleanser along with clean cloth or paper towels. You can also use antiseptic towelettes. Just make sure that you also wash your hands with soap and running water as soon as possible.

If your work area has any reasonable chance of exposure to bloodborne pathogens, you should never do any of the following while in that area:

- Eat
- Drink
- Apply cosmetics or lip balm
- Handle contact lenses
- Smoke

Universal Precautions were developed by the Center for Disease Control (CDC). OSHA 1910.1030(d)(1) provides for enforcement of Universal Precautions, stating that they "shall be observed to prevent contact with blood or other potentially infectious materials. Under circumstances in which differentiation between body fluid types is difficult or impossible, all body fluids shall be considered potentially infectious materials."

PERSONAL PROTECTIVE EQUIPMENT

Personal Protective Equipment (PPE) relating to medical procedures is special clothing and other equipment that is worn as a barrier of protection between an individual and any bloodborne pathogens in the workplace. Wearing appropriate medical-related PPE (and wearing it properly) should be your primary consideration when exposed to bloodborne pathogens. (Figure 4-21) For example, wearing latex gloves is a simple precaution for preventing blood or potentially infectious materials from contacting the skin.

Here are some guidelines to follow with PPE:

- Inspect your PPE thoroughly for wear and tear before putting it on.
- Make sure you know how to wear the PPE properly.
- Always remember to wear the appropriate PPE when exposed to bloodborne pathogens.
- Remove and replace any PPE that no longer forms an effective barrier between you and bloodborne pathogens.
- Remove your PPE before leaving your work area, and place it in appropriate bags or containers for decontamination or disposal (make sure you know where these bags or containers are before putting on the PPE).

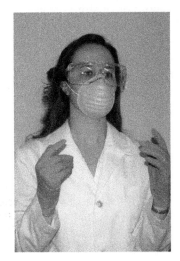

FIGURE 4-21 Proper PPE Should Always be Worn When Working with Blood or Other Body Fluids

Gloves are a vital form of PPE. Workers must wear gloves when their hands could come into contact with blood or other potentially infectious materials. Also, gloves must be worn when handling contaminated items (such as bloody bandages).

Gloves should be made of latex, nitrile, rubber, or any other material impervious to water. Latex or vinyl gloves are not completely impermeable to bloodborne pathogens, so it is acceptable to wear two sets of gloves if the material is thin or flimsy. This will provide an added layer of protection. If you have any cuts, sores, or damaged skin on your hands, be sure to bandage them before putting on any gloves.

Always inspect gloves for damage before use. You should remove contaminated gloves as soon as feasibly possible. When removing gloves, take care to avoid touching the outside of the glove with exposed skin. Immediately after removing your gloves, wash your hands thoroughly with soft, antibacterial soap. Dispose of contaminated gloves in a proper container.

Anytime there is a risk of blood or bodily fluids splashing, you should wear appropriate eye and face protection. Eye protection includes goggles and eyeglasses with solid side shields. Face protection includes face shields and masks. Many facilities provide prescription safety glasses (if not, you must provide your own), and also provide goggles or side shields. Goggles can prevent bloodborne pathogens from contacting your eyes and mucous membranes. Make sure the goggles fit securely against your face with no gaps. Eyeglasses with side shields do not provide the same protection as goggles.

A surgical mask can protect the nose and mouth, while a chin-length face shield protects the entire face. Make sure your mask or face shield is adjusted properly so it will not slide down or fall off. Eye protection such as glasses or goggles can be worn under a face shield for added protection.

Lab coats and aprons may be available to cover your clothing. If clothing becomes contaminated, remove it as soon as possible because fluids can seep through and come into contact with your skin. Contaminated laundry should be handled as little as possible, and it should be placed in an appropriately labeled bag or container until it is decontaminated, disposed of, or laundered.

Other PPE can include mouthpieces, resuscitation bags, pocket masks, or other ventilation devices. If you are required to perform CPR, make sure to use a disposable mouthpiece or similar device. (Figure 4-22) Most of this PPE is included in first aid kits at a facility.

Your facility should provide instructions on what PPE is required for first aid, how to properly wear it, where to place PPE when you leave the work area, and more.

If you ever experience a situation where you do not have appropriate PPE but must react quickly, you can improvise by using plastic, a towel, or other materials to add a barrier between you and the blood or bodily fluids.

Before entering a work environment that requires PPE you have never worn before, make sure you know how to wear it properly. Also, locate the bags or containers where you will place the PPE after its use. Placing PPE in bags or containers allows the plant to wash, decontaminate, dispose of, or store the gear properly.

EXPOSURE

Employers who can reasonably anticipate that employees may come into contact with blood are required under the OSHA Bloodborne Pathogen standard to create a written Exposure Control Plan. The plan is designed to eliminate or minimize employee exposure to bloodborne pathogens.

FIGURE 4-22 CPR Face Shields Help Prevent the Transmission of Bloodborne Pathogens Through Saliva

The Exposure Control Plan contains sections on the following:

- Exposure determination
- Methods of compliance
- Hepatitis B vaccination and post-exposure evaluation and follow-up
- Communication of Hazards to Employees
- Record keeping

You should be able to obtain a copy of a facility's Exposure Control Plan from the facility's designated Compliance Officer.

If you are exposed to blood or other potentially infectious materials, you should act as quickly as feasibly possible. Remove any PPE or garments that are soaked with fluids as soon as possible. If you get any bodily fluids on your skin, thoroughly wash the affected area with running water and soft, antibacterial soap. If you get blood or other fluids splashed in your eyes, flush your eyes with running water for at least 15 minutes. Flush splashes to the nose or mouth with running water as well.

After you have been exposed to blood or other potentially infectious materials, report the incident to your supervisor. This is called an exposure incident. According to the OSHA standard, an **exposure incident** is "a specific eye, mouth, other mucous membrane, non-intact skin, or parenteral (piercing mucous membranes or the skin barrier through such events as needle sticks, human bites, cuts, and abrasions) contact with blood or other potentially infectious materials that results from the performance of an worker's duties."

After reporting the incident, contact a healthcare professional. Your employer will ask you and the source individual to consent to blood testing as soon as possible after the exposure incident. You can also seek risk counseling.

DECONTAMINATION AND HOUSEKEEPING

All surfaces, tools, equipment, and other objects that come in contact with blood or potentially infectious materials must be decontaminated and sterilized. Also, any PPE that is soiled with blood and other infectious materials must be handled and cleaned or disposed of properly.

Some process technician jobs require basic housekeeping tasks such as cleanup. If you must perform these tasks as part of your job, make sure you familiarize yourself with the Bloodborne Pathogens standard sections that cover decontamination and housekeeping.

Summary

Biological hazards are hazards caused by living organisms, or parts or products of organisms that can generate an adverse response when they come into contact with the body.

The most common biological hazards a process technician might face are microorganisms (e.g., viruses, bacteria, and fungi), arthropods (e.g., spiders and insects), poisonous snakes (e.g., rattlesnakes, cottonmouths, copperheads, and coral snakes), plant allergens and toxins (e.g., pollen, poison ivy, and poison oak), and protein allergens (e.g., urine, feces, blood, hair, or dander from vertebrate animals).

Another hazard process technicians may face is bloodborne pathogens. The most common way for a process technician to be exposed to these types of pathogens is through the administration of first aid.

To prevent the transmission of bloodborne pathogens, technicians should always use universal precautions, the premise of which is to always treat blood or other bodily fluids or materials as if they are infectious, and avoid any direct contact with them.

In addition to avoiding contact with infectious materials, technicians should also wear proper personal protective equipment, wash their hands thoroughly, avoid eating, drinking, or smoking in areas where bloodborne pathogens might be present, and practice proper decontamination and sterilization procedures.

In the event of exposure, employees should always notify their supervisor and obtain immediate medical attention.

Checking Your Knowledge

1. Define the following terms:
   a. Allergen
   b. Arachnid
   c. Arthropod
   d. Biological hazard
   e. Bloodborne pathogen
   f. Exposure incident
   g. Microorganism
   h. Pathogen
   i. Protein allergen
   j. Toxin
   k. Universal precautions
   l. Venom

2. What type of biological hazards are viruses and bacteria?
   a. Arthropods
   b. Microorganisms
   c. Animal allergens
   d. Plant allergens

3. *(True or False)* Antibiotics do not cure viral infections.

4. Penicillin, a product of a microorganism, is made by a/an:
   a. Virus
   b. Bacteria
   c. Algae
   d. Fungus

5. The most unique feature of a brown recluse spider is a/an _____-shaped mark on its head.

6. Mosquitoes can carry which of the following diseases? (select all that apply)
   a. St. Louis encephalitis
   b. Cryptosporidium
   c. Anaphylaxis
   d. West Nile virus

7. Name the four types of poisonous snakes in the U.S.

8. *(True or False)* Dry and windy weather are conditions that promote the distribution of plant allergens.

9. List five types of protein allergens.

10. _____ gloves can be a source of a protein allergen.

11. The OSHA Standard 29 CFR 1910.1030 addresses _____ _____.

12. HBV affects which major organ in the body?

13. HIV can lead to _____.

14. *(True or False)* There is a vaccine to prevent AIDS.

15. Define the phrase *universal precaution*.

16. Which government agency created a set of universal precaution guidelines?

17. What is a simple, but critical, practice you can perform to prevent transmission of blood-borne pathogens?

Student Activities

1. Use the library or the Internet to research dangerous arachnids, insects, and snakes common to your region. In a two-page report, describe the organisms, their habitats, behaviors, and how they are dangerous to people.

2. List the causes, symptoms, and preventive measures of HIV and HBV.

3. Locate a copy of the OSHA Bloodborne Pathogens standard (such as from the OSHA Web site, www.osha.gov) and familiarize yourself with it. Pay close attention to the sections on methodology, labels and signs, and training. Write a one-page summary of the standard.

CHAPTER

5

Equipment and Energy Hazards

Objectives

This chapter provides an overview of equipment and energy hazards in the process industries that can affect people and the environment.

After completing this chapter, you will be able to do the following:

- Discuss the equipment and energy hazards posed by certain activities performed in the process industries:

  Working with moving or rotating equipment

  Working with equipment that is pressurized, has elevated temperatures, or emits radiation

  Working with energized equipment (powered by electricity or other power source)
- Describe government regulations and industry guidelines that address equipment and energy hazards.

Key Terms

Ampere (amp)—a unit of measure of the electrical current flow in a wire; similar to "gallons of water" flow in a pipe.

Arc—a spark that occurs when current flows between two points (contacts) that are not intentionally connected. See *spark*.

Conductor—a substance or body that allows a current of electricity to pass continuously along it.

Electricity—a flow of electrons from one point to another along a pathway, called a conductor.

Energy—the ability to do work.

Hydraulic—the use of liquid (hydraulic fluid) as the power source.

Insulator—a device made from a material that will not conduct electricity; the device is normally used to give mechanical support or to shield electrical wire or electronic components.

Ionizing radiation—radiation that contains enough energy to cause atoms to lose electrons and become ions.

Kinetic energy—energy associated with mass in motion.

Lockout/tagout—OSHA-mandated procedures for controlling hazardous energy.

Machine guard—a barrier that prevents a machine operator's hands or fingers from entering into the point of operation.

Nip point—a dangerous area where contact is made between two points on the equipment (e.g., a belt meeting a pulley or two gears intermeshing); also called a pinch point.

Non-ionizing radiation—low-frequency radiation that does not have enough energy to convert atoms to ions.

Ohm—a measurement of resistance in electrical circuits.

Pneumatic—the use of air pressure or a gas as the power source.

Point of operation—the area where the equipment actually performs its intended task (e.g., cutting, rotating, stamping).

Potential energy—stored energy; energy that has the potential to become kinetic.

Pressure—the force exerted on a surface divided by its area.

Radiation—the transfer of heat energy through electromagnetic waves.

Short circuit—a short circuit occurs when electrons in a current flow find additional unwanted paths outside of the intended circuit or conductor, and flow to it.

Spark—a single burst of electrical energy. See *arc*.

Static electricity—electricity "at rest"; an electrical charge caused by friction between two dissimilar materials.

Temperature—the degree of hotness or coldness that can be measured by a thermometer and a definite scale.

Volt—one volt is the electromotive force that will establish a current of one amp through a resistance of one ohm.

Voltage—the driving force needed to keep electrons flowing in a circuit.

Introduction

Process technicians work with many different types of equipment on a daily basis. Each type of equipment presents its own set of potential hazards. Not all hazards will apply to each type of equipment.

This chapter addresses some of the general hazards posed by different types of equipment. Electricity and the nature of electricity are also described so that process technicians can better understand it and its potential hazards.

Later chapters address the various types of controls (e.g., engineering, administrative, and Personal Protective Equipment) that can be used to reduce or eliminate these hazards.

Equipment and Energy Hazards Overview

Equipment can pose a variety of hazards. These hazards can be physical, chemical, ergonomic, or biological. Physical hazards may include rotating parts, cutting or punching motions, electricity, excessive noise, excessive vibrations, extreme **pressures,** extreme **temperatures,** or excessive radiation. Chemical hazards come from toxic, harmful, or corrosive substances such as cleaning solvents, lubricants, or process fumes. Ergonomic hazards are the result of repetitive or awkward motions during operations or maintenance. Biological hazards come from exposure to bacteria, fungi, and viruses on working surfaces, or in improperly treated cooling water.

Hazards presented by these types of equipment can be the result of any of the following:

- Incorrect design or construction of equipment
- Insufficient or missing guarding on moving or rotating parts
- Poor or improper maintenance or modifications
- Incorrect or unsafe operation
- Using the equipment for purposes or under conditions other than intended

Other potential equipment hazards include fire and explosion, work area and height hazards, and noise and hearing, although this is by no means a complete list. Other chapters in this textbook cover these types of potential hazards.

EQUIPMENT HAZARDS
Moving or Rotating Equipment Hazards

Many types of equipment in the process industries have moving or rotating parts. This equipment presents hazards that can cause injuries ranging from minor to life threatening. These injuries include the following:

- Pinches
- Strains and sprains
- Cuts, punctures, or tears of skin, from minor to severe
- Sheared or severed appendages (e.g. fingers, hands, arms, toes, legs, and feet)
- Crushed or broken appendages or body parts
- Blindness
- Strangulation

Hazardous equipment motions can include rotating, reciprocating, and cutting. Some equipment is in continuous motion, while other equipment operates intermittently. Also, some equipment may move accidentally.

The **point of operation** on a piece of equipment is the area where the equipment actually performs its intended task (e.g. cutting, rotating, and stamping). This area is extremely hazardous.

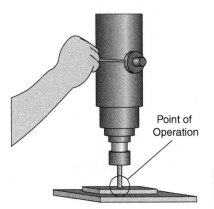

Point of
Operation

FIGURE 5-1 Point of
Operation

Power transmission is another potentially hazardous area, since it involves moving parts such as pulleys, gears, chains, and cranks.

Nip points (pinch points) are areas where contact is made between two points on the equipment, such as a belt meeting a pulley or two gears intermeshing. Nip points can crush fingers or other appendages, or potentially grab any loose articles of clothing or hair and pull the worker into the equipment. Other types of hazards can come from flying chips, sparks, or broken equipment parts.

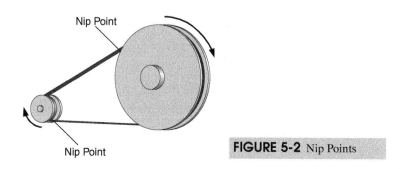

FIGURE 5-2 Nip Points

Various safeguards are used to protect process technicians from the hazards of moving or rotating equipment parts. One safeguard is called machine guarding (OSHA C.F.R. 1910 Subpart O). A **machine guard** is a barrier that prevents a machine operator's hands or fingers from entering into the point of operation.

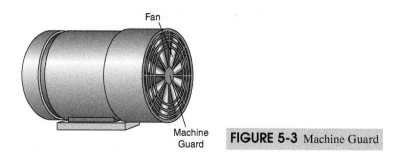

FIGURE 5-3 Machine Guard

OSHA requires that any machine part, function, or process that can cause injury must be safeguarded. If normal operations or accidental contact can cause injury to the equipment operator or others nearby, the hazard must be controlled or eliminated.

PRESSURE HAZARDS

Pressurized equipment presents potential hazards such as leaks, vibration, rupture, or explosion. Process technicians can be exposed to hazardous materials during a pressure-related accident. Vacuum, high pressure, compressed gases, runaway reactions, and so on can cause pressure hazards.

More detail about pressure and pressurized equipment hazards is provided in Chapter 7, *Pressure, Temperature, and Radiation Hazards*.

TEMPERATURE HAZARDS

Equipment can operate at extremely hot or cold temperatures. The human body reacts in a variety of ways to such extreme temperatures. For example, if the equipment generates a considerable amount of heat in an enclosed area, a process technician may develop heat stress after working around the equipment for too long. If the equipment operates under extreme cold, a process technician's skin can be severely damaged if it comes into contact with the surface of the equipment. Exposed skin can be burned by hot or cold surfaces.

Additional details on temperature and temperature-related equipment hazards are provided in Chapter 7, *Pressure, Temperature, and Radiation Hazards*.

RADIATION HAZARDS

Radiation is the process by which elements emit energy in the form of electromagnetic waves or small atomic particles. Government guidelines classify radiation into one of two categories: ionizing and non-ionizing.

Ionizing radiation is radiation that contains enough energy to cause atoms to lose electrons and become ions. As the ions are formed, the normal electrical balance in the atom changes and causes damage to living cells.

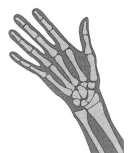

FIGURE 5-4 X-rays are a Form of Ionizing Radiation

Ionizing radiation takes different forms. Gamma rays are a highly penetrating form of ionizing radiation that can cause damage to body tissue. Alpha and beta particles, another form of ionizing radiation, can also cause damage to body tissues but do not travel as far as gamma rays. X-rays are another dangerous form of ionizing radiation. Ionizing radiation can cause headaches, increased risk of infection, fatigue, nausea/vomiting, sterility, mutations, cancer, and death.

Non-ionizing radiation is low-frequency radiation that does not have enough energy to convert atoms to ions. Non-ionizing radiation can take many forms, such as visible light, ultraviolet light, infrared light, microwaves, radio, and AC power. Although the effects of non-ionizing radiation are under debate, links between this type of radiation and cancer are being studied. However, non-ionizing radiation (depending on its form) can cause blisters, blindness, burns, heat stress, cataracts, skin cancer, and dry skin. The major potential hazards from non-ionizing radiation include electrical shock, fire, and explosion.

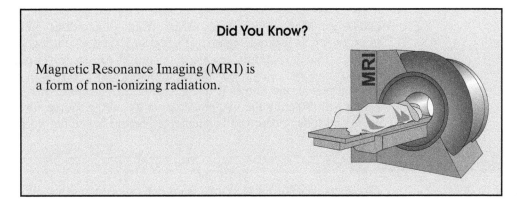

Did You Know?

Magnetic Resonance Imaging (MRI) is a form of non-ionizing radiation.

Some equipment that the process technician can encounter will emit radiation, either ionizing (e.g., radioactive sources used in instrumentation, X-ray devices, reactors) or non-ionizing (e.g. furnaces, flares, lasers, computer monitors). Exposure to radiation is regulated by government guidelines, including length of exposure, precautions, monitoring, warning signs/labels, and so on. Further details on radiation and radiation-related equipment hazards are provided in the *Pressure, Temperature, and Radiation Hazards* chapter.

ENERGY HAZARDS

Energy is defined as the ability to do work. Energy can take one of two major forms: kinetic and potential.

Kinetic energy is the term for energy in motion. Potential energy is the term for stored energy (i.e., energy that has the potential to become kinetic).

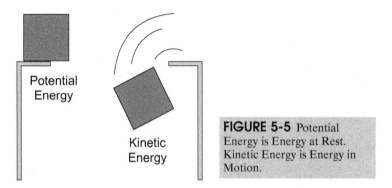

Potential Energy

Kinetic Energy

FIGURE 5-5 Potential Energy is Energy at Rest. Kinetic Energy is Energy in Motion.

Process technicians can face hazards from both kinetic and potential energy, especially when dealing with energy sources used to power equipment. Equipment can be powered in a variety of ways, using any of the following sources:

- **Electricity** (powered or operated by the flow of electrons)
- **Hydraulic** (powered or operated by fluid pressure)
- **Pneumatic** (powered or operated by air pressure)
- Steam (operated by stream pressure, such as a rotating steam turbine)
- Tension [powered or operated by a force (e.g., a spring) that pulls or stretches]
- Gravity (powered or operated by the force of attraction)
- Process (as in backflow through a valve into a rotating piece of equipment)

Energy sources can become hazardous if they are not controlled or maintained properly. Equipment hazards from energy sources are due to the force that the energy supplies. In other words, the energy puts the equipment's moving parts into motion. Energy sources can become hazardous if leaks or ruptures of hydraulics or pneumatics, or electrical shorts, occur. Also, maintenance or repair work can become hazardous if the energy source is not isolated from the equipment prior to the start of work.

GOVERNMENT REGULATIONS

OSHA established procedures for controlling hazardous energy, 29 CFR 1910.147 (also known as **lockout/tagout**). (Figure 5-6) The lockout/tagout procedure protects workers from the potential release of uncontrolled energy by isolating equipment from its energy source using locks, chains, and tags. This prevents the accidental startup of the equipment and allows it to be maintained, cleaned, or repaired when needed. Along with isolating the equipment from its energy source, any residual (remaining) energy must be released and documented before beginning work on the equipment.

Electrical Hazards

Electricity is a flow of electrons from one point to another along a pathway, called a conductor. **Conductors** are substances or materials that allow a current of electricity to pass continuously along them. The conducting path is called a circuit. Current is measured in units called **amperes**, which is a measure of the flow of electrons (similar to how you would measure liquid flow in gallons per minute). Electrical potential is the driving force needed to keep the electrons flowing in the circuit. This potential is called **voltage** and is similar to how you would describe the pressure of liquids flowing through pipes. A **volt** is the unit of electromotive force that will establish a current of one amp through a resistance of one ohm. An **ohm** is a measurement of resistance in electrical circuits. Within conductors, electrons can break free and flow more easily than other materials. Metals, as well as some types of hot gases (plasmas), and certain liquids (such as aqueous solutions), are good conductors. Materials that do not give up their electrons as easily are called **insulators**. Insulators are poor conductors, so they

FIGURE 5-6 Lockout/ Tagout Protects Workers from Uncontrolled Energy

can be used to maintain current flow in conductors. Air, rubber, and glass are examples of insulators.

Electrical circuits are designated by voltages and current capacity. Many circuits around the home or office environment are 120V, with some 240V circuits for heavier-duty electrical equipment. At process facilities, electrical systems can use voltages of 120V, 240V, and 480V. A 120V system can have a current capacity around 15-20 amps, while a 480V circuit can have a current capacity of 20 amps or higher (up to hundreds of amps).

Only a small amount of current is required to cause serious injury or even death. The amount of amps received during a shock determines the effects. As little as three amps can cause pain. Between three and ten amps causes pain, and the person may not be able to move. Twenty amps can cause a loss of muscle control and difficulty breathing. Over 50 amps can cause heart problems, nerve damage, muscular contractions, and death.

Another type of electricity is static electricity. **Static electricity** occurs when a number of electrons build up on the surface of a material, but have no positive charge nearby to attract them and cause them to flow. When the negatively charged surface comes into contact with (or comes near) a positively charged surface, current flows until the charges on each surface become equalized, sometimes creating a spark. Lightning and the shock that occurs from touching a doorknob after shuffling across the carpet are both good examples of static electricity. Static electricity is a hazard in the plant from the flow of non-conductive fluids and solids. Static electricity can result from equipment operation, forklift use, and other such sources.

Certain personal electronic devices are often banned in plants, such as cell phones, electronically charged music players, and so on. This is due to the potential spark or static electric discharge they might cause.

Electricity has the potential to be hazardous to people, causing injury or death. It can also damage equipment or upset processes, and can cause fires or explosions.

FIGURE 5-7 Frayed Electrical Wires can Short-Circuit and Cause Electrical Shock

CAUSES OF ELECTRICAL HAZARDS

Electrical hazards can be created by a variety of factors, including the following:

- Improper wiring or grounding
- Cracked or degraded insulation
- Improper operation of equipment
- Stored energy in capacitors or batteries
- Failure to follow lockout/tagout procedures
- Improper breaker operations
- Wet insulation
- Equipment failure
- Surges and overloads
- Downed power lines
- Static electricity buildup
- Lightning strikes

Short circuits are a common cause of electrical hazards. A short circuit occurs when electrons in a current flow find additional unwanted paths for current flow, outside of the intended circuit or conductor, and flow to them. For example, if the insulation is cracked on two wires in close proximity to each other, the electrons jump between the wires and create a "short" circuit (a shortcut outside the intended circuit). Sunlight, age, water, extreme heat or cold, and even animals (insects, birds) can cause cracks or splits in insulation materials.

Stray currents can develop when electrical lines "leak" or when batteries interact with conductive materials. Stray currents can corrode and destroy any conductive materials through which they flow. They can also cause arcs or shocks.

Water, because it is a good conductor, must also be considered when dealing with electricity. Water decreases the resistivity of materials to electricity. For example, dry

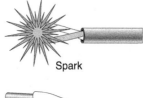

Spark

Arc

FIGURE 5-8 Illustration of a Spark vs. an Arc

skin has a resistance of up to 100,000 ohms. Wet skin, however, reduces resistance to as low as 450 ohms. Even sweat on the skin can decrease resistance.

SPARKS AND SHOCKS

Electricity presents two main types of hazards:

- Sparks that can lead to fire or an explosion
- Shock

A **spark** is a single burst of electrical energy. An **arc** is a spark that occurs when current flows between two points (contacts) that are not intentionally connected. The amount of current flowing and the distance traveled factor into the strength of the arc. Sparks and arcs can both cause flammable materials (such as fuel vapors) to ignite, resulting in a fire or explosion.

In hazardous environments, special wiring and equipment is required to prevent fires and explosions, since switches, lights, motors and even telephones produce a tiny arc as part of their normal operation. Wiring can also become overheated and cause an electrical fire.

An electrical shock occurs when a person comes into contact with electrical current. Electrical shock and other electrical hazards occur when a person contacts a conductor carrying electricity while also touching the ground or an object that has a conductive path to the ground. The person completes the circuit as current passes through the body.

Electrical shocks can occur when a person does any of the following:

- Comes in contact with a bare wire (either intentionally bare or as a result of cracked, worn, or damaged insulation)
- Uses improperly grounded electrical equipment
- Works with electrical equipment in a wet or damp environment, or while sweating heavily
- Works on electrical equipment without checking that the power source has been turned off and that all conductors have been tested to ensure they are de-energized
- Uses long metal equipment, such as cranes or ladders, that can come into contact with a power source

Electrical shock can cause burns, cardiac arrest, ventricular fibrillation (rapid, irregular contractions of the heart), muscle damage, cessation of breathing, or death. (Figure 5-9)

Three factors determine the seriousness of the shock and the extent of injury:

- Amount of the current (measured in amps), affected by the voltage and body resistance (wet versus dry skin)
- Path of the current, where the current enters and exits the body (the chest cavity and head are most sensitive)
- Length (duration) of time that the current flows

Other factors such as moisture (e.g., a damp environment or perspiration) and wounds or open cuts can increase the effects of electrical shock.

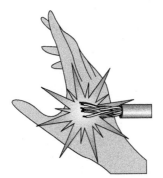

FIGURE 5-9 Electrical Shock can Result in Serious Injury or Death

STATIC ELECTRICITY

A common example of static electricity is walking on carpet and touching someone, creating a spark upon contact. Another example is inserting a key into the lock on a door and seeing a spark jump between the key and the door lock.

Static electricity can cause minor shocks. It can also damage sensitive electronics. Low humidity (dry conditions), the types of materials involved, and friction can increase the likelihood of static electricity buildup. If enough of a charge builds up, it can present a serious hazard by creating a spark. Sparks in a hazardous environment could cause a fire or explosion.

Even though static electricity buildup is best known from when solid materials interact, liquids (such as those used or created as part of a process) can also accumulate static electricity as they flow down a pipeline or hose, or are discharged from a nozzle into a container. As non-conductive fluids are poured, pumped, flowed, or allowed to free fall (like from the top of a tank or drum), these actions can charge the liquid with static electricity. Many fluids in the process industries are flammable, and a spark from static electricity can ignite a fire. Methods such as grounding, bonding, dip lines, and reduced turbulence can be used to prevent static electricity in these situations.

Dust and powders can also become charged with static electricity as they come into contact with each other and there is friction between the particles. Differences in particle makeup and size affect the buildup of the charge. Gases do not charge, but static electricity can build up around gases that are used in connection with the transport of liquids or solids. This can create a fire hazard if the materials are flammable.

Also, static shock hazards can occur when a person reacts to the shock, causing involuntary movement such as falling or jumping into the path of a moving device.

Static electricity can build up during different processing methods, such as the following:

- Blending
- Mixing
- Agitation
- Spraying
- Coating
- Filling/Flowing
- Power transmission
- Sand blasting
- Water blasting

Electrical Safety

Following are some general guidelines for ensuring electrical safety:

- Follow all lockout/tagout procedures thoroughly.
- Treat dead circuits as if they are live.
- Wear proper clothing and protective equipment including safety glasses or goggles and shoes with non-conductive (e.g., rubber) soles. See Chapter 24, *Personal Protective Equipment*, for more information.
- Verify there is no water near electrical equipment.
- Do not wear metal jewelry or rings around electrical equipment.
- Do not use metal ladders around electricity.
- Use insulated metal tools when working with electricity. In hazardous (e.g., flammable) locations, use intrinsically safe or explosion-proof tools and lights.
- Keep electrical equipment clean and free of dust, dirt, and oil; inspect electrical equipment regularly for wear.
- Make sure doors to electrical equipment are securely closed and locked.
- Never remove or tamper with grounding prongs on electrical equipment.
- Inspect power tools for wear or damage before use, and operate them properly (See Chapter 11, *Construction, Maintenance, and Tool Hazards*).

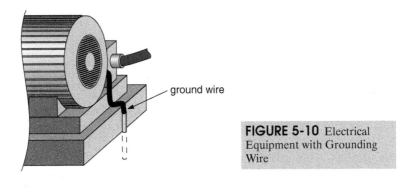

ground wire

FIGURE 5-10 Electrical Equipment with Grounding Wire

An understanding of fire and explosion hazards, types of fires, and fire extinguishers is also important. See Chapter 6, *Fire and Explosion Hazards,* for details.

GROUNDING, FUSES, AND CIRCUIT BREAKERS

Proper grounding, or establishing a connection between an electrical circuit and the ground (earth), is one of the most important ways to prevent electrical hazards. Sometimes, electrical equipment can become grounded unintentionally; this is called a ground fault. This can occur if insulation becomes damaged. A Ground Fault Circuit Interrupter (GFCI) provides protection against ground faults. Ground faults can either be built into the circuit, or plugged in. Electrical equipment is then plugged into the GFCI.

If a current exceeds the design amount of an electrical circuit, or an electrical circuit becomes damaged, the circuit can overload. When this occurs, fuses and circuit breakers break the circuit (open it) and cut off the flow of electricity. This reduces the risk of fire and explosion.

When a fuse blows or a circuit trips, this indicates that something is wrong, such as an overloaded circuit or short circuit. Do not start current flow again by replacing the fuse or resetting the circuit breaker until the problem is diagnosed and solved.

If a fuse needs to be replaced or a circuit breaker needs to be reset, follow these general tips:

- Make sure all power supplies to the equipment are turned off.
- When replacing a fuse, make sure to use the type and size recommended by the manufacturer.
- Examine circuit breakers carefully for damage before attempting to reset them.
- When opening or closing a circuit, stand to one side of the switch and operate it in one swift motion.

FIGURE 5-11 Electrical Breaker Box

• Visually inspect terminals and conductors for signs of damage (e.g., melting, discoloration).

Hazardous Environments

Some environments, such as those that contain flammable materials or hazardous gases, require special electrical safety measures. Electrical systems in such areas must be intrinsically safe or explosion proof, meaning that they generate only low levels of electrical energy. Some structures use explosion-proof enclosures to prevent the hazardous materials or atmospheres outside it from becoming ignited (sparks are restricted, heat sources are reduced, and explosions are contained). Intrinsically safe and explosion-proof tools and light sources (e.g., flashlights) must always be used in these hazardous environments.

Lightning

Lightning is an extremely destructive, naturally occurring electric current created when atmospheric electricity is discharged. Lightning, with its very high voltage and current, can result in damage to equipment (especially electrical equipment), and facilities. It can cause fires, or even explosions. If a person is struck by lightning, it can cause severe injuries or death. Chapter 13, *Natural Disasters and Inclement Weather*, discusses lightning in more detail.

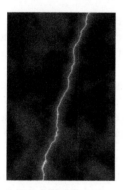

FIGURE 5-12 Lightning
Courtesy of the FEMA Photo Library

Government Regulations

All personnel working with electrical equipment must be properly trained and authorized for the particular type and level of work being performed. OSHA procedure 29 CFR 1910.332 covers electrical training for unqualified persons. Only qualified personnel can work on electrical systems. In most cases, process technicians are not required to perform work on high-voltage systems.

The National Fire Protection Association (NFPA), an international non-profit organization, publishes the National Electrical Code, a set of guidelines covering electrical installation, wiring requirements, and safety. In the process industries, the codes relating to electricity in flammable environments are extremely important. Electrical

FIGURE 5-13 Electrical Wires use a Standard Color Code to Differentiate Hot, Neutral, and Ground Wires

wires use standard color-coding to indicate hot (red or black), neutral (white or gray), and ground (green).

Summary

Process technicians work with many different types of equipment on a daily basis. When working with this equipment, technicians may be exposed to a variety of physical hazards, such as rotating parts, cutting or punching motions, excessive noise, excessive vibrations, extreme pressures, and extreme temperatures. Technicians may also be exposed to chemical, ergonomic, and biological hazards. Technicians must be aware of these hazards so they can prevent or eliminate exposure to them whenever possible through the use of engineering and administrative controls and personal protective equipment.

Checking Your Knowledge

1. Define the following terms:
   - a. Arc
   - b. Conductor
   - c. Insulator
   - d. Ionizing radiation
   - e. Kinetic energy
   - f. Lockout/tagout
   - g. Machine guarding
   - h. Nip point
   - i. Non-ionizing radiation
   - j. Physical hazard
   - k. Point of operation
   - l. Potential energy
   - m. Spark
   - n. Static electricity
2. List five hazards associated with rotating equipment.
3. A machine guard is used to:
   - a. Prevent technicians from damaging the blades on a machine.
   - b. Prevent a machine from being powered up during lockout/tagout.
   - c. Prevent a machine operator's hands or fingers from entering the point of operation.
4. Ionizing radiation is radiation that:
   - a. Contains enough energy to convert atoms to ions.
   - b. Does not contain enough energy to convert atoms to ions.
   - c. Poses no real threat to humans.
5. A rubber band stretched between two fingers is an example of:
   - a. Kinetic energy
   - b. Potential energy
6. Explain the purpose of lockout/tagout.
7. List two factors that determine the severity of an electric shock and the extent of the injuries that shock might cause.
8. List five processing methods that can produce static electricity.

Student Activities

1. All new cell phones come with an owner's manual and a warning label. Obtain a copy of one of these manuals and review the warning. Write a paragraph explaining what you learned.
2. Research the lockout/tagout process. Write a two to three-page paper explaining what you learned.
3. Use the Internet to research industrial accidents that have occurred due to energy hazards. Discuss the findings with your fellow students.

6

Fire and Explosion Hazards

Objectives

This chapter provides an overview of fire, explosion, and detonation hazards in the process industries, which can affect people and the environment.

After completing this chapter, you will be able to do the following:

■ Recognize specific physical hazards present in the process industries and explain the following potential safety, health, and environmental hazards:

Fire

Explosions

Detonation

■ Describe government regulations and industry guidelines that address fire and explosion hazards.

Key Terms

Chain reaction—a series of reactions in which each reaction is initiated by the energy produced in the preceding reaction.

Combustion—the process by which substances (fuel) combine with oxygen to release heat energy, through the act of burning (oxidation).

Combustion point—the ignition temperature at which a fuel can catch on fire.

Conduction—the transfer of heat through matter via vibrational motion.

Convection—the transfer of heat through the circulation or movement of a liquid or a gas.

Deflagration—a process of subsonic combustion that usually propagates through thermal conductivity (i.e., hot burning material heats the next layer of colder material and ignites it).

Detonation—a violent explosion that generates a supersonic shock wave and propagates through shock compression.

Explosion—a sudden increase in heat energy, released in a violent burst.

Fire—a type of combustion, resulting from a self-sustaining chemical reaction.

Fire point—the temperature at which burning is self-sustaining after removal of an ignition source.

Fire tetrahedron—the elements of a fire triangle (fuel, oxygen, and heat) combined with a fourth element, a chain reaction that keeps the fire buring.

Fire triangle—the three elements (fuel, oxygen, and heat) that must be present for a fire to start.

Flammable (inflammable)—the ability (inability) of a material to ignite and burn readily.

Fuel—any material that burns; can be a solid, liquid, or gas.

Volatility—the ability of a material to evaporate.

Introduction

Although fire, explosions, and detonations rarely occur in the process industries, the results can be devastating when they do happen. It is vital for process technicians to understand the characteristics, hazards, and prevention of fire, explosions, and detonations, along with the government regulations that address them.

This chapter provides an overview of fire, explosion, and detonation hazards that process technicians could encounter in the workplace. Later chapters describe related topics such as flame-resistant clothing and firefighting systems and equipment.

Fire and Explosion Hazards Overview

Fires and explosions are some of the most dangerous hazards process technicians can potentially face. The process industries deal with many hazardous materials that pose fire and explosion hazards. Although efforts are made to prevent fires and explosions at process facilities, process technicians must understand these potential hazards and the terminology used to describe them.

Combustion (or incineration) is the process by which substances (fuel) combine with oxygen to release heat energy, through the act of burning. During combustion, fuel is consumed to give off heat, light (flame), gases, and smoke. Controlled combustion is a key component in many processes. It is used to do such things as power engines, fire heaters and furnaces, generate steam, and cook food.

Fire is a type of combustion that results from a self-sustaining chemical reaction. During a fire a fuel in the presence of oxygen (air) is subjected to heat or some other energy source (ignition point) that results in combustion. Heat energy continues to release until all of the combustible fuel is consumed.

An **explosion** is a sudden increase in heat energy, released in a violent burst. Explosions are usually associated with high temperatures and the release of gases. Explosions cause pressure waves that travel outward. In technical terms, explosions are categorized as deflagrations if the pressure waves are subsonic (i.e., do not travel faster than sound) or detonations if they are supersonic (i.e., travel faster than the speed of sound).

Deflagration is a process of subsonic combustion that usually propagates through thermal conductivity (i.e., hot burning material heats the next layer of colder material and ignites it). Deflagrations burn with great heat and intense light. **Detonations** are violent explosions that generate a supersonic shock wave and propagate through shock compression. Detonations are extremely destructive and produce high-velocity (fast) pressure waves, called shockwaves. Detonations are characterized by their rapid reaction and the high pressure they produce.

FIGURE 6-1 Deflagration

Comparison of Detonation and Deflagration Pressure Levels

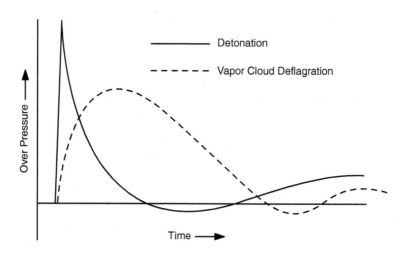

Fire Hazards

Fires pose many health hazards, including burns, suffocation, and exposure to smoke and toxic fumes. Most fire-related deaths are caused by suffocation or breathing smoke and toxic fumes. During combustion, materials are broken down into their basic elements that then form new compounds. Carbon is released when organic compounds are burned. The carbon produced during the fire then combines with oxygen to form carbon monoxide or carbon dioxide. If inhaled, these can deprive a body of oxygen. Carbon monoxide is the primary killer, causing suffocation. During a fire, lethal levels of carbon monoxide can be generated in a short amount of time, replacing the oxygen needed for respiration. Most fires produce more carbon monoxide than carbon dioxide. Fires also generate smoke, composed of gases, air, and suspended particles (due to incomplete combustion). All of these gases and toxic fumes present breathing hazards and create dangerous atmospheres that can cause injury or death.

Depending on the type of fuel burned by the fire, other dangerous by-products can be formed. Table 6-1 Provides some examples of some dangerous by-products and the substances that produce them.

Released gases can also present an explosion hazard if they come into contact with an ignition source. Burns from acids are also a potential hazard during a fire. For

TABLE 6-1   Dangerous By-Products and Their Sources

By-Product	Produced When the Following Fuels are Burned
Hydrogen sulfide	Compounds containing sulfur, such as rubber and crude oil
Hydrogen chloride	Polyvinyl chloride (PVC)
Hydrogen cyanide	Polyurethane
Sulfur dioxide	Sulfur containing compounds
Ammonia	Refrigerants, hydrogen-nitrogen compounds, and nylon
Carbon monoxide and carbon dioxide	Combusted carbon and oxygen

example, sulfur dioxide or nitrogen oxides can combine with water to create sulfuric acid or nitric acid.

Process technicians must understand the elements required for a fire to start, how to prevent fires and control them, and the combustible and **flammable** properties of the materials they are working with.

Flames and heat from fire pose hazards such as burns or death. Burns prevent the normal functioning of skin, hindering its ability to protect vital internal organs and disrupting the other functions. The severity to which a burn injures or damages the skin ranges from first degree (less severe) to third degree (most severe). These are discussed in greater detail elsewhere in the book.

CHARACTERISTICS OF FIRES

Fire is an exothermic oxidation process. Exothermic means that heat is generated, and oxidation is a process that occurs when oxygen is combined with another substance (usually carbon in a fire).

Along with heat, fires produce light (flames), smoke, toxic fumes, and other substances (such as ash and soot). Plumes of flame and smoke rise above the fire; the fire can spread based on wind direction, type and amount of fuel, temperatures, conditions (dry, wet), and other factors.

FIRE TRIANGLE AND FIRE TETRAHEDRON

As described previously, a fire starts when a substance (**fuel**), in the presence of air (oxygen), is heated to an ignition point (heat), resulting in combustion. Fire must have all of these elements (fuel, oxygen, heat) present to start.

Did You Know?

Many common devices (e.g., cars, gas appliances, wood stoves, and cigarettes) produce carbon monoxide as a by-product of combustion. If these devices are used improperly they can pose serious health risks, including death by suffocation.

Suffocation occurs because carbon monoxide (CO) binds with hemoglobin in the blood, which prevents oxygen from being transported to the cells.

Symptoms of low-level carbon monoxide poisoning are similar to those of the common flu.

Photo courtesy of CDC
Public Health Image Library
(PHIL)

The following three elements are referred to as a **fire triangle**:

1. **Fuel**—Fuel is any material that burns; it can be a solid, a liquid, or a gas.
2. **Oxygen**—Air is composed of 21% oxygen; generally, fire needs only 16% oxygen to ignite.
3. **Heat**—The energy or heat required by a fuel to produce ignition.

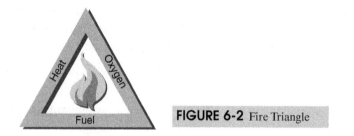

FIGURE 6-2 Fire Triangle

Fire must have all three elements (fuel, oxygen, heat) present to start. Removing one or more of these elements will extinguish a fire.

The fire triangle represents the elements necessary to create a fire. Once a fire has started, the fire tetrahedron represents the elements necessary to sustain combustion.

The **fire tetrahedron** consists of the components of the fire triangle and another component – the chain reaction.

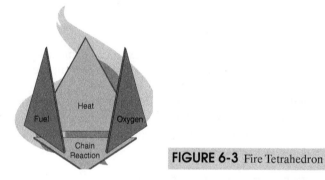

FIGURE 6-3 Fire Tetrahedron

A **chain reaction** is a series of reactions in which each reaction is initiated by the energy produced in the preceding reaction (e.g., when the first domino is knocked over it causes the second domino in the series to topple. This, in turn, makes the third domino topple, and so on). This type of reaction occurs when fuel, oxygen, and heat come together in proper amounts under certain conditions. Chain reactions are what cause fires to build on themselves and spread. For a fire to continue there must be a constant fuel, oxygen, and heat source.

To stop a fire, one of the four components of the fire tetrahedron must be removed. If fuel is removed, the fire will have nothing to sustain itself. If the atmosphere is rendered inert by removing the source of oxygen, the fire will cease. If the fire is cooled, the elevated temperatures will reduce and the chain reaction will stop. Some extinguishing agents stop a fire not by removing fuel, heat, or oxygen, but by preventing the chain reaction from occurring.

FUEL SOURCES

Fires need fuel in order to burn. Given a high-enough temperature, almost any substance can become fuel for a fire. The **combustion point** is the ignition temperature at which a fuel can catch on fire. Fuels for fires can be solids, liquids, or gases.

Many materials used in the process industries are extremely flammable or combustible.

Following are some examples of solid fuels:

- Wood
- Office furniture

Did You Know?

Static electricity and sparks can ignite hydrocarbon vapors. For example:

- Filling a portable gas can in the bed of a pickup truck can produce static electricity that can cause the container to ignite.
- Talking on a cell phone around gasoline can cause sparks which cause the gasoline vapors to ignite.
- Getting in and out of the car during fill-up can produce static electricity that can ignite gasoline vapors.

- Smoking a cigarette while working around a car can lead to fire and explosion.

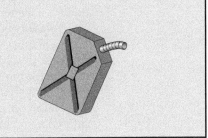

- Cardboard, paper, and packing materials
- Fabric, such as cotton and nylon
- Rubber
- Plastic

Following are some examples of liquid fuels:

- Gasoline
- Kerosene
- Benzene
- Isopropyl alcohol

The fuel for fire that comes from liquids is actually the vapors above the liquids, since liquids cannot burn. When the temperature of a flammable liquid rises, more vapors increase on its surface. Thus, it is critical for process technicians working with flammable liquids to understand the properties of the liquid and the temperatures at which the liquids are handled or stored.

Following are some examples of gas fuels:

- Propane
- Butane
- Methane

HEAT SOURCES AND IGNITION

A wide range of heat sources can produce the necessary ignition spark to start a fire. These include the following:

- Open flames (e.g., furnace, pilot light, or lit cigarette)
- Hot surfaces (e.g., catalytic converter)
- Electrical (e.g., frayed wiring, arcing, lightbulbs, and heating elements)
- Sparks (e.g., from welding or grinding)
- Friction (e.g., from brake pads)
- Static electricity (e.g., lightning)
- Chemical reactions (e.g., combining hydrochloric acid and caustic soda)

OXYGEN

Oxygen (O_2) is naturally present in most environments; air is composed of 21% oxygen. A fire needs only 16% oxygen to ignite. Since oxygen is usually readily available as a fire triangle component, the mishandling of fuel or a heat source typically starts fires. Other gases, such as chlorine or nitric oxide, can also substitute for oxygen as part of the fire triangle and fire tetrahedron.

FLAMMABLE AND COMBUSTIBLE SUBSTANCES

A number of flammable substances exist as either a liquid or a gas, depending on temperature and pressure. According to OSHA, a flashpoint is the "lowest temperature at which a flammable liquid will give off enough vapors to form an ignitable mixture with the air above the surface of the liquid or within its container." The flashpoint typically indicates the liquid's susceptibility to ignition. The **fire point** is the temperature at which burning is self-sustaining after removal of an ignition source. Normally, the fire point is a higher temperature than the flashpoint.

Volatility is how easily a liquid will evaporate, and liquids are said to have light (or high) volatility or heavy (or low) volatility. A liquid's boiling point can change due to pressure, so a 10% point is used to indicate the temperature at which 10% of the liquid has changed into a gas. These factors are used to understand the properties of liquids and classify them according to a scale.

Flashpoint, along with boiling point, is used to classify liquids. Liquids can be considered flammable or combustible. OSHA states:

"Flashpoint was selected as the basis for classification of flammable and combustible liquids because it is directly related to a liquid's ability to generate vapor, i.e., its volatility. Since it is the vapor of the liquid, not the liquid itself that burns, vapor generation becomes the primary factor in determining the fire hazard. The expression "low flash - high hazard" applies. Liquids having flashpoints below ambient storage temperatures generally display a rapid rate of flame spread over the surface of the liquid, since it is not necessary for the heat of the fire to expend its energy in heating the liquid to generate more vapor."

Flammable liquids are liquids with a flashpoint less than 100 degrees F. Combustible liquids are liquids with a flashpoint over 100 degrees F. Table 6-2 compares the differences between flammable and combustible liquids.

TABLE 6-2 Comparison of Flammable vs. Combustible Liquids

Substance Classification	Description
Flammable	Liquids with a flashpoint of less than 100 degrees F; designated as Class I liquids that fall into one of three subdivisions (IA, IB, and IC)
Class IA	Flashpoint less than 73 degrees F; boiling point less than 100 degrees F
Class IB	Flashpoint less than 73 degrees F; boiling point greater than or equal to 100 degrees F
Class IC	Flashpoint between 73 and 99 degrees F
Combustible	Liquids with a flashpoint over 100 degrees F; designated as Class II or Class III, with Class III being divided into two subdivisions (IIIA or IIIB)
Class II	Liquids with a Flashpoint between 100 and 139 degrees F
Class III	Liquids with a Flashpoint over 140 degrees F
Class IIIA	Flashpoint between 140 and 199 degrees F
Class IIIB	Flashpoint greater than or equal to 200 degrees F

Since many flammable liquids are lighter than water, water cannot be used to put out such fires. A flammable liquid will float on the water and spread.

Figure 6-4 illustrates the different classes of flammable and combustible liquids and provides examples of substances from each category.

The process of filling tanks or containers with flammable liquids can cause static electricity buildup, due to a phenomenon related to the flow of fluids. Precautions must be taken to prevent a static electricity discharge, which could ignite the liquid's vapors (e.g., hydrocarbons are good insulators and readily accumulate a static charge). When dispensing Class I liquids, an electrical bonding (grounding) can be used to connect the hose nozzle to the container; this reduces the chance of a static electricity discharge.

FIGURE 6-4 Classes of Flammable and Combustible Liquids

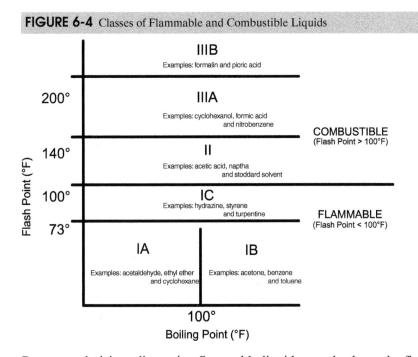

Process technicians dispensing flammable liquids can also keep the flow as smooth as possible by making sure the nozzle remains as close to the liquid's surface in the container as possible to avoid splashing. Always ensure that tanks and containers are grounded to prevent a buildup of static charge.

FLAMMABLE ATMOSPHERES

The vapors from flammable and combustible liquids, when combined with oxygen, have a flammability range (based on percent of fuel in the air) that is specific to each material. If the amount of fuel vapors falls below this range, then the mixture of oxygen and vapors is considered too lean to burn. If the amount of fuel vapors rises above this range, then the mixture is said to be too rich to burn.

OSHA states:

"When vapors of a flammable or combustible liquid are mixed with air in the proper proportions in the presence of a source of ignition, rapid combustion or an explosion can occur. The proper proportion is called the *flammable range* and is also often referred to as the *explosive range*. The flammable range includes all concentrations of flammable vapor or gas in air, in which a flash will occur or a flame will travel if the mixture is ignited. There is a minimum concentration of vapor or gas in air below which propagation of flame does not occur on contact with a source of ignition. There is also a maximum proportion of vapor in air above which propagation of flame does not occur. These boundary-line mixtures of vapor with air are known as the *lower* and *upper flammable limits* (LFL and UFL) respectively, and they are usually expressed in terms of percentage by volume of vapor in air. The LFL is also known as the lower explosive limit (LEL). The UFL is also known as the upper explosive limit (UEL)."

Confined spaces can prove dangerous if flammable gases or vapors are present. A common cause of explosions in confined spaces is improper ventilation or failure to remove flammable gases before any work is started in such areas.

STAGES OF FIRE

The growth of a fire proceeds in stages. Combustion begins with no flames or smoke and little heat (incipient). Smoldering starts when heat (flame) is applied to a combustible material (the fuel); the heat oxidizes the material's surface and turns it into

FIGURE 6-5 Comparison of Lower and Upper Explosive Limits (LEL and UEL)

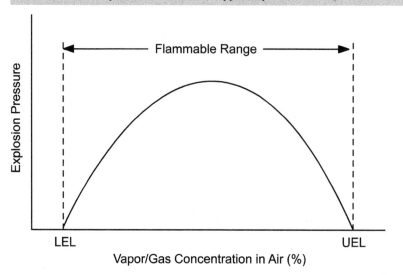

combustible gases. So, even if a fuel is solid, only the combustible gases from it actually burn (just like with a liquid fuel, where only the vapors burn). The oxidation process produces heat, which raises the temperature of the combustible materials. The availability of combustible materials and oxygen determines how quickly or slowly a fire proceeds to the next stage, which is free or open burning.

Free or open burning is when visible flames can be seen. Initially, the flames are limited to the area of origin. More heat is released during combustion, which raises the temperature of nearby objects to their ignition point. If the fire is contained inside a structure (such as a building or tank), the combustible gases rise because they are lighter than air. At this point, the fire either has sufficient oxygen to progress to the next stage (flashover) or does not have enough oxygen and returns to the smoldering stage.

If the fire has enough oxygen, it grows in size. Excessive heat, flame, smoke, and gases are generated. Combustible gases that have not yet been burned gather at the ceiling of the structure. Temperatures near the ceiling can reach 1,500 degrees F in this superheated gas layer. The gases continue to build, and then start to return to the floor. The gases heat all materials in the structure. If enough oxygen remains near the floor, flashover occurs and everything in the structure instantly bursts into open flame. Unprotected persons or living creatures in the room could not survive. The pressure, heat, and smoke can push out of the structure now, as the combustion process speeds up. The fire is harder to put out at this point, as it is fully developed.

FIGURE 6-6 Stages of Fire

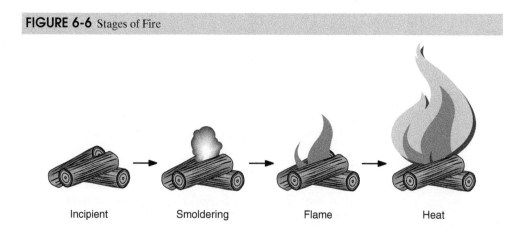

The post-flashover fire burns hot and moves fast. At this point, more resources are required to fight and contain the fire, and more firefighting materials must be used to reduce the ignition temperature (whatever is being used to put out the fire, such as water or foam). Search and rescue operations become more dangerous and difficult at this stage.

Finally, the fire reaches the decay stage. The fuel has been consumed and the temperatures start to decline.

HEAT TRANSFER

Once a fire has started, it can generate three different types of heat transfer:

- Radiation—the transfer of heat through electromagnetic waves
- **Convection**—the transfer of heat through the circulation or movement of a liquid or a gas
- **Conduction**—the transfer of heat through matter via vibrational motion

FIGURE 6-7 Conduction, Convection, and Radiation

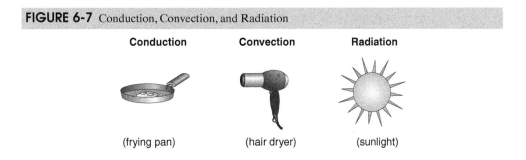

Conduction   Convection   Radiation

(frying pan)   (hair dryer)   (sunlight)

Radiant heat provides the energy from the flames to the fuel, which affects the burn rate. Radiant heat is what causes flames to spread.

CLASSES OF FIRE

Fires are classified according to their properties, which are typically determined by the type of fuel being burned. The type of fuel best determines the way that the fire is fought. Fires are grouped according to classes, based on fuel:

- Class A—Combustible materials such as wood, paper, and plastic
- Class B—Grease or combustible and flammable gases or liquids
- Class C—Fire involving live electrical equipment
- Class D—Combustible metals (e.g., aluminum, sodium, potassium, and magnesium)
- Class K—Cooking oil, fat, grease, or other kitchen fires; intended to supplement a fire suppression system (like in a commercial kitchen)

Firefighting equipment, such as extinguishers and systems, are usually designed to fight specific classes of fire. These types of equipment are discussed in more detail elsewhere in this textbook.

FIGURE 6-8 Fire Extinguisher

VESSELS

Vessels, including tanks and other containers, can pose fire and/or explosion hazards. Vapors and gases can build up in enclosed spaces and can catch fire if heat or another ignition source comes into contact with them. This fire can build rapidly, and potentially cause an explosion.

Tanks can be built to vent vapors to the atmosphere, or use a floating roof that rises and falls with the liquid level inside the tank (and thus allows for expansion of vapors from the liquid). Foam firefighting equipment can also be installed on a tank ceiling. Dikes and other containment systems can be built around tanks to contain any fires involving flammable liquids in the event the tank bursts or fails.

Accurate inventory records must be kept for Class I liquids, so that potentially dangerous leaks can be detected and corrected.

OSHA and EPA regulations address other safety issues related to tanks, such as monitoring leaks, automatic shutoffs for pressurized systems, tank and pipe construction, and spill and overfill protection.

NATIONAL FIRE CODE SECTION 704

The National Fire Protection Association created a National Fire Code that includes a Section 704, which is a system for identifying the hazards of materials. Section 704 allows people to easily visually identify the hazards associated with a substance when it burns. NFPA 704 uses a color-coded diamond along with a numbering system. Products and packaging must be clearly marked with labels using this diamond.

The white section of the diamond displays special information that varies based on who created that particular diamond. NFC Section 704 specifies only two symbols, W and OX explained in Table 6-6. Some commonly used types of special information are also listed in the Tables 6-3, 6-4, and 6-5. The field can be left blank if no special hazards are present.

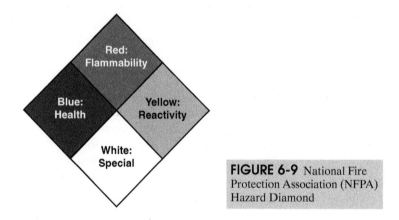

FIGURE 6-9 National Fire Protection Association (NFPA) Hazard Diamond

TABLE 6-3   Red: Flammability

Rating	Description
0	No hazard; the material will not burn (e.g., water)
1	Slight hazard; the material must be pre-heated before ignition can occur (e.g., corn oil)
2	Moderate hazard; the material must be moderately heated or exposed to relatively high ambient temperature before ignition can occur (e.g., diesel fuel)
3	Extreme fire hazard; liquids and solids that can be ignited under almost all ambient temperature conditions (e.g., gasoline)
4	Extremely flammable gases or liquids with very low flashpoints; materials that will rapidly or completely vaporize at atmospheric pressure and normal ambient temperature, or that are readily dispersed in air and that will burn readily (e.g., propane gas)

TABLE 6-4   Blue: Health

Rating	Description
0	Material that on exposure under fire conditions would offer no hazard beyond that of ordinary combustible material (e.g., peanut oil)
1	Material that on exposure would cause irritation but only minor residual injury (e.g., turpentine)
2	Material that on intense or continued but not chronic exposure could cause temporary incapacitation or possible residual injury (e.g., ammonia gas)
3	Material that on short exposure could cause serious temporary or residual injury (e.g., chlorine gas)
4	Material that on very short exposure could cause death or major residual injury (e.g., hydrogen cyanide)

TABLE 6-5   Yellow: Reactivity

Rating	Description
0	Material that is normally stable, even under fire exposure conditions, and is not reactive with water (e.g., liquid nitrogen).
1	Material that in itself is normally stable, but which can become unstable at elevated temperatures and pressures (e.g., phosphorous)
2	Material that on intense or continued but not chronic exposure could cause temporary incapacitation or possible residual injury (e.g., calcium metal)
3	Material that on short exposure could cause serious temporary or residual injury (e.g., fluorine gas)
4	Material that on very short exposure could cause death or major residual injury (e.g., trinitrotoluene — TNT)

TABLE 6-6   White: Special Information

Rating	Description
W	Material has an unusual reactivity with water; do not use water (e.g., magnesium metal).
OX	Material possesses oxidizing properties (e.g., ammonium nitrate).
ACID	Material is an acid.
ALK	Material is a base (alkaline).
COR	Material is corrosive.
☢	Material is radioactive.

Other types of labeling systems can also be used, depending on the manufacturer.

DANGEROUS GOODS CLASSIFICATION AND LABELING

The United Nations uses a globally recognized system of classifying and labeling dangerous goods, which the U.S. Department of Transportation recognizes. These classification labels are discussed in more detail in Chapter 3, *Recognizing Chemical Hazards*.

Explosion Hazards

The sudden burst of heat from an explosion generates energy waves that can travel rapidly outward, up to a few feet per second. If an explosion occurs in a container, such as a vessel, the explosion could be contained. The container walls could be designed to stretch and absorb the pressure wave. However, if the internal pressure exceeds the container's pressure capacity, the container can rupture.

Explosions can cause a variety of direct and indirect safety and health hazards to workers:

- Fatalities
- Flying or falling debris and shrapnel
- Collapsed structures, buildings, and equipment (which can trap or injure personnel)
- Concussion
- Ruptured eardrums
- Flames and burns
- Suffocation (due to smoke inhalation or oxygen-deficient atmospheres)
- Toxic fume inhalation
- Injuries due to being knocked down or thrown by the shockwave
- Product exposure from hazardous spills and gas releases

Explosions can cause extensive damage to property and equipment, result in upsets or interruptions to processes, and harm the environment (through destruction and release of hazardous materials).

EXPLOSIVES

Most process technicians do not handle explosive materials as part of their jobs. However, some jobs might require handling and storing explosives (such as those in mining). Explosives are categorized by their degree of hazard.

The U.S. Department of Transportation uses a classification system that was developed based on an explosive material's sensitivity to heat, impact, friction and shock waves, and by measuring the resulting violence of an explosion.

- Class A—the most hazardous; highly explosive materials (usually considered to be high explosive bombs), including TNT and nitroglycerine
- Class B—propellants, some fireworks, and pyrotechnic signal devices
- Class C—items with explosives in restricted quantities that pose a minimum hazard, such as smokeless powders, blasting caps, and small-arms ammunition

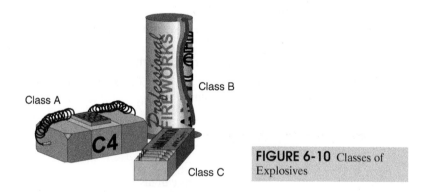

Class A

Class B

Class C

FIGURE 6-10 Classes of Explosives

Storage magazines for explosives are divided into two groups, depending on the amount (weight) of explosives stored:

- I: More than 50 pounds
- II: 50 pounds or less

TYPES OF CHEMICAL EXPLOSIONS

A chemical explosion occurs when chemical reactions generate and discharge high-pressure gas. Chemical explosions are categorized into five types:

- Combustion—explosions involving vapor clouds, gas, mist, dust, or backdraft
- Thermal—when two or more chemical compounds rapidly react together and result in an explosion (common to oil/gas and chemical facilities)

- Condensed Phase—when rapid chemical reactions occur between solids or liquids to cause an explosion (typical of high explosives)
- Nuclear—the rapid release of energy caused by an uncontrolled nuclear reaction (see Chapter 8 for details on nuclear reactions)
- BLEVE—when a boiling liquid (BL) creates expanding vapor (EV), which then explodes (E)

Detonation Hazards

A detonation is an explosion that generates a supersonic combustion shock wave. The shock of the detonation compresses fuel and causes temperatures to reach the ignition point. The ignited fuel burns behind the shock wave and releases energy that supports the shock wave and sustains it. Detonations can be extremely violent, due to the high-velocity pressure waves (between five to seven times the speed of sound) that accompany them.

Detonations present all the hazards of other types of explosions, but are much more violent and dangerous. The accompanying shock waves can cause extensive damage to structures and equipment.

High-speed detonations usually occur in long containers, such as pipe runs, which allow shock waves from the detonation to increase in speed and intensity as they reflect along the container length. Spherical detonations grow in all directions and the shock waves travel more rapidly and with greater force.

DETONATION CHARACTERISTICS

Detonations have the following characteristics, which set them apart from other types of explosions:

- The entire detonation process takes between a fraction of a second up to two seconds.
- The initial pressure produced can be 60 to 100 times greater than other types of explosions.
- Detonation shock waves are supersonic (traveling faster than sound), and can reach up to 8,000 feet per second (for some hydrocarbon mixtures).
- Very few containers are designed to withstand detonations.
- Shock waves produced by detonation can travel easily against gas flow (for example, traveling back up pipes into an upstream process vessel).
- Shock waves can build in intensity each time they reflect off a container wall, causing pressure to increase.
- The high velocity and pressure causes high-impact stress, which can cause containers to break before the container can expand or absorb any of the impact.

Hazard Protection

In the process industries, companies use engineering to reduce the risk of fire and explosion hazards. Hazard protection can also be passive, meaning that sufficient heat sources are not present to cause ignition. Some engineering solutions to reduce these hazards include the following:

- Fire monitors, detection systems, and alarms
- Firefighting systems (e.g., sprinklers, hydrants, hose stations, deluge systems, halon systems, and CO_2 systems)
- Low-heat lighting (lighting designed to put off a minimal amount of heat)
- Firewalls
- Fire-retardant materials
- Leak monitoring and detection
- Vessel design

Also, the following administrative programs are used to reduce fire and explosion hazards:

- Fire-safety training
- Fire extinguisher placement and training
- Vessel loading and unloading training
- Electrical safety training
- Evacuation plans
- Written procedures
- Signs and markings
- Formation of fire brigades
- Material handling and storage
- Hazard communication and documentation (e.g., Material Safety Data Sheets, labeling)
- Respirator training
- Confined-space entry training
- Hot-work permits

Personal Protective Equipment can be used to protect against some hazardous conditions:

- Fire-retardant clothing protects against burns.
- Respirators and rescue air prevent suffocation and inhalation of harmful substances.
- Gloves protect against hot surfaces.

It is the responsibility of the process technician to be aware of potential fire hazards and watch for the elements that can produce fires. Immediate and appropriate responses to fires can mean the difference between controlling and putting out the fire, versus having it spread and cause wide damage, injuries, and even fatalities. Process technicians should know how to report fires at their facilities and be familiar with the location and use of fire extinguishers. They should also know how to properly handle and store flammable materials and understand the properties and hazards of those materials.

Government Regulations

There are numerous government regulations relating either directly or indirectly to fire and explosion hazards as Table 6-7 shows.

TABLE 6-7 Government Regulations Relating to Fire and/or Explosion Hazards

Regulation	Description
OSHA 1910.155 and 156	Fire protection scope, applications, and definitions; fire brigades
OSHA 1910.157 and 158	Portable fire suppression equipment including fire extinguishers and standpipe/hose systems
OSHA 1910.159-163	Fixed fire suppression equipment (sprinkler systems, dry chemical, gaseous agent, water spray, foam)
OSHA 1910.164 and 165	Fire detection systems and employee alarm systems
OSHA 1910.106	Flammable and combustible liquids
OSHA 1926.152	Handling and storage of flammable and combustible liquids
OSHA 1910.1200	Hazard communication
OSHA 1910.134	Respiratory protection
OSHA 1910.119	Process Safety Management (PSM), including hot-work permits
OSHA 1910.307	Electrical for hazardous locations, such as electric equipment and wiring in locations which are classified depending on the properties of the flammable vapors, liquids or gases, or combustible dusts or fibers (Note: the National Electrical Code also addresses this hazard.)
OSHA 1910.109	Explosives and blasting agents
OSHA 1910.146	Permit-required confined spaces

The National Fire Protection Association (NFPA), American Petroleum Institute (API), American Society for Testing and Materials (ASTM), and other organizations also provide information, guidelines, and recommendations for fire and/or explosion safety and prevention.

Summary

Although fire, explosions, and detonations rarely occur in the process industries, when they do occur the results can be devastating. Fire and explosions are some of the most dangerous hazards a process technician may face.

The heat from a fire can cause toxic fumes, burns, and suffocation. Fire can also spark a chain reaction in other areas of the process. Explosions and deflagrations produce violent bursts of energy and pressure that can cause serious damage, injury, or death.

In order for a fire to start, three elements must be present: fuel, oxygen, and heat. These three elements together are called the fire triangle. Removing any one element from the triangle will extinguish the fire. Another factor in the propagation of a fire is a chain reaction (a series of reactions in which each reaction is initiated by the energy produced in the preceding reaction). When a chain reaction is added to the components of the fire triangle, it is referred to as a fire tetrahedron. Like the fire triangle, removing any one element from the tetrahedron will extinguish the fire.

Heat from a fire or some other heat source can be transferred in three ways: convection, conduction, and radiation. Radiation is the transfer of heat through electromagnetic waves. Convection is the transfer of heat through the circulation, or movement, of a liquid or gas. Conduction is the transfer of heat through matter via vibrational motion.

It is important for process technicians to be familiar with the characteristics of flammable substances and the best methods for extinguishing the different types of fires.

Process technicians should be familiar with the different classes of fire (A, B, C, D, K); understand the characteristics, hazards, and prevention of fires, explosions, and detonations; and the governmental regulations that address all these issues.

Checking Your Knowledge

1. Define the following terms:
   a. Combustion
   b. Deflagration
   c. Detonation
   d. Explosion
   e. Fire
   f. Fire point
   g. Fire tetrahedron
   h. Fire triangle
   i. Flammable
   j. Flashpoint
   k. Fuel
   l. Volatility

2. Which of the following is the primary killer during fires?
   a. Carbon monoxide
   b. Carbon dioxide
   c. Atmospheric oxygen

3. How much oxygen does a fire need to ignite?

4. List and describe the four components of the fire tetrahedron.

5. *(True or False)* Liquid burns during a fire.

6. _____ is the temperature at which a fire on the liquid's surface is sustained.

7. What is the term for the temperature at which vapors are produced in sufficient concentrations for flammable or combustible liquids to flash?
   a. Combustion
   b. Convection
   c. Fire point
   d. Flashpoint

8. What does Section 704 of the National Fire Code specify?
   a. Classes of fire
   b. Classes of flammable liquids
   c. Types of fire extinguishers
   d. Hazards of materials

9. The three different methods of heat transfer are:
   a. Radiation, combustion, and conduction
   b. Convection, fusion, and flame
   c. Radiation, convection, and conduction
   d. Flashpoint, fire point, and convection

10. What type of fuel is involved in a Class B fire?
    a. Wood
    b. Electrical equipment
    c. Metals
    d. Flammable liquids or gases

Match the explosion description with its type:

11. Combustion explosion
12. Thermal explosion
13. BLEVE
14. Nuclear explosion
15. Condensed phase explosion

a. Produced when boiling liquid creates expanding vapor, which then explodes
b. Generated when two or more chemical compounds rapidly react together and explode
c. Created by dust, gas, mist, or backdrafts
d. Produced when rapid chemical reactions occur between solids or liquids
e. Occurs due to uncontrolled fusion or fission reaction

16. *(True or False)* Water can be used to put out a flammable liquid fire, if the liquid is lighter than water.

Student Activities

1. Inspect your home or work area for potential fire hazards, then make a list of them. Describe how each fire hazard can be reduced.
2. Research the Texas City explosion of 1947, involving a cargo ship loaded with ammonium nitrate fertilizer that caught fire, then exploded. Write a three-page paper describing the event, the devastation it caused, and the impact on the people and the community.
3. Select a government regulation relating to fire or explosion, then research the regulation. Present your research to your fellow students. Describe the hazard, how the regulation addresses or reduces the hazard, and how it is enforced.

7

Pressure, Temperature, and Radiation Hazards

Objectives

This chapter provides an overview of pressure, temperature, and radiation hazards in the process industries, which can affect people and the environment.

After completing this chapter, you will be able to do the following:

■ Name specific hazards associated with pressure and pressurized equipment used in the process industries and discuss the specific hazards posed by this equipment:

Vacuum

High pressure

Compressed gases

Pressure vessels (runaway reactions)

■ Recognize the hazards of heat and temperature in the working environment.

■ Understand the effects of ionizing and non-ionizing radiation.

■ Describe government regulations and industry guidelines that address pressure, heat, and radiation hazards.

Key Terms

Atmospheric pressure—the pressure at Earth's surface (14.7 psi at sea level).

Dose—the amount of a substance taken into or absorbed by the body.

Endothermic—a chemical reaction that requires the addition or absorption of energy.

Exothermic—a chemical reaction that releases energy.

Rad—radiation absorbed dose; the unit of ionizing radiation absorbed by a material, such as human tissue.

Rem—roentgen equivalent man; a unit of measure of the dose of radiation deposited in body tissue, averaged over the body.

Runaway reaction—a reaction that is out of control; can be either endothermic or exothermic.

Vacuum—any pressure below atmospheric pressure (14.7 psi).

Introduction

In the process industries, the physical hazards associated with temperature, pressure, and radiation can cause injury, illness, or death if these hazards are not handled properly. They can also have short or long-term effects on the environment.

Government agencies use regulations, and the process industries use engineering, administrative controls, and Personal Protective Equipment to protect or limit personnel exposure to temperatures, pressures, and radiation hazards. Process technicians are

---

Did You Know?

The pressure and wind speed from an explosion is powerful enough to rupture a person's lungs, or even knock down a building.

Consider, for example, the bomb (code named "Little Boy") dropped on Hiroshima during World War II. The explosion produced by this bomb created estimated wind speeds of 980 mph on the ground directly beneath the explosion, and generated a pressure equivalent to 8,600 lbs per square foot (60 psi)!

A mile away from the blast, the wind speed was still 190 mph with a pressure equivalent of 1,180 lbs per square foot (about 8 psi)! A force of this magnitude would still be capable of bringing down even the sturdiest of buildings today.

Photo courtesy of the U.S. Department of Energy

required to share the responsibility to protect themselves, their co-workers, and the community from such hazards.

Process technicians must be able to recognize and understand the temperature, pressure, and radiation hazards associated with their operating facilities and how these hazards can impact local communities.

This chapter provides an overview of the temperature, pressure, and radiation workplace hazards that process technicians might encounter. Each hazard is briefly described, along with its potential effects.

Pressure Hazards

We are constantly surrounded by air that pushes on every surface it touches. Pressure is the force exerted on a surface divided by its area.

$$P = F/A$$

At sea level, air exerts a pressure on the surface of Earth of 14.7 pounds per square inch (psi). We call this **atmospheric pressure**. Any pressure below 14.7 psi is referred to as **vacuum**.

It is important for process technicians to understand the concept of pressure and its impact on processes, equipment, and the human body. For example, pressure is used frequently in the process industries to move fluids through pipes, hoses, and equipment. Our bodies also use pressure to circulate our blood, operate our lungs, and control other functions such as balance and hearing.

Process technicians must familiarize themselves with the proper operation and maintenance of pressure-related processes and equipment, and learn to respect the hazards of compressed gases and air.

While the specific details of pressured equipment operations and maintenance are outside the scope of this textbook, technicians can do several things to reduce pressure-related hazards:

- Perform proper maintenance.
- Conduct periodic inspections.
- Ensure proper storage or placement of pressurized equipment and storage vessels (e.g., store in locations away from heat or cold, if possible).
- Ensure that lines and hoses are firmly clamped.
- Release pressure before working on pressurized equipment.
- Use shielding around high-pressure systems.
- Limit access to areas near high-pressure systems.

PRESSURE IN THE PROCESS INDUSTRIES

OSHA defines pressure vessels as storage tanks or vessels designed to operate at pressures above 15 psi. The process industries are full of pressure vessels and pressurized equipment, so most process technicians will encounter them on a daily basis. Because of this, it is important for technicians to know that this equipment can pose significantly hazardous conditions if not handled properly.

Hazards posed by pressurized equipment include ruptures, fires, explosions, and leaks, as well as pulsation or vibration of equipment and lines. All of these factors pose potential hazards to technicians, equipment, the environment, and the community. For example, a leak can expose technicians and the surrounding community to potentially hazardous substances. A fire can expose individuals to hazardous substances and combustion products, and can even trigger a chain reaction (a self-sustaining series of reactions) that leads to an explosion. Explosions can produce violent releases of energy that cause extreme changes in air pressure, flying debris, flames, and smoke, as well as exposure to hazardous materials that can cause poisoning, suffocation, or other effects.

The most significant sources of pressure hazards are posed by the following situations:

- Vacuum
- High pressure

FIGURE 7-1 Imploded Railcar
Courtesy of the U.S. Department of Energy

- Compressed gases
- Runaway reactions in pressure vessels

Vacuum

A vacuum occurs when the pressure inside equipment is less than the air pressure outside of the equipment. If the equipment is not designed for use with a vacuum, it could implode and cause significant damage and/or potentially cause the release of toxic substances. (Figure 7-1) Less-than-atmospheric pressures and changes in atmospheric conditions can also be caused by weather phenomena such as hurricanes and tornadoes.

High Pressure

High pressure can occur due to thermal expansion of a fluid inside its container. Take water, for example. When water is heated to its boiling point (212°F), it expands up to 1,600 times its original volume when converted from a liquid to a gas! This type of thermal expansion can cause pipeline ruptures, fractures, and other equipment damage. That is why pressurized systems are typically designed to handle expansion, and include safety relief valves that reduce the pressure if it reaches a dangerous level.

Compressed Gases

Compressed gases can be extremely dangerous if suddenly released. For example, if a cylinder of compressed gas (e.g., a scuba tank or a tank of helium used to blow up balloons) develops a substantial leak, or if the tank falls over and the valve on the top of tank breaks off, the cylinder may immediately be propelled like a rocket since the energy being released from the leak can have more force than the weight of the cylinder. (Figure 7-2) This can cause serious injury to personnel and equipment. That is why it is extremely important for process technicians to inspect compressed gas cylinders periodically, and report any that appear dangerous or are not secured properly.

Did You Know?

The S.S. *Grandcamp,* the ship that caused the Texas City explosion of 1947, was loaded with ammonium nitrate (a substance used in fertilizers and demolition bombs) when a small fire broke out.

When responders failed to extinguish the fire with jugs of drinking water and a portable fire extinguisher, the captain ordered the hatches on the ship be closed, and the steam fire suppression system activated.

This deadly combination of pressure and steam caused the small fire to turn into a runaway reaction that eventually killed 600 people and caused millions of dollars in damage.

Photo courtesy of: http://www. locall259iaff.org/disaster.html

FIGURE 7-2 Compressed Gas Cylinders Secured to a Dolly

Another hazard of compressed gases is they can replace oxygen and potentially cause a hazardous atmosphere if they are released in a small or confined space, or if flammable, can rapidly release a cloud of gas or toxic material.

Compressed air can also be dangerous if it is not handled properly. For example, air lines can whip around or cause flying debris that can damage the eyes or other parts of the body, or damage equipment. In extreme cases, compressed gases or steam can be powerful enough to cut through the skin and cause serious physical injury or death. Because of this, process technicians should never point air lines directly at a person, use them for horseplay, or use them to clean any parts or clothing on the body.

Runaway Reactions

A **runaway reaction** is a reaction that is out of control. These types of reactions can be extremely dangerous and difficult to stop. Reactions can be **exothermic** (generating heat) or **endothermic** (consuming heat). If an exothermic reaction gets out of control, the reaction vessel can become over-pressurized due to the uncontrolled release of heat or gas generation, which increases the pressure. If an endothermic reaction gets out of control, a vacuum can form inside the vessel as a result of pressure drops due to a drop in temperature.

Pressure Leaks

Pressure leaks can be difficult to determine. If enough fluid leaks out, equilibrium with the surrounding atmosphere may be reached and the leak may stop.

Leaks can be caused by a variety of factors, including the following:

- Dirty or damaged valves, threads, gaskets, and other devices used to control flow
- Vessel over-pressurization
- Thermal expansion of joints and materials due to temperature fluctuations

---

Did You Know?

Nitrogen, which cannot be detected by smell, is frequently used for purging tanks because it has inert (non-reactive, non-flammable) properties.

However, in high concentrations, nitrogen can be lethal because it displaces oxygen in the body.

When oxygen levels in the body decrease, the brain triggers the breathing rate to increase. This causes a further build-up of nitrogen in the body.

When oxygen levels fall to 4-6% percent, the victim will fall into a coma in less than 40 seconds. Once this happens, chances of survival are minimal.

It is important to detect leaks whenever possible to prevent product loss, environmental contamination, injury, or death. Some gases (e.g., the natural gas used in homes) have scents added to them to make them detectable by smell. Other gases are odorless and must be detected using leak detectors or "sniffers." Following are some other methods for detecting leaks:

- Monitoring pressure gauges
- Listening for the sounds of escaping gas
- Checking cloth streamers for movement
- Looking for corrosion

PRESSURE AND OUR BODIES

It is vital that process technicians understand pressure, since it can have an impact on safety. For example, pressure affects the process of breathing, so in order for supplied-air respirators to work, adequate pressure is required. Failing to wear a respirator, wearing a respirator that is incorrect for the environmental conditions, or wearing a respirator that does not fit properly can result in serious injury or death.

In the process industries, technicians may encounter environments that are not optimum for our bodies and our health. For example, some environments may be oxygen-rich, while others are oxygen-deficient. In these environments, the pressure changes that regulate breathing can be impacted.

Temperature Hazards

Temperature is the degree of hotness or coldness that can be measured by a thermometer and a definite scale. Extremely hot or cold temperatures can affect how people work and their performance.

Process technicians face temperature hazards from equipment and processes (e.g., fired equipment, unfired pressure vessels, refrigeration, and cryogenics) that produce extremely hot or cold working environments.

HEAT

Temperature and heat are not the same. Heat is the transfer of energy from one object to another as a result of a temperature difference between the two objects. Heat energy moves from hot to cold, and it cannot be created or destroyed; it can only be transferred from object to another. Heat can be transmitted in three different ways: conduction, convection, or radiation. (Figure 7-3)

Conduction is the transfer of heat through matter via vibrational motion from one object touching another, such as from a frying pan on a stove burner to an egg in the pan.

Convection is the transfer of heat energy through the circulation or movement of a fluid (liquid or gas), such as warm air being circulated through a heating system.

FIGURE 7-3 Heat can be Transferred Through Conduction, Convection, or Radiation

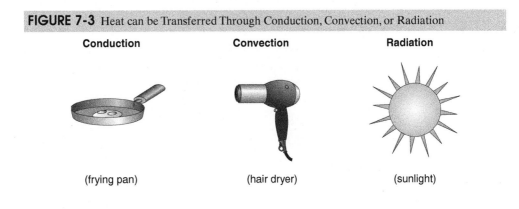

Conduction	Convection	Radiation
(frying pan)	(hair dryer)	(sunlight)

Radiation is the transfer of heat energy through electromagnetic waves in space, without matter moving in that space, such as the warmth from the Sun or an open flame.

How the Body Responds to Heat

A person's body constantly generates heat called metabolic heat. Since humans are "warm-blooded," the body maintains a consistent, constant temperature, despite being exposed to a wide range of environmental temperatures. In order to maintain a constant temperature, the body must rid itself of excess heat to keep its internal temperature in safe limits, usually somewhere between 96°F and 99°F. The body also varies the rate and amount of blood circulated to help regulate temperature.

The body's primary method for removing excessive heat is by sweating. As sweat evaporates, the skin is cooled and excess heat is eliminated. A person at rest, with no stress conditions, generates approximately one liter of sweat per day. A person performing extensive work, working in hot conditions, or feeling stress can produce up to four liters of sweat in as little as four hours. Since sweat consists of water and salt, a person must replace these in order to remain healthy.

Environmental heat is heat generated outside of the body. This heat could be generated by the Sun, a car, an oven, a heater, or many other things. In the workplace, this heat can be generated by equipment and processes (e.g., high air temperatures, radiant heat sources, high humidity).

As environmental heat approaches the body's normal temperature and humidity increases, it is more difficult for the body to cool itself. When the body cannot cool itself, less blood is circulated to muscles, the brain, and other internal organs. When this happens, the person loses strength and fatigue sets in. The person is also not as alert and mental capacity can be affected.

In hot working environments, sweaty palms can affect a person's ability to grip tools. Safety glasses can become fogged. A person can become dizzy more easily. Hot surfaces can also lead to burns. Hot environments can also impact a person's mental capabilities and physical performance. Emotionally, people can become irritable or angry, causing them to overlook safety procedures or become distracted during hazardous tasks.

Heat Rash

Heat rash is a series of raised bumps or blisters on the skin that feel "prickly" to the person. This condition occurs when sweat does not evaporate properly, such as in high humidity environments.

Heat fatigue and heat collapse are other possible heat stress conditions. Sunburn is another potential heat-related hazard.

Did You Know?

There are many ways to prevent heat-related hazards.

- Acclimating to the environment.
- Drinking plenty of fluids (avoid alcohol and caffeine)
- Adding salt to food to help replace salt lost through sweating
- Avoiding working outdoors during the hottest part of the day
- Resting between work periods
- Wearing appropriate clothing and/or PPE
- Using sunscreen

- Getting enough sleep and eating light meals
- Paying attention to warning signs of heat stress
- Understanding how some types of PPE (e.g., protective suits and respirators) can reduce the body's ability to tolerate heat
- Avoiding contact between skin and extremely hot surfaces

Heat Cramps

Heat cramps are muscle spasms in the arms, legs, or abdomen caused when salt and potassium are depleted as a result of heavy sweating. This problem is compounded if the person is drinking a lot of water but not replacing the lost salt and potassium. Drinking specially designed liquids with added salt and potassium (such as sports drinks) can help alleviate the cramps.

Heat Exhaustion

Heat exhaustion can also be caused by sustained physical exertion in a hot environment, along with inadequate replacement of water and salt lost through sweating. Heat exhaustion occurs when the body's water and/or salt levels become lower than normal. As the body becomes dehydrated, the amount of circulating blood is reduced. Following are common symptoms of heat exhaustion:

- Clammy, damp skin
- Pale or flushed skin
- Fainting/dizziness
- Fatigue
- Nausea
- Headache

If any of these symptoms occur, emergency personnel should be notified and the person should be moved to a cool place where they can drink plenty of liquids (avoid alcohol and caffeine, however).

Heat Stress

Heat stress is a common hazard associated with excessive temperatures. Heat stroke (sometimes referred to as "sun stroke") is one type of heat stress that happens when the body's ability to sweat becomes impaired or breaks down, causing the core temperature to rise rapidly.

Prolonged physical activity in a hot environment can cause heat stress. Other factors that can contribute to heat stress are obesity, poor physical shape, and cardiovascular disease. A person experiencing heat stress must be cooled down immediately. The first step is to contact emergency personnel, then move the person into a cool environment and cover them with a damp sheet or spray them with cool water and fan them with a newspaper or fan.

Heat Stroke

Heat stroke is one of the most serious heat-related conditions. It can occur even in people who are not exercising if temperatures are hot enough. Individuals suffering from heat stroke have warm, flushed skin and do not sweat. In most cases, heat stroke victims also have a very high temperature (106 degrees F or higher), and may become delirious, lose consciousness, or have seizures.

These individuals need to have their temperature reduced quickly, and must be given IV fluids for rehydration. Emergency medical treatment should be sought as quickly as possible. A hospital stay for observation may be required following heat stroke since many different body organs can fail as a result of the excessive temperatures and their physiological effects.

Reducing the Impact of Heat-Related Hazards

Process technicians can use a variety of techniques to counter heat-related hazards. These include the following:

- Acclimate to the environment by gradually increasing exposure to the hot work environment.
- Drink plenty of fluids, including water and sports drinks (you should drink 5 to 7 ounces of fluids, preferably cooled slightly or room temperature, every 15 to 20 minutes; do not depend on your body to signal thirst).

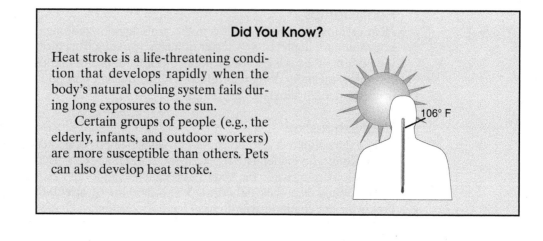

Did You Know?

Heat stroke is a life-threatening condition that develops rapidly when the body's natural cooling system fails during long exposures to the sun.

Certain groups of people (e.g., the elderly, infants, and outdoor workers) are more susceptible than others. Pets can also develop heat stroke.

- Add salt to food, if necessary, to replenish salt lost due to sweating (salt tablets should not be used).
- Avoid working outdoors between noon and 2 p.m. if possible, or at least limit your exposure.
- Wear appropriate clothing and Personal Protective Equipment.
- Use a sunscreen (the Sun Protection Factor, or SPF, indicates the length of sun exposure; for example, SPF 15 indicates 15 times longer sun exposure than without sunscreen).
- Get enough sleep and eat light meals.
- Pay attention to the warning signs of heat stress hazards, for yourself and your co-workers.
- Understand how certain gear such as respirators and protective suits can reduce the body's ability to tolerate heat.
- Avoid contact between skin and extremely hot surfaces.

Companies can also reduce heat-related hazards by limiting non-essential tasks in hot environments or adding extra workers. Following are some other controls:

- Work-rest periods
- Climate-controlled facilities (air conditioning, fans, ventilation)
- Heat shielding and insulation
- Heat hazard and heat acclimation training
- Worker monitoring
- Personal Protective Equipment (e.g., reflective clothing, insulated suits and gloves, face shields, ice vests, air cooled garments, vortex coolers, and water-cooled garments)

COLD

Process technicians can also be exposed to cold conditions in the workplace. For example, technicians may have to work outside in extremely cold conditions, or they may experience cold due to process requirements or equipment. When the body is unable to warm itself properly, serious cold-related injuries can result. Permanent tissue damage or even death can occur.

Low temperatures, combined with wind speed and moisture, can result in cold-related injuries. Wind speed can make the body sense coldness lower than the thermometer reading. This is called the wind-chill factor. Temperature and wind-chill factor combine to determine a temperature colder than what a thermometer may indicate. Along with moisture, wind can increase the hazard posed by cold temperatures considerably.

Cold-related injuries are considered to be either generalized (affecting the whole body) or localized (affecting a particular body part, such as the feet).

Did You Know?

The temperatures in remote regions of Alaska can reach 80 degrees below zero!

Process technicians working in those environments must always wear proper protective clothing and minimize their exposure to the outside air, since temperatures that low can burn the skin, and cause the tears in your eyes to freeze instantly.

Hypothermia

Hypothermia is a general hazard, during which the core body temperature drops to dangerous levels, below 95° F. Symptoms of hypothermia include the following:

- Fatigue/drowsiness
- Uncontrollable shivering
- Cool, blue-colored skin
- Slurred speech
- Uncoordinated movements
- Irritable, confused, or irrational behavior

If not treated, hypothermia can result in death. Therefore, if hypothermia is suspected, immediately contact emergency personnel, then move the person to a warm, dry area. Remove any wet clothing and replace with dry clothing, then wrap the person in blankets. Have the person drink warm, sweet drinks such as sugar water or sports drinks (avoid caffeinated beverages). Have the person move the extremities to create muscle heat. If this isn't possible, place warm water bottles or hot packs on the arm pits, groin, neck, and head. DO NOT rub the body or place the person in warm water.

Frostbite

Frostbite is a localized hazard that generally affects the hands, feet, fingers, toes, ears, and the nose. Frostbite involves a freezing of deep layers of skin and tissue. Following are symptoms of frostbite:

- Sensation of cold
- Hard, numb skin
- Tingling or stinging feeling
- Cramps
- Pale, waxy white skin color

If frostbite is suspected, contact emergency personnel, then move the person to a warm, dry area. Remove any tight or wet clothing that affects the frostbitten area. Place the affected area in a warm water bath (105 degrees F) to slowly warm the area. DO NOT rub the affected area, or pour warm water directly on the affected area. If there is a chance the affected area could get cold again, do not warm it until the person can stay in a warm area. If the affected area is warmed then becomes cold again, tissue damage can occur. After 25-40 minutes of warming, dry the affected area and wrap it to keep it warm.

Frostnip

Frostnip is a mild case of frostbite. If not treated properly, frostnip can become frostbite. Trench foot is a frostnip-type condition caused by cold, wet environments (but not freezing). Symptoms include foot pain, tingling, swelling, and itching. If not treated, trenchfoot can result in tissue death and blistering, permanently damaging the foot.

Process technicians can use a variety of ways to counter cold-related hazards, such as the following:

- Acclimate to the environment by gradually increasing exposure to the cold work environment.
- Drink plenty of warm drinks, including water and sports drinks, while avoiding caffeinated drinks.
- Avoid working outdoors during the coldest part of the day, if possible, or at least limit your exposure.
- Wear appropriate clothing in layers and Personal Protective Equipment.
- Use the buddy system.
- Keep feet and extremities dry.
- Take frequent, short breaks in a warm, dry shelter.
- Get enough sleep and eat warm, high-calorie meals (like pasta).
- Pay attention to the warning signs of cold hazards, for yourself and your co-workers.
- Avoid contact between skin and extremely cold surfaces.

BURNS

The skin, the body's largest organ, is composed of a tough outer layer called the epidermis and an inner layer called the dermis. The skin provides these important functions:

- Protection
- Secretion
- Respiration
- Sense

Burns prevent the normal functioning of skin, hindering its ability to protect vital internal organs and disrupting the other functions listed above. The severity to which a burn injures or damages the skin, and how deep the damage goes, is based on the following (Figure 7-4):

- First-degree burns affect the outermost layer of skin, and result in a mild skin inflammation (such as sunburn); the person may feel some pain and the skin can feel sensitive to touch.
- Second-degree burns affect the outermost and underlying layer of skin, and result in blisters, pain, redness, and swelling. The person feels more intense pain and the skin is sensitive to touch.
- Third-degree burns extend into deep tissues under the skin and result in blackened, charred skin or white skin that can be numb. Moist burns from steam or hot liquids produce white skin, while dry burns from fire or hot surfaces cause blackened, charred skin. Third degree burns can be fatal.

Along with the degree of burn, the surface area of the body burned is critical. Burns covering more than 75 percent of the body's surface area can be fatal. Burns are classified as minor, moderate, or critical based on the degree of burn and the area burned.

Both hot and cold objects can produce burns. Hot objects transfer heat to the skin, while cold objects take heat from the skin. Cold burns can result from prolonged exposure to moderately cold objects (such as snow) or brief contact with extremely cold objects (such as dry ice or liquid nitrogen).

Chemicals can also cause burns if they come into contact with the skin. Chemical burns are covered in the *Chemical Hazards* chapter.

RADIATION HAZARDS

Radiation is defined as energy traveling through space, either as rays or waves of particles. Radiation is a part of our daily lives, in sunlight, cell phones, power lines,

FIGURE 7-4 Comparison of First, Second, and Third Degree Burns

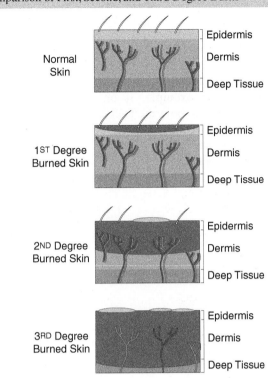

microwaves, radios, televisions, X-rays, and so on. We are exposed to background radiation on a daily basis.

There are two types of radiation:

- Ionizing
- Non-ionizing

Process technicians may encounter both types of radiation in the workplace. Various radiation sources can be found in a wide range of industrial settings. If radiation is not properly controlled, it can prove hazardous to the health of workers. For example, in the process industries radiation can be generated by processes, instrumentation, testing, and other sources.

Ionizing radiation is radiation that is powerful enough to causes atoms to release electrons and become ions.

Examples of ionizing radiation include X-rays and gamma rays. Ionizing radiation can cause damage to the tissues and organs of the body. Ionizing radiation can also cause mutations which could lead to cancer or birth defects. Ionizing radiation takes two forms:

- Particulate (alpha, beta, neutrons)
- Electromagnetic (X-rays, gamma rays, high-speed electrons, and high-speed protons)

Non-ionizing radiation is low-energy radiation such as radio and television waves. Non-ionizing radiation falls into one of the following categories, based on frequency (hertz, or cycles per second):

- Extremely Low Frequency (ELF), including AC current fields
- Very Low Frequency (VLF)
- Radio Frequency (RF)
- Microwaves (MW)
- Infrared (IR)

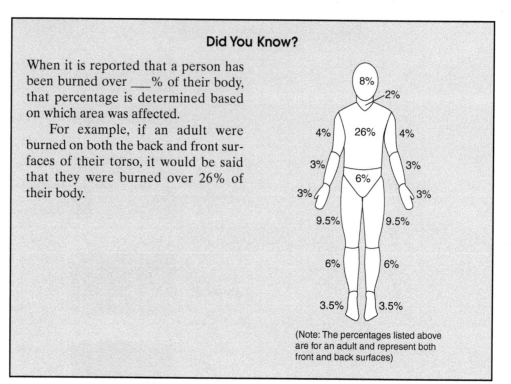

Did You Know?

When it is reported that a person has been burned over ___% of their body, that percentage is determined based on which area was affected.

For example, if an adult were burned on both the back and front surfaces of their torso, it would be said that they were burned over 26% of their body.

(Note: The percentages listed above are for an adult and represent both front and back surfaces)

- Ultraviolet (UV)
- Visible Light

Radiation at these low frequencies does not have enough energy to ionize atoms, and thus does not damage tissue and organs as readily as ionizing radiation. The higher frequencies of electromagnetic radiation, X-rays, and gamma rays, do have enough energy to ionize atoms and are considered ionizing radiation. Non-ionizing radiation can still cause damage though, as it can heat the skin and cause blisters or damage vision. (Figure 7-5)

Ionizing Radiation

A **dose** is the amount of a substance taken into or absorbed by the body. A dose of radiation is the amount of ionizing radiation absorbed by the body, based on the unit of mass for the body part that received the dose or the entire body. A dose is measured in rads, the amount of ionizing radiation energy absorbed by body tissue related to the mass of the tissue. A **rad** is equal to the absorption of 100 ergs per gram of tissue (an erg is an extremely small unit of energy). A **rem** is a measure of ionizing radiation dosage, based on the biological effect of that radiation on body tissue, compared to an X-ray dosage. A millirem is one-thousandth of a rem.

FIGURE 7-5 Radiation Symbol

Ionizing radiation exposure can produce the following effects:

- Radiation poisoning
- Nausea
- Loss of hair
- Cancer
- Neurological damage
- Reproductive problems
- Internal organ damage
- Death

Worker exposure to radiation must be carefully controlled and monitored. OSHA regulates the amount of exposure to radiation that a worker can receive. Companies must ensure that these exposure limits are not exceeded.

Maximum rem dosage over a calendar year for the following body parts is as follows:

- Entire body 1.25
- Head/trunk 1.25
- Blood forming organs 1.25
- Eye lens 1.25
- Reproductive organs 1.25
- Hands/forearms 18.75
- Feet/ankles 18.75
- Skin 7.5

Exposure for workers under 18 years old cannot exceed 10 percent of the maximum rems stated above.

OSHA does allow exceptions to the maximum exposure for a worker in a restricted area, as long as the following criteria are met:

- The maximum dosage to the entire body over the calendar quarter does not exceed three rems.
- The entire body dosage, when added to the accumulated occupation dosage to the entire body, does not exceed 5(N-18) rems, where N is the person's age.
- The company maintains updated records that show the worker's past and current exposure, to indicate that the addition of a dosage will not exceed the worker's specified dose.

The Nuclear Regulatory Commission (NRC) states that the total internal and external dose for a worker cannot exceed five rems per year. Pregnant workers must not exceed a dosage of 0.5 rems over the term of the pregnancy.

Personal monitoring devices must be used where appropriate. These devices include dosimeters, film badges, and other similar devices worn or carried by a worker, to measure the radiation dosage received. (Figure 7-6) Special Personal Protective Equipment (PPE) can also be worn when working in areas with radiation hazards (e.g., shielded aprons, gloves, and radiation suits).

Access to areas with radiation hazards can be restricted. A radiation area is any accessible area where a radiation hazard can deliver a dose of five millirem in one hour to a major portion of the body, or 100 millirem over five days to a major portion of the body. A high-radiation area is any accessible area where a radiation hazard can deliver a dose of more than 100 millirem in one hour. Caution signs must be used to mark radiation areas, high-radiation areas, areas where radioactive materials are stored, airborne radioactive areas, and containers used to store radioactive materials. Companies that produce, use, store, or transport radioactive materials may also have an auditory warning signal in the event the facility must be evacuated.

The best protection from radiation exposure is to distance yourself from the source. The reduction in intensity is inversely proportional to the square of the distance from the source. In other words, the intensity drops off rapidly as you move away from the source.

Did You Know?

In the 1950s and 1960s, before the hazards of radiation were completely understood, many shoe stores contained special X-ray machines that customers could use to determine proper shoe size and fit.

These machines were often out of adjustment and were constructed so radiation leaked into the surrounding area. By 1970, 33 states had banned their use, and the other 17 states had restrictions in place that made their use impractical.

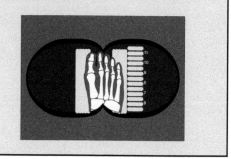

Radiation source shields can melt or be destroyed by fire and the source unaffected. Because of this, process technicians should always follow post-fire procedures for entering areas with radiation sources.

Ionizing radiation used in pipe and vessel X-rays can cause instrumentation to produce false readings where nuclear gauging is used and also cause flame detectors in boilers and furnaces to fail. These instrument failures could be hazardous, so process technicians need to be aware of their possibilities. All radiation-related problems must be reported to the designated site-radiation safety representative.

Companies are required to notify OSHA of incidents and keep records and reports of overexposure, along with cumulative exposure totals.

Various regulations apply to storage and disposal of radioactive materials, including hazardous waste operations (HAZWOPER), which regulate hazardous material cleanup.

Radioactive materials must also be protected from unauthorized removal, so companies that handle radioactive materials above certain regulatory quantities must be prepared to deal with terrorist attacks or other attempts to gain control of these materials.

Non-Ionizing Radiation

Sunlight is the most common form of electromagnetic energy, consisting primarily of infrared and ultraviolet light. Ultraviolet radiation from sunlight can cause eye damage, premature aging of the skin, and skin cancers (such as melanoma). Workers can protect themselves from these hazards by using sunscreen, and wearing appropriate clothing and eyewear when outdoors.

FIGURE 7-6 Dosimeter Attached to Employee Badge

Ultraviolet light sources also include sunlight, welding, UV lamps, and lasers. Infrared light sources include high-temperature processes which can produce heat that leads to heat stress hazards, dry skin and eyes, and other health hazards (e.g., skin cancer and cataracts).

Visible radiation can also cause distortion of light sources, which affect individuals who are color deficient ("color-blind"). This can pose a safety issue if someone needs to visually interpret color-coded warning signs, symbols, and alarms. Good lighting can improve safety and production, while poor lighting can potentially lead to hazards.

The potential hazards of other types of non-ionizing radiation, such as that generated by cell phones, power lines, and computer monitors, are being researched. Conflicting studies have been reported, with some stating that such sources can cause cancer, neurological problems, and reproductive effects, while others debate the findings. It is known, however, that radio frequency and microwaves can interfere with medical devices such as pacemakers.

Government Regulations

While OSHA does not have specific standards that address temperature in the workplace it does publish fact sheets for working in hot and cold environments.

For pressure-related hazards, OSHA does not have one specific standard for pressure vessels. Some OSHA standards require a pressure vessel to be built in accordance with the industry codes and standards. Compressed gases are covered in 1910.101.

For radiation in the workplace, the following government regulations apply:

- Non-ionizing radiation OSHA 1910.97
- Ionizing radiation OSHA 1910.1096
- Hazardous Waste Operations (HAZWOPER) 1910.120 and 1926.65
- Occupational radiation protection Department of Energy 10 CFR 835
- Standards for protection against radiation Nuclear Regulatory Commission 10 CFR 20

Summary

In the process industries, the physical hazards associated with temperature, pressure, and radiation can cause injury, illness, or death if these hazards are not handled properly. They can also have short or long-term effects on the environment.

Government agencies use regulations, and the process industries use engineering, administrative controls, and Personal Protective Equipment to protect or limit personnel exposure to temperatures, pressures, and radiation hazards. Process technicians share the responsibility to protect themselves, co-workers, and the community from such hazards.

Process technicians must be able to recognize and understand the temperature, pressure, and radiation hazards associated with their operating facilities and how these hazards can impact local communities.

Checking Your Knowledge

1. Define the following terms:
   a. Atmospheric pressure
   b. Conduction
   c. Convection
   d. Endothermic
   e. Exothermic
   f. Heat
   g. Ionizing radiation
   h. Non-ionizing radiation
   i. Pressure
   j. Radiation
   k. Temperature
   l. Vacuum
2. When the pressure inside equipment (such as a vessel) is less than the air pressure outside, what occurs?
   a. Explosion
   b. Toxic flashover
   c. Implosion
   d. High pressure reversal

3. _____ is the transfer of energy from one object to another as a result of a temperature difference between the two objects.
    a. Temperature
    b. Pressure
    c. Radiation
    d. Heat
4. Define convection, conduction, and radiation.
5. How many liters of sweat can a person lose if working extremely hard or in hot conditions?
6. What is the serious medical condition that occurs when the body's ability to sweat becomes impaired or breaks down, and the core temperature rises rapidly?
    a. Heat stroke
    b. Heat cramp
    c. Heat stress
    d. Metabolic heat
7. Which of the following is NOT a good recommendation for protecting yourself against heat stress?
    a. Drinking fluids regularly
    b. Eating light meals
    c. Drinking caffeinated beverages
    d. Avoiding work outdoors between noon and 2 p.m.
8. *(True or False)* Hypothermia is a localized cold-related injury, only affecting extremities such as the hands, feet, or nose.
9. _____ can prevent the normal functioning of skin, hindering its ability to protect vital internal organs.
    a. Frostbite
    b. Hypoxia
    c. Narcosis
    d. Burns
10. Which of the following is NOT a type of ionizing radiation?
    a. X-ray
    b. Alpha particle
    c. Gamma ray
    d. Microwave
    e. High speed electron
11. *(True or False)* Sunlight is a form of electromagnetic radiation.

Student Activities

1. Look on the OSHA Web site for information on pressurized vessels or compressed gases. Find out what the hazards are and how they can be prevented (controls). Make a list of hazards and potential controls.
2. Research a heat stress or cold stress-related medical condition and write a one-page paper describing the condition, potential effects (e.g. tissue damage, death), symptoms and treatment.
3. Use the Internet or library to research the short and long terms effects of radiation on people and the environment, from either Hiroshima/Nagasaki or Chernobyl. Then, report the findings to your fellow classmates.

CHAPTER 8

Hazardous Atmospheres and Respirator Hazards

Objectives

This chapter provides an overview of respiration hazards in the process industries, which are caused by hazardous atmospheres.

After completing this chapter, you will be able to do the following:

■ Name specific hazards associated with hazardous atmospheres, ventilation, and other respiratory-related issues.

■ Understand the effect of hazardous atmospheres on respiration.

■ Describe government regulations and industry guidelines that address hazardous atmospheres and respiration hazards.

Key Terms

Absorption—the complete uptake of a contaminant into a liquid or a solid.

Adsorption—the adhesion of a contaminant to the outer surface of a solid body or a liquid.

Air—a layer of gases surrounding Earth; composed mainly of nitrogen (79%) and oxygen (21%).

Air-purifying respirator—a type of PPE that usually covers a wearer's nose and mouth, using a filter or cartridge to remove any contaminants before they enter the wearer's lungs.

Air-supplying respirator—a type of Personal Protective Equipment (PPE) that covers a wearer's face with a mask, providing breathable air through a hose that connects the mask to a clean-air source (usually a compressed air tank or compressor).

Atmosphere—the air space or environment in which the process technician is working.

Confined space—a work area, not designed for continuous employee occupancy, that restricts the activities of employees who enter, work inside, and exit the area, and provides a limited means of egress.

Contaminant—a substance not naturally present in the atmosphere or present in unnaturally high concentrations; also called an impurity. Can be a physical, chemical, biological, or radiological substance.

Hazardous atmosphere—an atmosphere that can cause death, illness, or injury if people are exposed to it. Examples of hazardous atmospheres are flammable, oxygen deficient/enriched, toxic, or irritating/corrosive environments.

HVAC—ventilation systems used to control workplace environmental factors such as temperature, humidity, and odors. The acronym is short for heating, ventilating, and air conditioning.

Immediately Dangerous to Life and Health (IDLH)—any condition that presents an immediate threat to a person's life or causes permanent health problems. This usually refers to an airborne concentration that is immediately dangerous to life and health or can impair a person's ability to escape the atmosphere.

Oxygen-deficient—an atmosphere in which the oxygen concentration is less than 19.5%.

Oxygen-enriched—an atmosphere in which the oxygen concentration is greater than 23.5%.

Respiration—the bodily process of taking oxygen from air breathed in (inhalation) and giving off carbon dioxide (exhalation); also called breathing.

Introduction

Many process facilities expose workers to various potentially hazardous atmospheres, biological hazards, airborne contaminants, and other similar health threats. These hazardous atmospheres include insufficient oxygen levels, or harmful contaminants in the air such as dust, particulates, smoke, gases, vapors, mists, and sprays. Respiratory protection is critical in these situations and requires the use of ventilation systems and respirators.

Hazardous atmospheres can cause short-term effects, such as irritations to the lungs or mucous membranes, coughing, shortness of breath, light-headedness, or mental or physical impairment. Some can also result in long-term effects such as cancer or lung disease. In some cases, these hazards can even result in death. The lungs can serve as a route of entry for hazardous chemicals. If you inhale them, they can act on other parts of the body such as the brain or liver.

Process technicians must be aware of the dangers that hazardous atmospheres pose, especially to respiration (breathing), along with controls that can reduce or remove the hazard. These controls include ventilation, limited exposure, training, respirators, and other protective devices. Process technicians must also know how to select, fit, wear, use, clean, maintain, and store respirators. Respirators can be used in everyday work situations, such as performing a maintenance task in a confined space with a hazardous atmosphere or in an emergency situation such as a release, spill, fire, or terrorist attack.

This chapter provides an overview of hazardous atmospheres and respiratory hazards that process technicians might encounter.

Respiration

Oxygen is vital for life, and required for the fundamental processes that keep us alive and healthy. **Respiration,** or breathing, is the bodily process of taking oxygen from air breathed in (inhalation) and giving off carbon dioxide (exhalation).

The purpose of the respiratory system is gas exchange: It absorbs oxygen from the air into the body and expels the waste gas carbon dioxide to the environment. Oxygen comes from the **air** around us, a layer of gases surrounding Earth that is typically made up of about 21% oxygen and 79% nitrogen. Carbon dioxide is produced as the body uses energy.

The respiratory system starts at the nose and ends in the tissue of the lungs, called alveoli. (Figure 8-1) The major parts are the nose, mouth, throat (pharynx), voice box (larynx), wind-pipe (trachea), and lungs. The windpipe branches into two sections (bronchus), going to the left and the right lung. In each lung, the bronchus branches into smaller tubes, bronchioles, which then branch into small clusters of alveoli, or air sacs.

Alveoli, only two cells thick, handle the actual gas exchange. Because alveoli are so thin, oxygen can pass directly from the lungs to the blood (through small blood vessels called capillaries). The oxygen is then circulated to other parts of the body, where it is used as part of the body's energy process. The capillaries carry the waste gas carbon dioxide from this energy process to the alveoli, where it is removed (exhaled) through the lungs.

As mentioned in Chapter 2, *Types of Hazards and Their Effects*, inhalation is one possible way for hazardous materials to enter the body from the **atmosphere**, the air space or environment in which the process technician is working. Inhalation is the most common route of entry for a hazardous substance and occurs when an airborne hazard (e.g., dust, mist, or vapor) is inhaled. Inhalation hazards occur when substances enter the body through the mouth or nose and are then transported into the lungs. Hazardous chemical or biological agents, in forms such as dust, mist, or vapors, can be inhaled and cause problems in the nose, throat, or lungs (or be absorbed into the body through the lungs). These can cause short and long-term health effects. Even a nonhazardous substance can affect breathing if enough of it builds up in the lungs.

The respiratory system has a variety of defensive mechanisms to protect the body from hazardous chemicals:

* Nose hairs act as a filter to trap large particles, preventing them from going farther into the respiratory system.

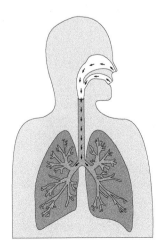

FIGURE 8-1 The Respiratory System Starts at the Nose and Ends in the Tissue of the Lungs

- A mucous layer coats the respiratory system, including the inside of the nose, to trap particles and absorb gases or vapors. This mucous can be expelled from the body by sneezing, blowing the nose, or coughing. Although this mucus can also be swallowed, it is advisable to remove it from the body.
- The windpipe (trachea) is covered with tiny hair-like cells called cilia, which moves mucus and any trapped particles it contains to the throat so it can be coughed out.

Some examples of health effects that can result from inhalation of hazardous substances include the following:

- Direct effect on the respiratory system tissues, causing irritation, shortness of breath, difficulty breathing, and/or fluid buildup in the lungs (edema), can be caused by irritating or corrosive chemicals (see the Irritant and Corrosive Atmosphere section later in this chapter). Edema is life threatening because it can fill the alveoli with fluid and prevent gas exchange, resulting in suffocation.
- Scarring of lung tissue can be caused by particles (dust and fibers) building up in the lungs. Scarred lung tissue becomes thickened and has a harder time expanding. This makes the gas exchange process more difficult. The body cannot inhale as much oxygen as necessary, and the inhaled oxygen does not pass into the blood in adequate amounts. Some particulates (e.g., asbestos) can cause lung cancer.
- Organ problems, including the brain, liver, kidneys, and bone tissue, can be caused by chemical gases or vapors being inhaled and absorbed into the blood and being carried to different organs. For example, solvent vapors can be inhaled and carried by the blood to the brain, where they can cause disorientation and dizziness.

Hazardous Atmospheres

Process technicians can encounter a wide variety of hazardous atmospheres in the workplace. A **hazardous atmosphere** is a concentration of any substance in the air that can cause death, incapacitation, impaired abilities (including self-rescue), injury, or illness. Hazardous atmospheres are any atmospheric conditions that exceed the Threshold Limit Value (TLV) for that substance (see Chapter 2, *Types of Hazards and Their Effects*, for details). Knowing TLVs is important when selecting the proper respirator to use when exposed to a hazardous atmosphere. (Figure 8-2) When a condition presents an immediate threat to a person's life or could cause permanent health problems, it becomes **Immediately Dangerous to Life and Health (IDLH)**. This term usually refers to an airborne concentration that can impair a person's ability to escape the atmosphere.

Benzene is one example of a substance that can contribute to a hazardous atmosphere. Benzene is commonly used in the process industries as a solvent. Benzene acts as an irritant or as a depressant on the central nervous system. Some studies have linked leukemia to benzene exposure (a systemic poison). People can be exposed to benzene from exposure to non-industrial sources, such as smoking, gasoline pumps (contributing to the use of vapor recovery systems), glue, or other common products.

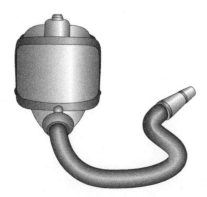

FIGURE 8-2 Proper Protective Gear can Help Prevent Health Effects that can Result From Inhalation of Hazardous Substances

Asbestos is another example. Asbestos was a commonly used building material (typically in insulation or fireproofing) that is now banned by the EPA, since it can release fibers into the air that can be inhaled. These fibers have been linked to health problems such as lung cancer, lung scarring (called asbestosis), and cancer of the chest or abdominal lining (mesothelioma).

The following are descriptions of the major types of hazardous atmospheres.

Oxygen-Deficient or Enriched Atmospheres and Asphyxiants

Typically, oxygen accounts for a little over 20% of a healthy atmosphere. An **oxygen-deficient** atmosphere occurs when the oxygen level of an atmosphere drops below 19.5%. Oxygen-deficient atmospheres deprive the body of the needed amount of oxygen, which can lead to blackouts, brain damage, or death. (Figure 8-3) Oxygen-rich atmospheres can also cause problems such as oxygen intoxication and fire hazards (some substances will ignite and burn more quickly).

Two main factors can create oxygen-deficient atmospheres: consumption of oxygen, such as during a fire, or displacement of oxygen, when another gas forces oxygen or air out of the area. Other oxygen-deficient atmospheres include welding operations and chemical reactions that consume oxygen. Displacement can occur during operations such as "inerting" or purging a vessel, by placing a gas such as nitrogen in it. The atmosphere inside of process equipment is often intentionally made inert with nitrogen inside a vessel or other process containment systems in order to remove hydrocarbons and oxygen so that the atmosphere inside is not flammable.

An **oxygen-enriched** atmosphere occurs when the oxygen level rises above 23.5%. This can happen when oxygen is added to an atmosphere, replacing other gases. Oxygen-enriched environments can pose fire hazards. Some materials not usually considered fire hazards will ignite and burn rapidly in an oxygen-rich environment.

Both oxygen-deficient and oxygen-enriched atmospheres can cause a variety of health problems, including short and long-term effects, or even death.

Asphyxiants are substances that can result in suffocation. A simple asphyxiant displaces oxygen to the point where the body cannot satisfy its oxygen requirements. Types of simple asphyxiants include carbon dioxide, nitrogen, helium, hydrogen, ethane, and methane. Chemical asphyxiants interfere with the body's ability to process oxygen, through chemical reactions. Types of chemical asphyxiants include carbon monoxide, hydrogen sulfide, and hydrogen cyanide (used in gas chambers, but also an industrial insecticide).

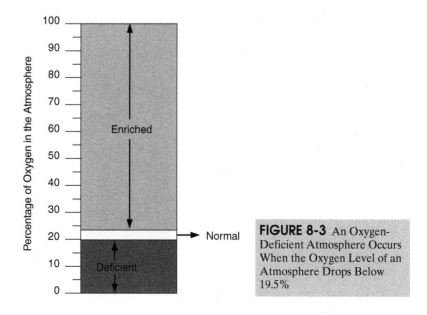

FIGURE 8-3 An Oxygen-Deficient Atmosphere Occurs When the Oxygen Level of an Atmosphere Drops Below 19.5%

Flammable Atmospheres

A flammable atmosphere exists when the air in a specific area contains enough of a gas, vapor, or mist to cause it to exceed a lower flammable limit; or, when a combustible dust reaches a concentration in the atmosphere meeting or exceeding the lower flammable limit. (Figure 8-4)

Often, a flammable atmosphere is the result of an oxygen-enriched atmosphere. Flammable atmospheres can also result from vaporization of a flammable liquid, chemical reaction, work by-product, extreme concentrations of combustible dust, or a release of trapped substances from the lining of a confined space (such as propane being removed from a tank).

For an atmosphere to become flammable, the ratio of oxygen to combustible material must reach a point where it is neither too lean nor too rich for combustion to take place (see Chapter 6, *Fire and Explosion Hazards*). Flammable gases, vapors, or mists can accumulate when there is improper ventilation in an area, such as a confined space. Combustible dust is usually created during loading or unloading operations involving certain materials (e.g., coal, grain, fertilizers). Static electricity can spark these materials, especially during low humidity. Flammable atmospheres pose hazards such as fires, explosions, and respiratory problems.

FIGURE 8-4 A Flammable Atmosphere Exists When the Air in a Specific Area Contains Enough of a Gas, Vapor, Mist or Dust to Exceed a Lower Flammable Limit

Toxic Atmospheres

A toxic atmosphere is air that is poisonous. Toxic atmospheres can be created of gases, dust, particulates, or vapors from the process itself, a stored product, tasks performed in a confined space (e.g., welding), leaks, or spills. Such atmospheres can also result from routine handling of chemicals during such tasks as sampling and equipment maintenance. Examples of toxic atmospheres include carbon monoxide and hydrogen sulfide (H_2S).

Some types of toxic atmospheres include systemic poisons. Systemic poisons are chronic hazards that attack the body's vital organs and systems. Mercury is a commonly known systemic poison that can be vaporized at room temperature.

Irritant or Corrosive Atmospheres

Irritant or corrosive atmospheres contain substances that cause irritation to the skin, eyes, upper-respiratory tract, nose, mouth, and throat. (Figure 8-5) Irritants can cause temporary discomfort, while corrosives can cause serious damage to tissues. Irritant or corrosive atmospheres are divided into two groups: primary and secondary. A primary irritant shows no systemic toxic effects (effects on the entire body), but directly affects the respiratory system. Some common primary irritants are chlorine, hydrochloric acid, sulfuric acid, and ammonia. A secondary irritant produces systemic toxic effects along with surface irritation. Examples of secondary irritants are benzene, carbon tetrachloride, ethyl chloride, and dusts (such as coal dust, silica, and grain dust), which can all cause long-term health problems.

FIGURE 8-5 Irritant or Corrosive Atmospheres Contain Substances that Cause Irritation to the Skin, Eyes, Upper Respiratory Tract, Nose, Mouth, and Throat

Prolonged exposure to irritant or corrosive concentrations can produce "little or no evidence of irritation," according to OSHA. "This may result in a general weakening of the defense reflexes from changes in sensitivity. The danger in this situation is that the worker is usually not aware of any increase in his or her exposure to toxic substances."

Depressant, Anesthetic, or Narcotic Atmospheres

Depressant, anesthetic, or narcotic atmospheres contain substances that can affect the body's nervous system. The effects can be temporary and cause no lasting damage. However, in high concentrations they can cause asphyxiation and result in loss of consciousness or even death. They can also make a person unable to handle emergency situations or remove themselves from dangerous environments (self-rescue), or impair a person's ability to concentrate (which can lead to any number of hazards in the workplace). Symptoms include dizziness, staggering, slow reflexes, and personality changes.

In carefully controlled amounts, anesthetics can be used to put a patient "under" for medical procedures and controlling pain (e.g., ether or chloroform for dentistry, surgery, etc.).

One familiar depressant, ethyl alcohol, is familiar to everyone as drinking alcohol. In the process industries, it is typically referred to as ethanol. When vaporized and inhaled, ethanol creates effects similar to drunkeness.

A substance such as methyl alcohol (or methanol), is both a depressant and a systemic poison; it also poses a fire and explosion hazard. Benzene is a depressant, irritant, and systemic poison, as well as a fire and explosion hazard.

Hazardous Atmosphere Controls

Hazardous atmospheres can be addressed through the use of engineering controls, administrative controls, or Personal Protective Equipment (PPE). Engineering controls include ventilation or using closed processes so workers are not exposed to hazardous atmospheres. Administrative controls include education and training, written procedures, emergency plans, and other similar efforts. To prevent workers from entering spaces with hazardous atmospheres, appropriate warning signs and barricades can be used.

OSHA requires employers to develop a written respiratory protection program to protect workers from hazardous environments. Workers must be trained on the respiratory protection program.

Finally, if engineering controls and administrative controls do not sufficiently reduce the hazard, PPE must be used. The primary type of PPE used to address respiration hazards is a device called a respirator.

The following are explanations of various engineering, administrative, and PPE controls that are used to deal with hazardous atmospheres:

- Ventilation
- Monitoring
- Confined space permits
- Medical evaluation
- Respirators

VENTILATION

Ventilation is a common engineering control used to address hazardous atmospheres. (Figure 8-6) Exhaust (local) ventilation is used to trap and remove hazardous atmospheres from a source. Examples are vent hoods (e.g., welding hoods, lab hoods), similar to those found in kitchens. Dilution ventilation involves moving air in and out of an area, room, or building to reduce (dilute) the concentration of a potentially hazardous atmosphere.

Other workplace environmental factors, such as temperature, humidity, and odors, are controlled with ventilation systems collectively called **Heating, Ventilating, and Air Conditioning (HVAC)**.

In areas where ventilation does not totally remove a hazardous atmosphere or is not feasible, workers must wear approved respirators.

FIGURE 8-6 Ventilation is a Common Engineering Control Used to Reduce or Eliminate Hazardous Atmospheres

MONITORING

Monitoring can be used to detect the type and measure the concentration of airborne contaminants. This information can determine what type of respirator to wear for that atmosphere. If the type and/or concentration of an atmosphere is unknown, then the highest level of respiratory protection should be worn (most often a supplied-air system).

Various devices can be used to detect and measure hazardous atmospheres, which can alert workers to potentially dangerous changes in the environment. These devices can either be permanently affixed to an area and tied into an alarm system, or be a portable device. These devices sample the air and analyze it for various contaminants and concentrations. Monitoring involves both engineering controls (the devices) and administrative controls (procedures for using them). See Chapter 23, *Monitoring Equipment,* for more details.

CONFINED-SPACE PERMITS

Confined-space permits are an administrative control. (Figure 8-7) **Confined spaces** are areas with restricted or limited access, large enough for a person to enter but not designed for ongoing occupancy. Examples of confined spaces include tanks/vessels, furnaces, boilers, reactors, and excavations. Confined-space procedures take into account hazardous atmosphere monitoring, testing, and PPE.

FIGURE 8-7 Confined-Space Permits are Administrative Controls Used to Restrict Entry to Hazardous Areas

Ventilation is a key control that can be used to reduce hazardous atmospheres before workers enter the confined space, and remove any contaminants produced during the work in the space. The atmosphere must be checked before any workers enter the confined space. Conditions can also be monitored for changes using devices that can detect and measure hazardous atmospheres.

Persons entering a space containing a hazardous atmosphere must be instructed in the nature of the hazard, precautions to be taken, and the use of protective and emergency equipment. Standby observers, similarly equipped and instructed, must continually monitor the activity of employees working within the confined space.

Confined spaces are covered in more detail in Chapter 9, *Working Area and Height Hazards,* and Chapter 21, *Permitting Systems.*

MEDICAL EVALUATION

Medical evaluation is an administrative control. The OSHA respiratory protection standard requires that a physician or other licensed health care professional must perform a medical evaluation on each worker who will be exposed to hazardous atmospheres. The determination must be made that the worker will be able to perform work and use the equipment. (Figure 8-8)

The evaluation consists of a medical examination and/or questionnaire. Certain conditions can affect whether a worker can wear a respirator, such as asthma, emphysema, epilepsy, diabetes, or claustrophobia.

The health care professional must provide a written recommendation to the employer before the employee can be fit tested or allowed to use a respirator.

If the employee's health condition changes, an additional medical evaluation is required. Other reasons for an additional medical evaluation are a supervisory recommendation or workplace changes.

FIGURE 8-8 Medical Evaluation is an Administrative Control Used to Protect Workers from Hazardous Atmospheres

OTHER CONTROLS

The following are some other controls that can be used to address hazardous atmospheres:

- Changing a process to use a material that is not hazardous or less hazardous (engineering)
- Creating a closed process that does not expose workers to the hazardous atmosphere (engineering)
- Limiting exposure through breaks and temporary reassignments (administrative)
- Performing housekeeping tasks to keep a work area clean and orderly (administrative)

Respirators

Respirators are a type of Personal Protective Equipment (PPE) that protects workers from exposure to hazardous atmospheres that can result in acute or chronic health hazards. (Figure 8-9)

FIGURE 8-9 Respirators Protect Workers from Exposure to Hazardous Atmospheres

Respiratory protection training can cover topics such as the following:

- Hazardous atmospheres specific to a work area or unit
- Purpose of respirators and the consequences of improper fit and use
- Limitations of respirators
- Fit testing, or the process for determining the correct size of face mask that will give a good seal. NOTE: Workers must be fit tested for every type of respirator that might be used
- Types of respirators available for an area or unit
- How to properly select, put on, and remove a respirator
- How to inspect, clean, maintain, and repair respirators
- Medical warning signs and symptoms that can hinder the use of respirators
- How to use respiratory protection in an emergency
- How to store respirators after use

Employees are required to be fit tested for a respirator before initial use, and be re-tested annually. (Figure 8-10) The fit test determines that the respirator fits properly and has a good seal. Workers must use respirators that have a good seal, with no leakage. The fit test is administered using an OSHA-approved seal test. Workers must not have facial hair or facial conditions that will interfere with the respirator's seal or valve function. A good seal can be affected by facial hair (e.g., beards), scarring, dental changes, glasses, cosmetic surgery, or a significant change in body weight.

FIGURE 8-10 Employees are Required to Complete a Computerized Respirator Fit Test Before Initial Use

Employees must clean and disinfect respirators, following manufacturer recommendations and company policy. If the respirator is to be used by more than one employee, it must be cleaned and disinfected before each use or after each fit testing and training use.

Before entering a hazardous atmosphere, employees should inspect the respirator for maintenance needs. All maintenance needs must be addressed before the respirator can be used. The wearer should then perform a fit check after donning the respirator. After each use, the respirator should be inspected again.

TYPES OF RESPIRATORS

Respirators fall into one of two types:

- **Air purifying**—one type filters contaminants out of the air, protecting the wearer from particles as small as 0.3 microns (e.g. a HEPA filter). Another type uses adsorbing cartridges to trap chemical vapors. Air purifying respirators cannot be used in atmospheres with less than 19.5% oxygen (O_2) content. Examples include dust masks, half or full face masks, and gas masks.

- **Air supplying**—provide breathable air to the wearer through a mask and hose connected to a clean air source. Examples include breathing air hose and a Self-Contained Breathing Apparatus (SCBA).

Employers must provide respirators, certified by the stringent guidelines set by the National Institute for Occupational Safety and Health (NIOSH), to all workers who need them. Employers must also identify and evaluate the respirator hazards in the workplace. These hazards must be categorized as Immediately Dangerous to Life and Health (IDLH) or non-IDLH.

IDLH is any condition (e.g., oxygen-deficient) that presents an immediate threat to a person's life; it can also cause permanent health problems. Another term used is IDL, or Immediately Dangerous to Life. Employers must select the appropriate types of respirators to use for IDLH and non-IDLH atmospheres, and train employees on their selection, fit, use, and other important information.

For protection against gases or vapors, the respirator must be equipped with an end-of-service-life indicator and be certified by NIOSH for a specific gas or vapor (i.e., if an air-purifying respirator is used and the filter is saturated, the wearer may not be able to detect the presence of the gas leaking into the respirator; this is why the end-of-life indicator is required).

For protection against very small particulates, the employer must provide either an air-supplying respirator or an air-purifying respirator equipped with High-Efficiency Particulate Air (HEPA) filters.

The type of hazard usually determines the type of respirator that should be used. (Figure 8-11) Selecting the proper respirator is based on answering the following questions:

- What are the contaminants and hazards that the worker could encounter? If unknown, the highest level of respiratory protection must be used, typically an air-supplying respirator.
- What are the physical, chemical, and/or biological characteristics of the contaminants?
- Is the contaminant organic or inorganic?
- Is the contaminant an acid or a base?
- What is the concentration of the contaminant?
- Are the contaminants airborne?
- What is the boiling point and vapor pressure? This can help determine airborne levels.
- Is the atmosphere oxygen deficient (less than 19.5% oxygen) or oxygen enriched (greater than 23.5%)? If so, an air-supplying respirator must be used.
- Does the hazardous atmosphere contain a gas that has no distinguishable warning properties? If so, an air-supply respirator must be used.
- What is the exposure-time limit?
- What are the physiological effects on the body? If the contaminant is an eye irritant, a full face mask must be worn. If it is an asphyxiant, an air-supplying

FIGURE 8-11 The Type of Hazard Usually Determines the Type of Respirator to Select and Use

respirator must be worn. If the contaminant can be absorbed through the skin, a full face mask must be worn.

- What specific tasks will be performed around the contaminant, and how long will they take?
- What other Personal Protective Equipment is required (i.e., can the hazard also be absorbed through the skin)?
- What is the current health condition of the worker?
- Has the worker been fit tested to wear specific types of respirators?
- What types of respirators are currently available?

The most important question to consider is if the respirator will adequately protect the worker from the hazard.

Before entering a hazardous atmosphere, other factors should also be considered:

- What approvals must be obtained first?
- How will conditions be monitored?
- What permits might be required?
- How will emergencies be dealt with?
- How will communications be handled?
- Have all the entrants familiarized themselves with the MSDS for possible contaminant?

AIR PURIFYING RESPIRATORS

Air-purifying respirators are designed to remove particulates and gases/vapors. Air-purifying respirators are designed to handle specific **contaminants,** protecting the wearer's respiratory system from substances not naturally present in the atmosphere or that are present in unnaturally high concentrations. (Figure 8-12) Companies or manufacturers mark the respirator with the type of contaminant for which it is designed to handle. Filter types use a mechanical method to remove particulates (dust, fumes, fibers, or mists). Air passes through the filter, which removes any contaminants by trapping them. Cartridge types contain materials (such as activated charcoal or silica gel), which remove chemical gases and vapors from the air through a chemical reaction (either through **absorption,** the complete uptake of a contaminant into a liquid or a solid; or **adsorption,** the adhesion of a contaminant to the outer surface of a solid body or a liquid).

The following are descriptions of various air-purifying respirators:

- Dust masks can handle particulates (solids in the air), such as irritant dusts. Dust masks are often misused, and can result in up to 20% leakage due to fit. Dust masks can be disposable.
- Half masks cover the face from under the chin to the bridge of the nose. Elastic or rubber is used to secure the mask to the head. A variation is a quarter mask, which does not cover the chin.
- Full masks cover the entire face, with the filtering chamber attached to the face mask.
- Gas masks also cover the entire face. However, the filtering chamber is too heavy to attach to the face mask and is typically suspended by a harness; the chamber is attached to the face mask using a flexible hose.

FIGURE 8-12 Air Purifying Respirators Use Canisters Designed to Protect Against Specific Contaminants

Air-purifying respirators are typically designed to protect the wearer from between five to 50 times the exposure limit for contaminants. In some conditions, air-purifying respirators can be overpowered by the amount of contaminant and exceed the respirator's design limitations. The wearer will notice that breathing becomes difficult or the contaminant is entering the respirator. The wearer should immediately leave the area.

Air-purifying respirators and the filters they use are color coded to match. Filters from one manufacturer should never be used on another manufacturer's respirator.

Filters are divided into three classes:

- Class N (not oil resistant) cannot be used in atmospheres with oil-based particulates present.
- Class R (oil resistant) can be used in any atmosphere; however, the filters must be changed after every shift when used in atmospheres with oil-based particulates.
- Class P (oil proof) can be used in any atmosphere, regardless of the particulates present.

Each class has three different efficiency ratings: 95%, 99%, and 100%. High-Efficiency Particulate Air (HEPA) filters are useful for extremely small size particles, such as asbestos.

Filters must be replaced as necessary, based on how long they have been in use and the concentration (or amount) of contaminants to which they have been exposed. Otherwise, the filter can become clogged or contaminants can break through. When the wearer has difficulty breathing normally, the filter likely needs to be changed.

There are different types of cartridges based on the type of chemical contaminant, which are also color coded:

- Organic vapor: black
- Acid gas: white
- Ammonia/amine: green
- Combined organic vapor/acid gas: yellow

Cartridges must be replaced when the wearer can smell the chemicals inside the mask. Therefore, these types of respirators should never be used around chemicals that have no distinguishing warning properties (such as smell or taste). When the cartridge becomes saturated, and thus no longer effective, the wearer cannot detect that the gas or vapors are leaking into the respirator.

Cartridge respirators can be used only with chemicals that have odor thresholds below the TLV (safe exposure limit). If the wearer cannot detect the chemical until it is at levels above the TLV, then the danger is that the wearer will be overexposed to the chemical if there is a leak.

Workers should check the protection factor for a cartridge before using it. By knowing the TLV, a worker can check it against the cartridge protection factor, typically described as X times the TLV, where X is a number, such as 5 x the TLV. The cartridge will have a maximum use concentration printed on it, indicating the level of concentration that should not be exceeded.

Limitations of cartridges are that they cannot be used under the following conditions:

- If the hazard is unknown, because the correct cartridge cannot be selected
- In oxygen-deficient atmospheres or IDLH atmospheres
- With substances that have poor warning properties for the wearer (smell, taste, reaction)

Organic vapor cartridges are not acceptable for use with organic vapors. For example, methanol cannot be readily handled by a cartridge respirator, so only air-supplying respirators can effectively protect the wearer. Because of the complex nature of chemicals, organic vapor respirators cannot be labeled with all the organic compounds that they can be effective against. Workers should check with the cartridge manufacturer for a full list of substances that the cartridge will handle.

AIR-SUPPLYING RESPIRATORS

Air-supplying respirators are used in environments with toxic concentrations of contaminants or oxygen-deficient atmospheres. (Figure 8-13)

Following are descriptions of various air-supplying respirators:

- **Air-line respirators** use a small-diameter hose to connect an air-supply source to a face mask or hood. Air can be supplied by cylinder (or bank of cylinders) or compressors rated for breathing air. The air supply must be properly filtered and meet OSHA guidelines for breathing air. The air line typically should not exceed 300 feet in length.
- There are three different types of air line regulators:
  - Continuous flow, in which the wearer receives air automatically; the respirator forces the air to the wearer. Because of the continuous flow, positive pressure builds up and prevents hazardous atmospheres from entering any leak points. However, the large amount of air used limits the use of cylinders as an air supply (since they can quickly become depleted).
  - Demand flow, in which the wearer inhales and produces a negative airflow that opens a valve to supply air. When the user exhales (positive pressure), the valve is closed. This type of regulator uses less air than a continuous flow type, so cylinders can be used. The process of inhalation required to open the air valve can draw in hazardous atmospheres through leak points. This type of regulator is becoming obsolete, because a steady leak in the mask can cause positive pressure, thus preventing the air valve from opening.
  - Pressure demand, in which a preset exhalation valve maintains a positive pressure.
- Hose masks are a crude type of air-line respirator which uses a larger-diameter hose. This allows air to be inhaled using normal lung power, although a blower can be used to assist air flow. Hose masks are decreasing in use.
- SCBA, or Self-Contained Breathing Apparatus, which requires the wearer to carry all breathing equipment. Air is supplied using a cylinder attached to a harness, which can be worn (typically on the back). SCBA respirators allow the wearers more mobility than air-line or hose respirators and there is no air-supply line trailing behind the wearer, which can get crushed, damaged, or tangled. One drawback is that the cylinder may restrict the wearer from entering some spaces. At more than 30 pounds, SCBA also weighs more than air-line or hose-mask respirators. Following are two types of SCBA:
  - Open-circuit, with exhaled air venting to the atmosphere
  - Closed-circuit, with exhaled breath recycled to keep up the amount of oxygen. Closed-circuit systems can be smaller and lighter than open-circuit systems.

For compressor-provided air, use only a compressor approved for breathing air. Never use a mechanical tool compressor. A potential problem with using compressors to supply air is that the compressor can have mechanical problems and fail. When the compressor shuts down, then the air stops flowing to air-line or hose-mask wearers. Breathing air compressors should have an alarm that indicates impending failure.

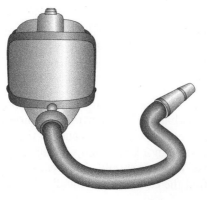

FIGURE 8-13 Air-Supplying Respirators are Used in Environments with Toxic Concentrations of Contaminants or Oxygen-Deficient Atmospheres

Some other types have an automatic shutdown procedure to keep wearers from breathing carbon monoxide created by an overheated compressor. Air-line or hose-mask wearers must carry an escape unit respirator (like a mini-SCBA unit), which typically provides about five minutes of emergency air.

A drawback with cylinders is that they are limited on the amount of air they can supply (potentially less than 30 minutes). Several factors can further reduce the amount of time available:

- Wearer's exertion
- Wearer's physical condition
- Wearer's body size
- Wearer's emotional state (panic, fear)
- Total cylinder charge
- Leaks
- Atmospheric pressure

For air-line or hose-mask respirators, wearers do not have to carry a cylinder and can work for a long time. However, wearers must be aware of the hose and its limitations. Its weight can affect movement. The length of the hose limits the distance the wearer can travel. The hose must not get tangled, crushed, or rubbed against sharp edges.

Often, SCBA use is reserved for emergency situations, such as rescue operations and fires. SCBA users might be required to don the equipment in one minute or less, so practice is essential. SCBA also changes the wearer's center of gravity, so a worker must get used to wearing it. Process facilities often hold monthly training sessions on SCBA gear use. SCBA should have a low-pressure warning device, which should be tested regularly.

EMERGENCY RESPIRATORS

Other types of respirators are called emergency escape respirators. These are intended for use to escape from an area where a hazardous leak or spill has happened. One type of escape respirator is called a mouthpiece respirator. The wearer clamps it in the mouth. A plug is used to prevent air intake through the nose, so the wearer must breathe through the mouth. Since it is difficult to use and form a proper seal, it is intended for use only in emergencies. Another type is called an escape pack or egress bag. It consists of a clear plastic bag the wearer dons, along with a limited-supply air bottle.

RESPIRATOR FIT, USE, AND CARE

Air-purifying respirators are less expensive and easier to use than air-supplying systems, making them the best option if they sufficiently address the hazard. However, some hazardous atmospheres cannot be sufficiently reduced by using air-purifying respirators. Air-supplying respirators must be used in these situations and in oxygen-deficient atmospheres. If the contaminant type and/or concentration of an atmosphere is unknown, then the highest level of respiratory protection should be worn.

When doing fit testing, the wearer should ensure there are no leaks and that a good seal is established. After being fitted to a particular make and model of respirator, a worker should be able to use all respirators of that type, unless a change occurs that affects facial structure (e.g., weight change, scars).

Fit tests typically involve the person donning the respirator, then being exposed to a "challenge" atmosphere. A challenge atmosphere is one to which a substance is added (such as banana oil, saccharin, or smoke). The wearer goes through some exercises that match work-related movements (e.g., moving the head around, talking, and jogging). If the wearer detects the challenge atmosphere (by smelling, tasting, or reacting), this means that the seal is not proper. Some tests use an electronic probe placed inside the mask to check the fit.

Your respirator can be checked two ways:

- Negative testing, where the user dons the respirator and ensures a good seal by covering the inlet(s), typically with a hand, to see if a vacuum is created inside the respirator. The user should cover the inlet for at least 10 seconds to properly check the seal.
- Positive testing, where the user dons the respirator and ensures a good seal by covering the exhaust ports, typically with a hand, to see if pressure is created inside the respirator. The user should cover the exhaust ports for at least 10 seconds to properly check the seal.

Companies can perform other types of tests as their policies require.
For respirators with face masks, some hazards include the following:

- Reduced peripheral vision
- Limited vision, if the mask fogs (due to heat and humidity)
- Muffled communication
- Potential overheating of the wearer

Following are some general recommendations for respirators:

1. Inspect, clean, and maintain respirators properly (see the following section).
2. Respirators must be placed in areas where they are easily accessible and in plain sight.
3. Often, a respirator must be worn with other PPE. Make sure to properly fit all PPE along with the respirator.
4. Wearing glasses or contact lenses can affect respirator use. Check with a health care professional or safety personnel about this issue.
5. Understand the limitations of respirators.
6. Familiarize yourself with emergency procedures, such as switching to emergency air if a compressor failure occurs when using an air-supplying respirator.
7. If you have any physical changes that can affect respirator fit (e.g., weight changes), make sure you perform another fit test.

After a respirator has been used, follow these steps:

1. Inspect the respirator for damage and make any necessary repairs. If the respirator is damaged beyond repair, dispose of it and replace it.
2. Clean and disinfect the respirator using company-provided cleaning materials and procedures.
3. Dispose of any used filters/cartridges properly; if the filters are contaminated, you may need to follow special procedures for disposal.
4. Make sure that the face piece seal is not bent or folded.
5. Store the respirator in a plastic bag to keep it clean and free of dirt.
6. Return the respirator to its proper location, so it is ready for the next use.

Government Regulations

There are a variety of government regulations relating either directly or indirectly to hazardous atmospheres and respiratory protection:

- OSHA General Industry Standards
  - 1910 Subpart G: Health and Environmental Controls
  - 1910.94 Ventilation
  - 1910 Subpart H: Hazardous Materials
  - 1910 Subpart I: Personal Protective Equipment
  - 1910.134 Respiratory Protection
  - 1910 Subpart J: General Environment Controls
  - 1910.146 Permit-Required Confined Space
  - 1910 Subpart L: Fire Protection
- 1910 Subpart Z: Toxic and Hazardous Substances

- NIOSH 42 CFR Part 84
- Environmental Protection Agency (EPA) Toxic Substance Control Act (TSCA)

The OSHA Respiratory Protection standard addresses the following:

- Employee training (respiratory hazards, care and use of respirators)
- Medical certification prior to respirator use
- Fit testing
- Objects that can interfere with a good seal (e.g., facial hair, glasses)
- Written program

Summary

Many process facilities expose workers to various potentially hazardous atmospheres, biological hazards, airborne contaminants, and other similar hazards. These hazardous atmospheres include insufficient oxygen levels, or harmful contaminants in the air such as dust, particulates, smoke, gases, vapors, mists, and sprays. Respiratory protection is critical in these situations and requires the use of ventilation systems and respirators.

Hazardous atmospheres can cause short-term effects, such as irritations to the lungs or mucous membranes, coughing, shortness of breath, light-headedness, or mental or physical impairment. Some can also result in long-term effects such as cancer or lung disease. In some cases, these hazards can even result in death. The lungs can serve as a route of entry for hazardous chemicals. If you inhale them, they can act on other parts of the body such as the brain or liver.

Process technicians must be aware of the dangers that hazardous atmospheres pose, especially to respiration (breathing), along with controls that can reduce or remove the hazard. These controls include ventilation, limited exposure, training, respirators, and other protective devices. Process technicians must also know how to select, fit, wear, use, clean, maintain, and store respirators. Respirators can be used in everyday work situations, such as performing a maintenance task in a confined space with a hazardous atmosphere or in an emergency situation such as a release, spill, fire, or terrorist attack.

Checking Your Knowledge

1. Define the following terms:
   a. Absorption
   b. Adsorption
   c. Confined space
   d. Contaminant
   e. Hazardous atmosphere
   f. HVAC
   g. Oxygen enriched
2. The percentage of oxygen in an oxygen-deficient atmosphere is _____ or less.
   a. 20%
   b. 23.5%
   c. 21.8%
   d. 19.5%
3. _____, or breathing, is the bodily process of taking oxygen from air breathed in (inhalation) and giving off carbon dioxide (exhalation).
4. *(True or False)* Inhalation is the most common route of entry for hazardous substances.
5. _____ is a common engineering control used to address hazardous atmospheres.
   a. Ventilation
   b. PPE
   c. Training
   d. Medical evaluation
6. Describe three potential hazards of confined spaces.
7. What is the purpose of a medical evaluation in regard to respirator use?
8. What is fit testing?
   a. Monitoring hazardous atmospheres for leaks
   b. Checking that ventilation works properly
   c. A process to determine what type of respirator forms a good seal on a process technician's face
   d. An emergency procedure to follow if an air-line respirator fails

9. *(True or False)* Facial hair does not interfere with the seal of a respirator.
10. Which of the following are administrative controls for hazardous atmospheres? (select all that apply)
    a. Changing a process to use a material that is not hazardous or less hazardous
    b. Creating a closed process that does not expose workers to the hazardous atmosphere
    c. Limiting exposure through breaks and temporary reassignments
    d. Performing housekeeping tasks to keep a work area clean and orderly
11. Name the two types of respirators.
12. Air-purifying respirators use _____ to remove any contaminants.
    a. Ventilators
    b. Filters
    c. Air hoses
    d. Fans
13. A _____ pressure air-line respirator allows the wearer to receive air automatically; the respirator forces the air to the wearer.
    a. Positive
    b. Negative
    c. Hose
    d. Filter
14. What does the acronym SCBA represent?
15. A process technician can perform which types of respirator fit checks?
    a. Negative and charged
    b. Positive and atmospheric
    c. Mechanical and chemical
    d. Negative and positive
16. *(True or False)* It is not necessary to clean a respirator after use.

Student Activities

1. Review the OSHA Web site and the regulations on respiratory protection. Select one aspect of the regulation (as outlined in the Government Regulations section of this chapter) and write a two-page summary.
2. Using the Internet or other resources, look up different manufacturer's recommended care and maintenance requirements of their respirators. Discuss your findings with your fellow students.
3. Research the process of respiration along with potential respiratory hazards that the process industries pose. List five hazards, along with how they affect respiration.
4. Test fit a respirator (if available), and then check the seal. Practice donning the respirator a few times, then have someone time you. Attempt to don the respirator in less than one minute.

Working Area and Height Hazards

Objectives

This chapter provides an overview of working area and height hazards in the process industries, which can affect the safety and health of workers.

After completing this chapter, you will be able to do the following:

■ Name specific hazards associated with work areas, such as the following:

Working surfaces

Means of egress

Heights

Confined spaces

■ Describe government regulations and industry guidelines that address working surfaces, means of egress, height, and confined-space hazards.

Key Terms

4-to-1 rule—a safety rule for using straight or extension ladders.

Dockboard—a temporary platform used during the loading operations of cargo vehicles.

Engulfment—the state of being surrounded or completely covered by materials or products within a confined space.

Fall protection—a system designed to minimize injury from falling when the work height is six feet or greater (above or below grade).

Guardrail—a rail secured to uprights and erected along the exposed sides and ends of platforms (OSHA).

Handrail—a single bar or pipe supported on brackets from a wall or partition, as on a stairway or ramp, to furnish persons with a handhold in case of tripping (OSHA).

HAZWOPER—the acronym for the OSHA standard for Hazardous Waste Operations and Emergency Response.

Means of egress—an exit or way to evacuate a building or facility during an emergency.

Three-point contact—a safety practice in which both feet and at least one hand are used when ascending or descending stairs.

Toeboard—a vertical barrier at floor level erected along exposed edges of a floor opening, wall opening, platform, runway, or ramp to prevent falls of materials (OSHA).

Walking and working surfaces—how OSHA refers to floors, walkways, passageways, corridors, platforms, and other similar surfaces.

Introduction

Many accidents in the workplace result from slips or falls. These can be caused by someone tripping over an obstacle in a walkway, slipping on a spill, falling off a ladder, stumbling off a loading dock, or a wide variety of other actions.

Other workplace hazards include blocked exits during an emergency, falls from heights, falling objects, and confined spaces with potentially hazardous atmospheres.

The process technician must be aware of these hazards and the safety measures that can be used to reduce their incidence. Also, they should perform housekeeping duties to keep work areas clean and free of clutter.

This chapter provides an overview of hazards that the process technician can encounter in general work areas, including **walking and working surfaces** (e.g., floors, walkways, passageways, corridors, and platforms), ladders, exits, heights (falls and falling objects) and confined spaces.

Working Areas

Process facilities consist of many common workplace elements, which can seem harmless but actually result in a number of workplace accidents. Following are some of these common elements:

- Floors and walking surfaces
- Lighting
- Doors, exits, and aisles
- Stairs, ladders, and elevated walkways

Federal, state, and local codes address a variety of issues relating to safety in building design, construction, and maintenance. Despite these codes, hazards can still exist, since the work environment at a process facility is dynamic and changes daily. For example, parts from a piece of equipment could be left in a walkway during maintenance, causing a tripping hazard. So even though the walkway itself might meet the building code, a hazard can still exist. Below are other examples of hazards:

- New security measures that result in an exit being locked
- Burned-out lights
- A cable lying across a walkway
- A spill creating a slippery walking surface
- Materials being moved into a warehouse, blocking an aisle
- Hoses left unrolled or not stored properly after use

Also, facilities are not always designed and built for convenient access during maintenance tasks. These types of tasks can put workers at risk as they clean, repair, and perform general maintenance on equipment or the facility itself.

Walking and Working Surfaces

Floors are referred to in government regulations as walking and working surfaces. These surfaces include the following:

- Walkways and aisles
- Stairs
- Platforms
- Catwalks
- Scaffolds
- Ramps
- Docks

Some of the most common injuries in the workplace occur due to falls or slips on walking and working surfaces. Various factors can result in falls or slips:

- Foreign objects on the walking surface (typically, objects that are not a permanent part of the walking surface, such as cables, boxes, and tools)
- Slippery or slick walking surfaces (spills, water, smooth concrete or metal) (Figure 9-1)
- Improper lighting (e.g., dim lighting, no lighting, or extremely bright light)
- Incorrectly designed or built walking surfaces (no traction, low areas where water can collect, steep inclines, narrow stair treads)
- Irregular or uneven surfaces

A person's physical condition can also result in falls. People with disabilities might have difficulties moving around in a poorly designed or constructed work area. Tired or distracted workers can trip and fall, as can people carrying bulky or awkward loads.

Falls from heights, such as ladders and walkways, are covered in a later section.

Falls and slips can result in the following:

- Cuts, scrapes, and bruises
- Sprains
- Broken bones (especially legs, arms and ribs).
- Concussion
- Death

FIGURE 9-1 Non-Slip Walking Surface

When OSHA inspectors review walking and working surfaces at a process facility, they look for how well the surface is maintained (i.e., is it "clean and orderly"?). To help lessen hazards associated with walking surfaces, the process technician must contribute to general "housekeeping" duties, such as these:

- Make sure permanent aisles are marked with signs, and areas leading to exits are kept clear of hazards.
- Check that fall guards are in working order.
- Make sure walkways are kept clear of foreign objects.
- Roll up hoses or store them properly after use.
- Clean up spills promptly.
- Tidy up work areas.
- Barricade and identify areas where hazardous conditions cannot be immediately corrected.
- Notify a supervisor of any hazards that the process technician cannot correct.

Following are some safety tips:

- Wear the proper footgear with non-skid soles at all times. (Figure 9-2)
- Do not allow yourself to be distracted when walking around the facility.
- Remember that the nature of process facility work includes equipment maintenance (e.g., parts may be placed out in walkways during this work) and that materials are often moved around. Liquids could spill or solids could spill out of their containers. A hazard that was not there yesterday could be there today.
- Keep a watchful eye for potential hazards.
- Tape off or block hazardous areas.
- Do not overload yourself when carrying objects. You should always be able to extend your arms and catch yourself if you begin to fall.
- Do not walk up or down stairs too quickly. Use the **handrail** that is attached to the wall to prevent tripping.

Areas with wet processes should include drainage along with false floors, platforms, mats, or other dry surfaces for workers to stand or walk on.

Another important issue regarding floors is load limits. Some floors are marked with floor-load marking plates to indicate the load that it can withstand. Process technicians must be aware of floor load limits and help follow that limit.

An important note about aisles and walking surfaces is that process technicians must be aware that vehicles such as forklifts and hoists could be in use. Also, working with **dockboards** (temporary platforms used during loading operations of cargo vehicles) can prove dangerous if they shift during use. Refer to Chapter 12, *Vehicle and Transportation Hazards*, for more details on these hazards.

Guardrails erected along the exposed sides and ends of platforms or secured to uprights, and covers are used to protect open floors, platforms, and runways when areas are four feet or more off the adjacent floor or ground level. Stairs have landings, tread width, height, guardrails, handrails, **toeboards** (vertical barriers placed along exposed edges of floor openings, platforms, runways, or ramps to prevent materials from falling on someone below), and other safeguards to protect workers. Do not climb or lean over railings. Climb stairs carefully; do not hurry when going up or down them. Use caution if you must carry anything on stairs. Have both feet and at least one hand available when ascending and descending stairs, a safety practice known as **three-point contact**.

FIGURE 9-2 Technicians should Always Wear Proper Footgear with Non-Skid Soles

Ladders

Process technicians might be required to work with ladders, either portable (e.g. step, straight, or extension) or fixed. (Figure 9-3) **Fall protection** measures must be followed whenever the work height is six feet or greater to minimize injury.

To be safe while using a portable ladder, inspect it before use with the following procedures:

- Read the manufacturer's label on the ladder. Familiarize yourself with the ladder's weight capacity and applications, and select the correct ladder for the task. Select a ladder that is the proper height for the job.
- Make sure the ladder is not cracked, has no loose rungs or braces, or any damaged connections.
- Look for corrosion or weather damage (from heat or cold).
- If using a wooden ladder, inspect it for moisture (which can conduct electricity).
- Pay extra attention when inspecting painted wooden ladders, as paint can hide cracks. If available, use ladders with varnish or clear lacquer instead of painted ones.
- If using a metal ladder, inspect it for sharp edges.
- If using a fiberglass ladder, make sure the fiberglass is not deteriorating.
- If the ladder has safety shoes (rubber or other non-skid materials on the feet), make sure they are secure and in good shape.
- Never use a makeshift ladder.
- If working around electricity, remember that a metal ladder can conduct electricity (rubber or non-conductive feet might reduce the hazard, but the hazard is still present).
- Inspect the rungs to make sure they are not slippery.
- Check the latch and rope on extension ladders.

Once you have inspected a portable ladder, follow these recommendations for its use:

- Make sure the ladder is properly opened and placed on a firm, dry base.
- Do not set a ladder on top of any loose objects or raise its height by placing it on other objects.
- Allow only one person at a time on the ladder.
- When climbing up or down a ladder, always face toward the ladder.
- If the ladder is near an entrance, place a barricade around the ladder.
- Do not have tools in your hand when climbing the ladder. Use a tool belt, have someone hand you the tools, or find some other way of getting tools to the raised work area.
- Do not lean too far to either side when working on a ladder. Climb down, reposition the ladder, and then resume the work.
- If reaching up, do not let your waist extend above the top rung.

FIGURE 9-3 Stepladder

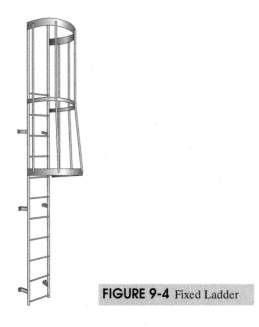

FIGURE 9-4 Fixed Ladder

- Follow the manufacturer's warnings.

Follow these rules for straight or extension ladders:

- When leaning a ladder against a surface, make sure the surface is not slippery or fragile.
- Use the **4-to-1 rule**: Put the base one foot away from the wall for every 4 feet of height between the base and the support point (where you lean the ladder). For example, if the support point is 12 feet from the ground, place the base at least 3 feet away from the wall.
- The ladder should extend about three feet beyond the top of the support point.
- Tie off the ladder at the top with a rope when in place. Always have someone hold the ladder at the base if it is not tied.

For fixed ladders, check that rungs and side rails (if present) are properly secured and not slippery. (Figure 9-4) If a ladder safety device is used instead of a cage for long fixed ladders, inspect the harness for wear and tear, then make sure you properly secure it to yourself and the fixed rail. Check that the braking system (if available) is functioning properly.

In some situations, safety net systems or guardrails can be used instead of fall protection equipment.

Exit Routes (Means of Egress)

Means of egress refers to an exit or other way to escape in case of an emergency (such as fire). Workers must be able to exit a building or work area quickly and safely. The location and types of doors, lighting, aisles, stairways, limited-access areas, locked or blocked interior doors, obstructions, and other factors all affect exits and how quickly workers can escape.

The number of exits is determined by various factors, such as the size of the facility, total occupancy, workplace arrangement, and number of employees. Often, two or more exits are required at process facilities. Signs must be posted along the exit route to indicate where to find the nearest exit, and all exit signs must be illuminated. Exits must open up to a street, walkway, refuge area, public way, or open space with access to the outside and these areas must be large enough to accommodate the facility occupants using that route.

Even areas outside exits can affect escape. For example, landscaping or fences might restrict escape efforts. In addition, efforts to improve security at process facilities can also impact exits and safety. As access is restricted, so are the potential exits during

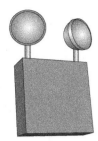

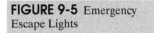
FIGURE 9-5 Emergency
Escape Lights

FIGURE 9-6 Emergency
Exit Sign

an emergency. Exits should include a way to open the door from inside while preventing outside access (e.g., a device such as a "panic bar" on a door).

Lighting is important for a quick and safe exit. Emergency light sources can help guide workers out of dim or dark situations, or situations where vision is obscured (e.g., smoke). (Figure 9-5) Some areas where emergency lights are installed include stairs, aisles, ramps, and passageways, especially those that lead to exits. (Figure 9-6)

Process technicians should follow these recommendations:

- Know the different ways that you will be notified in an emergency (lights, horns, announcements, etc.) and how to respond to each one.
- Make sure you know where exits are located and the emergency plan for your work area. Go immediately to any designated assembly points and report to the designated emergency coordinator.
- Remember that emergency exits may have an alarm that sets off, so use them only in an actual emergency.
- Follow housekeeping practices that help exits and aisles remain clear and unobstructed.
- Make sure flashlights or other light sources used in your work area always have a sufficient charge.

Heights and Fall Protection

Process technicians may be required to work at a variety of heights. Job tasks may require them to climb ladders, work on raised platforms or catwalks, (Figure 9-7) walk on top of tanks, climb towers, and so forth. A surprising number of deaths from falls occur in the eight-foot range, so OSHA mandates that employees use fall protection if working at elevations of six feet or greater (measured from the walking or standing surface to the next lower level) unless proper protection already exists (e.g., railing and cages).

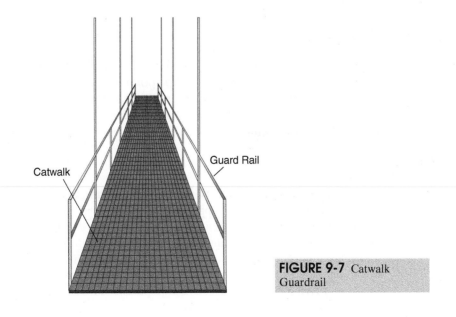

Catwalk

Guard Rail

FIGURE 9-7 Catwalk
Guardrail

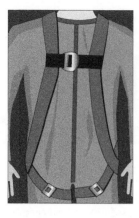

FIGURE 9-8 Fall Protection Harness

For fall protection (or fall arrest), process technicians must wear a full-body harness. (Figure 9-8) This harness is designed to evenly distribute the forces of a fall to muscle groups that can better absorb the forces than other body parts. In a fall-arrest system a lanyard is used to attach the harness to an anchor point. The lanyard and anchor point must be able to support at least 5,000 lbs., and the lanyard length must not allow a fall of greater than six feet. The following can help you remember the ABCs of fall protection equipment:

- <u>A</u>nchor point
- <u>B</u>ody harness
- <u>C</u>onnecting device (lanyard)

Before performing a task that requires a fall protection system, make sure you understand how to properly adjust and wear the harness, as well as remove it (your facility should provide training on this).

Never re-use fall protection equipment that has been worn during a fall, unless the manufacturer states that re-use of such equipment is okay. The impact from a fall can weaken equipment; most often, this equipment is removed from service. Your company might have a policy against re-using fall equipment worn during a fall, no matter what the manufacturer states.

In addition, you should also do the following before working at heights:

1. Inspect the harness for wear or damage (frayed or torn material), paying special attention to the buckles.
2. Make sure the lanyard has no knots on it, and that the D-ring on the lanyard is not damaged (check for sharp edges, dents, or corrosion).
3. Check snap hooks for proper operation.
4. Put on the harness, adjust it properly, and attach the lanyard.

When you reach the desired height, immediately attach the lanyard (already attached to the harness) to the anchor point. Once the job is complete, return to ground level, remove the harness, and return it to the proper location. If any fall protection equipment must be cleaned or repaired, notify the person responsible for the maintenance of this equipment.

Falling Objects

Falls and slips involve impacts between a person and a surface. Falling objects also involve impacts between the dropped object and either a surface or a person. Other accidents can also involve impacts, such as vehicle accidents or a flung object (e.g., a part that has broken off a piece of rotating equipment). This section focuses on dropped objects.

Falling objects typically pose a hazard to the head (including the face and eyes), but can also involve legs, feet, arms, hands and even the torso. Protection from falling objects involves Personal Protective Equipment (PPE) [e.g., hard hats, safety shoes,

FIGURE 9-9 Hardhats Help Prevent Head Injuries Caused by Falling Objects

and safety goggles (or glasses) and in some cases, drop nets (mesh arranged to catch falling objects), screens, or toeboards (vertical barriers, approximately 4 inches high, placed across the bottom of railings]. (Figure 9-9) Refer to Chapter 22, *Personal Protective Equipment and First Aid*, for more details on the different types of PPE.

If you are working at heights, make sure tools are not placed on the edge or in walkways where they can be accidentally knocked off. Wipe any moisture off your hands, or use gloves to maintain a good grip on tools. Also, watch your step to make sure you do not trip and drop any objects you are carrying.

Confined Spaces

Confined spaces are areas with restricted or limited access that are large enough for a person to enter but are not designed for ongoing occupancy.

Examples of confined spaces include the following:

- Tanks/vessels
- Reactors
- Silos
- Furnaces
- Boilers
- Columns
- Sewers
- Excavations
- Pits

Entry into confined spaces is controlled because the area could contain hazardous atmospheres (e.g., oxygen-deficient or flammable), toxic chemicals, powered equipment, extreme temperatures, heights, and so on. Confined spaces can cause potential hazards such as asphyxiation, **engulfment** (being trapped in materials or products in a confined space), heat stress, flooding, electrocution, exposure to toxic chemicals, entrapment, falls, and contact with rotating equipment. In addition, confined spaces do not have easy egress, meaning they can be difficult to exit in case of emergency. (Figure 9-10)

Because of such hazards, OSHA regulations require workers to obtain a permit before entering a confined space. Confined spaces may also include dangerous equipment, piping, and other hazards with which a worker will come into close proximity.

FIGURE 9-10 Extreme Caution should be Used in Confined Spaces, as They may Contain Hazardous Atmospheres and may be Difficult to Exit in the Event of an Emergency

The OSHA Permit-Required Confined Spaces regulation defines confined-space entry, stating that companies must develop a confined-space entry program. Such programs should address the hazards associated with confined-space entry, who is involved with the confined-space procedures, and equipment needed for testing, monitoring, communications, ventilation, and retrieval (safety line). Workers must be trained on the confined-space program and its procedures.

Companies determine what constitutes confined spaces in their facility. All confined spaces should be marked with warning signs, and efforts should be made to prevent unauthorized entry. In order to monitor and control confined spaces, a permitting process must be established, and workers are required to obtain a permit before entering the confined space.

Confined-space permits typically require the following pieces of information:

- Authorization (authorized signatures, time period covered)
- Location of the confined space
- Confined-space entry team (including the name of the person entering the space)
- Potential hazards, including atmospheric levels
- Personal Protective Equipment (PPE) required
- Communication procedures
- Rescue procedures
- Equipment checklist

Each company determines the exact information required for a permit. Confined-space permits are separate from other types of permits, such as lockout and tagout or hot work (see Chapter 21, *Permitting Systems*, for more details). Any additional permits must be obtained before a worker enters the confined space. (Figure 9-11)

For safety reasons, teams are often used for work in confined spaces. Teams typically consist of at least three people: an entry supervisor, the person entering the confined space (entrant), and a standby (or attendant). An entry supervisor oversees the operation and checks that conditions are safe and constantly monitors any changes. There can be one or more entrants.

The entrants perform the work in the confined space, communicating status and conditions with the team. An attendant or standby communicates with the entrants, monitors conditions, watches for hazards, keeps unauthorized workers out of the confined space, coordinates rescues if required, and remains on-station outside the confined space until the entrant leaves the space.

CONFINED SPACE ENTRY PERMIT 867534

Date and Time Issued: ______ Job Site / Space ID: ______ Equipment to be worked on: ______
Date and Time Expires: ______ Job Supervisor: ______ Work to be performed: ______

Entry Signature IN: ______ ______ ______
Entry Signature OUT: ______ ______ ______

Stand-by personnel: ______ ______ ______

Back-up personnel: ______ ______ ______

Atmospheric Check Requirements: Time: ____ Oxygen %: ____ Explosive %: ____ Toxic PPM ____
Tester's Signature: ______

Ventilation Required: Yes ____ No ____
Type: Blower ___ Fan ___ Ambient ___ Other ______

Atmospheric Check After Iso & Vent: Time: ____ Oxygen %: ____ Explosive %: ____ Toxic PPM ____
Tester's Signature: ______

Entry, standby, and back up persons currently trained: Yes ____ No ____

Equipment:	N/A	Yes	No
Breathing Apparatus			
Communications			
Gas Monitor			
Hoisting Equipment			
Protective Clothing			
Safety Harness			
Other:			

Periodic Atmospheric Check :
Time: ____ Oxygen %: ____ Explosive %: ____ Toxic PPM ____
Time: ____ Oxygen %: ____ Explosive %: ____ Toxic PPM ____
Time: ____ Oxygen %: ____ Explosive %: ____ Toxic PPM ____
Time: ____ Oxygen %: ____ Explosive %: ____ Toxic PPM ____
Time: ____ Oxygen %: ____ Explosive %: ____ Toxic PPM ____

Permit prepared by: ______ Approved by: ______

FIGURE 9-11 Sample Confined-Space Permit

It is critical that the name of the person entering the confined space (entrant) is listed on the permit, and that the person be tracked upon entering or exiting the confined space (such as a sign-in/sign-out process).

Before anyone enters the confined space, the following tasks should take place:

- The permit is posted near the confined space.
- The confined space is isolated from the process using blinds or shunts.
- Any chemicals are purged and the area is cleaned.
- All hazardous energy is controlled.
- Ventilation is used to remove any potentially hazardous atmospheres.
- Access to the space is restricted with barriers and signs.
- Proper PPE is selected, tested, and the fit adjusted.
- A safety review is conducted, and the confined space is monitored for any hazardous atmospheres or chemicals that might still be present.

OSHA defines entering a confined space as when any part of a person's body breaks the plane of the space's opening. It is extremely dangerous to put even a part of the body into a confined space, especially the head or even a hand, without following proper confined-space entry procedures.

Workers have died or been injured by just briefly looking into a confined space. Imagine an invisible wall at the perimeter of the confined space, and never put any part of your body across that wall, or "plane," under any circumstances.

Process technicians must remember these vital recommendations about confined spaces:

- Never enter any confined space, or break the plane of such a space with any body part, without a permit.
- Observe all signs relating to confined spaces.
- Follow all confined-space program requirements exactly.
- Obtain all other required permits before entering a confined space.

    If working in a confined space remember the following:

- Check that all confined-space entry requirements have been met.
- Make sure that you properly wear all required PPE.
- Be aware of your own physical conditions, such as excessive sweating, difficulty breathing, increased heart rate, blurred vision, ringing in the ears, and mental fogginess.
- Constantly update your team on your status, including work and physical condition changes.
- Understand that conditions can change in a confined space. Be ready to exit immediately if the situation requires it.

Welding in confined spaces presents additional hazards. Welding is discussed in Chapter 11, *Construction, Maintenance, and Tool Hazards*.

Hazardous atmospheres and confined spaces are covered in Chapter 8, *Hazardous Atmospheres and Respiratory Hazards*. Chapter 21 *Permitting Systems*, also covers confined-space permits.

Government Regulations and Industry Standards

A variety of government regulations relate either directly or indirectly to working areas, fall protection, and confined spaces:

- OSHA General Industry Standards
  - 1910 Subpart D: Walking working surfaces
  - 1910 Subpart E: Exit routes, emergency action plans, and fire prevention plans
  - 1910 Subpart F: Powered platforms, manlifts, and vehicle-mounted work platforms

- 1910 Subpart I: Personal Protective Equipment
- 1910 Subpart R: Special industries
- OHSA 1910.146 covers permit-required confined spaces

OSHA designates illumination levels in the workplace through its Hazardous Waste Operations and Emergency Response **(HAZWOPER)** standard. It sets illumination standards for areas such as corridors, hallways, exit ways, warehouses, mechanical and electrical rooms, storerooms, living quarters, locker rooms, indoor restrooms, workrooms, and other similar areas.

The National Fire Protection Association publishes a Life Safety Code, which includes recommendations on exits and evacuations. The American National Standards Institute (ANSI) publishes a Fall Protection Standard.

Summary

There are many potential hazards in the workplace. Slips and falls are just two of these hazards. Slips and falls can occur when walkways are slippery, improperly built, littered with foreign objects, or when improper footwear is worn. They can also occur as a result of improper ladder usage.

Another process industry hazard is confined spaces. When working in confined spaces, process technicians must always be aware of exit locations, the potential for hazardous atmospheres, and any other hazards that could be associated with a confined space. Technicians must also be familiar with the permits and procedures associated with confined-space entry.

While federal, state, and local codes are in place to reduce the likelihood of workplace injury, hazards and injuries can still occur. In order to prevent these injuries, process technicians should always be safety conscious and aware of their surroundings, follow proper procedures, and use appropriate personal protective equipment when required. In addition, technicians should take actions to prevent accidents or injury whenever possible (e.g., cleaning up a spill, or removing hazardous obstructions from walkways).

Checking Your Knowledge

1. Define the following terms:
   a. Dockboards
   b. Engulfment
   c. Guardrail
   d. HAZWOPER
   e. Three-point contact
   f. Toeboard
2. *(True or False)* Falls and slips are one of the most common cause of injuries in the workplace.
3. Select the statement that is NOT true for working with ladders:
   a. Inspect the ladder before using it.
   b. Do not place any objects under the ladder to raise its height.
   c. Never extend your waist above the top rung.
   d. Face away from the ladder when climbing down it.
4. Means of egress refers to _____.
5. What is the 4-to-1 rule when leaning a ladder against a support position?
6. A surprising number of deaths occur because of falls from which height?
   a. 4 feet
   b. 8 feet
   c. 12 feet
   d. 20 feet
7. According to OSHA, employees need fall protection when working at heights of ____ feet or higher.

8. The length of a lanyard on a fall protection system must not allow a fall of greater than how many feet?
   a. 2 feet
   b. 4 feet
   c. 6 feet
   d. 8 feet
9. A(n) _____ point is the name of the object that a lanyard is attached to when a worker is using a fall-protection system.
10. Name three types of Personal Protective Equipment that can protect against falling objects.
11. *(True or False)* A permit must be obtained before entering a confined space.
12. What is the name of the OSHA regulation covering confined spaces?

Student Activities

1. Inspect your home or work area for potential fall or slip hazards, then make a list of them. Describe how each hazard can be reduced.
2. Using your school or work area, locate the escape route for fires or emergencies. Make sure you know how to exit the facility in case of evacuation. Think about potential obstacles that you must overcome, or how you would handle different scenarios (such as the lights going out).
3. Research the OSHA confined-space entry regulations, and write a three-page paper on what the regulation is, what it protects against, what it requires companies to do, and how it is enforced.
4. If available, practice inspecting and properly using a portable ladder.

10

Hearing and Noise Hazards

Objectives

This chapter provides an overview of hazards from noise in the process industries, which can affect people and the environment.

After completing this chapter, you will be able to do the following:

- Name specific hazards associated with noise generated in a process industry environment.
- Describe how these variables can impact hearing:

  Volume of noise

  Length of exposure

- Describe government regulations and industry guidelines that address noise and hearing protection.

Key Terms

Amplitude—the measurement used to describe the intensity of sound.

Decibel (dB)—the measurement of the intensity of a sound, based on the human ear's perception. A unit that is used to measure sound-level intensity (how loud a sound is).

Frequency—the number of sound vibrations per second (peaks of pressure in a sound wave).

Hertz (Hz)—a measurement used to describe frequency. One hertz is one cycle per second.

Intensity—the loudness of a sound (pressure-peak intensity of a sound wave).

Noise—any unwanted or excessive sound.

Sound—a form of vibrational energy conducted through a medium (e.g., solid, liquid or gas) that creates an audible sensation that can be detected by the ear.

Sound wave—a pressure wave that moves through the air and is audible to the human ear.

Vibration—the rapid movement of an object back and forth along its radial or horizontal axis, in a periodic motion.

Wavelength—the distance between successive points of equal amplitude on a sound wave.

Introduction

The process industries can be noisy workplaces, with much of the noise created by the equipment used in operational processes and flow of materials through pipes. Process technicians must recognize the hazards that noise presents, understand how to reduce the risk of noise-induced hearing loss, and be familiar with the government regulations that address noise and hearing conservation.

This chapter explains what sound and hearing are, how noise is defined, the impact of noise on hearing, vibration and its effects, and the government regulations that address noise and hearing conservation.

Noise and Hearing Hazards Overview

Noise is any unwanted or excessive sound. In the workplace, excessive noise can result in a variety of potential hazards. These include the following:

- Hearing loss or reduction (either temporary or permanent)
- Ringing in the ears (tinnitus)
- Equilibrium problems (dizziness, disorientation)
- Speech problems
- Distraction (which could lead to an accident)
- Interference with communication or the ability to understand communication (e.g., verbal instructions or alarms)
- Psychological effects (e.g., stress)
- Fatigue

The effects of noise hazards can be caused by long-term exposure (e.g., repetitive noise over an extended period of time) or a single acute exposure (e.g., an explosion).

The next section explains more about noise by discussing the characteristics of sound and how hearing works.

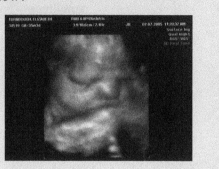

Did You Know?

Ultrasound equipment uses high-frequency sound waves to create an image of internal organs.

The frequency of these sound waves (approximately 20,000 to 10 billion cycles per second) is beyond the range of what the human ear can hear.

The Basics of Sound and Noise

Sound is a form of energy conducted through a medium (e.g., solid, liquid, or gas) that creates an audible sensation that can be detected by the ear. Sound starts with a vibration that travels in waves (called **sound waves**). Sound waves can vary in length and amplitude.

Wavelength is the distance between successive points of equal amplitude on a sound wave. **Amplitude** is the distance from the midpoint to the top (crest) of a sound wave or from the midpoint to the bottom (trough) of the wave. Figure 10-1 illustrates wavelength and amplitude.

Vibration is the rapid back-and-forth movement of an object along its radial or horizontal axis, in a periodic motion. This vibration disturbs the surrounding medium (e.g., air) causing it to vibrate. As the atoms in the medium vibrate, they transfer energy (pressure) outward from the source (similar to ripples in a pond). When the pressure wave reaches our ear, it is received and then converted into a signal that our brains interpret as sound.

The ear is made up of three main sections: the outer ear, the middle ear, and the inner ear. The outer ear is the external part that "gathers" sound waves and directs

FIGURE 10-1 Illustration of Wavelength and Amplitude

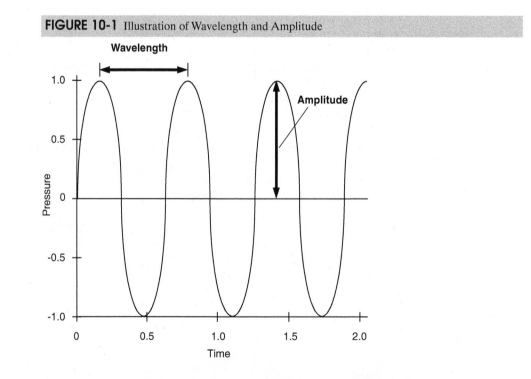

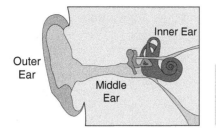

FIGURE 10-2 The Human Ear Gathers Sound Waves and Converts them to Signals the Brain Interprets as Hearing

them toward the eardrum. When a sound wave is received, the eardrum vibrates. As the eardrum vibrates, the bones and muscles in the middle ear begin to move and conduct sound from the outer ear to the inner ear. The inner ear then converts sound to nerve impulses, which are sent to the brain for interpretation.

Two terms are important to understand about sound: frequency and intensity. **Frequency** is the number of sound vibrations per second. The measurement used to describe frequency is called **hertz (Hz)**. The human ear has the ability to hear frequencies of approximately 20 Hz to 20,000 Hz.

Intensity is the loudness of a sound. The concept of loudness is subjective. Whether a sound is considered to be "soft" or "loud" depends on the person experiencing the sound.

The measurement used to describe intensity is called amplitude. In addition to intensity, the human ear can sense frequency (pitch). The body can also feel loud or low frequency sounds (think of a stereo with the bass turned up). Figure 10-3 illustrates frequency and amplitude.

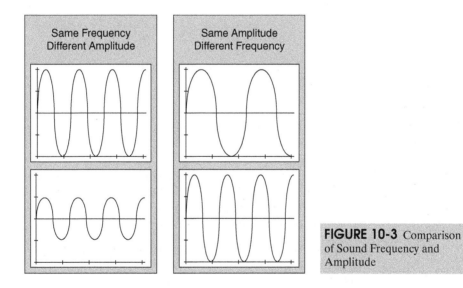

FIGURE 10-3 Comparison of Sound Frequency and Amplitude

Sound is measured in decibels. A **decibel (dB)** is the smallest difference in the level of sound that the human ear can perceive. The human ear is sensitive to a wide range of sound frequencies. However, the ear has been found to be more sensitive to frequencies in the upper middle range. Sound-level measurement takes this into account, using an international standard adjustment called the "A weighted scale." Readings from this scale are listed as dBA, instead of just dB.

The decibel scale does not follow a linear route; it is more of a rising curve (called a logarithmic scale). According to this scale, noise intensity doubles with each 3-dB increase. This means that a sound that is 88 dB is twice as loud as a sound that is 85 dB, and a 30-dBA sound is more than 1,000 times greater than a 1 dBA sound.

The threshold of hearing (the weakest sound a human ear can hear in a quiet environment) is 1-10 dBA. Discomfort can occur during sounds in the 85-95 dBA range. The threshold of pain (the maximum level before the sound becomes painful to the ear) is 120-140 dBA. Table 10-1 lists some common sounds and their approximate dBA values.

TABLE 10-1   Common Sounds and their Corresponding Decibel (dB) Values

Sound	Loudness (dB)
Gunshot (peak level)	140 to 170
Jet takeoff	140
Rock concert	110 to 120
Chain saw	110 to 120
Diesel locomotive	110 to 120
Stereo headphones	110 to 120
Motorcycle	90
Lawn mower	90
OSHA level for required hearing protection	**85***
Conversation	60
Quiet room	50
Whisper	30 to 40

* 8-hour time-conservation program weighted average

150
145
140 — Jet Taking Off
135
130
125
120
115
110
105
100
95
90 — Heavy Truck
85
80
75
70
65
60
55
50
45
40 — Quiet Office
35
30
25
20
15
10 — Leaves Rustling
5
0 — Threshold of hearing

Often, the way a sound is perceived by a person determines whether the sound is considered to be noise or not (e.g., loud rock music may be entertainment to one person, and painful noise to another).

In the process industries, noise is sound that is excessive or can cause harm. Government regulators have established guidelines for hearing conservation and managing noise hazards. A later section in this chapter will describe these guidelines.

Noises can be described by the following terms:

- Continuous noise—steady, unchanging noise; typically generated by equipment in operation, such as pumps, compressors, turbines, and fans
- Impulse noise—noise that has a sharp rise followed by a rapid decline in sound levels (less than one second); can be repetitive or occur just once. Some pumps or piston-driven equipment can make this type of noise
- Impact noise—noise generated by an object striking another surface. This is typically a noise that lasts less than a second in duration, but may produce a delayed sound such as a ringing. Some pneumatic tools (e.g., jackhammers and nail guns) can cause this kind of noise
- Intermittent noise—noise in which the level is interrupted by periods of low sound levels
- Varying noise—noise in which the levels change substantially
- Ambient noise—background noise associated with a given environment (e.g., heating or air conditioning noise or traffic noise). Localized ambient noise in the process industries can often exceed 85 dB

Noises can consist of sounds from a wide range of frequencies (e.g., many types of process operations environments) or a narrow range of frequencies (e.g., power tools).

> **Did You Know?**
>
> Because sound is the vibration of matter, it does not travel through a vacuum or in outer space.
>
> In movies and TV shows, the sounds of spaceships exploding are added for dramatic effect. In reality, you would see the explosion but would not be able to hear it.
>
>

Noise-Related Hazards

Hearing loss is the primary hazard associated with noise. While noise-induced hearing loss can be prevented, once hearing loss starts it is permanent (i.e., damage cannot be reversed). Noise-induced hearing loss usually occurs following long-term exposure to excessive noise, but it can also be caused instantly by a quick, intense sound such as a blast or some other trauma.

Gradual hearing loss occurs over time and may not be noticeable for 20 years or more when normal sounds begin to seem muffled or garbled. Usually a person feels no pain during gradual hearing loss.

Hearing loss can result from damage to just about any part of the ear. However, noise-induced hearing loss specifically targets hair cells in the inner ear. As sounds are received, these hairs wave back and forth, helping to transmit sound vibrations to the audiometric nerve (the nerve that transmits hearing to the brain). When the ears are subjected to constant repetitive noise, the hair cells in the ear become worn from continuous movement. This leads to permanent damage of the hair cells (similar to how walking on grass repeatedly will wear a path of damaged or dead patches). If the noise is excessively loud and sudden, the hairs can actually be sheared off. Unlike the hairs on the head, hair cells in the ear do not repair themselves once they are damaged.

Depending on conditions, hearing loss can be short or long-term. For example, a person in a noisy environment (e.g., a rock concert), or who has heard a loud noise (e.g., a gunshot), may experience short-term hearing problems. Once away from the noise, however, normal hearing may return. However, with gradual hearing loss, normal hearing does not return.

High-frequency hearing loss occurs first (above around 2,000 Hz), affecting the ability to hear speech. Once this loss occurs, low-frequency hearing loss can follow, impacting the ability to hear other sounds as well.

Some individuals with hearing loss may also experience tinnitus, a condition which results in a ringing or roaring sound in the ear.

If a person experiences a permanent hearing loss across a variety of frequencies, this can be considered a handicap. Such a handicap can impact employment, family relationships, social interactions, and more. A hearing handicap can negatively impact the person's ability to communicate with others.

Noise can cause workers to be distracted and lose focus on their task and safety, which can lead to accidents. Noise can also drown out important sounds such as communications, alarms or warnings, and how running equipment sounds.

Noise in the workplace can produce other effects in addition to hearing loss:

- Interference with communication or the ability to understand communication such as verbal instructions or alarms (this is probably the second most common and serious hazard in the workplace)
- Disorientation or dizziness (related to equilibrium and caused by inner ear problems)
- Speech problems (related to hearing loss)

Did You Know?

The decibel was named after Alexander Graham Bell, inventor of the telephone and founder of the Bell Telephone Company.

Bell was also known for his work educating the deaf.

- Pain and nausea
- Muscle tension
- Stress or other psychological issues
- Increased pulse and elevated blood pressure (related to stress)
- Fatigue

Acoustic trauma is a single event that results in abrupt hearing loss. For example, a loud explosion can cause acoustic trauma. But other incidents, such as a blow to the head or flying objects (e.g., chemical splashes, welding sparks, or rotating equipment throwing a part) that hit the ear or ear drum can also result in acoustic trauma.

Exposure to some chemicals, such as solvents and heavy metals, can potentially contribute to hearing loss. Other factors, such as a person's age and health, also contribute to hearing problems.

Noise Exposure and Duration

Hearing loss can occur quickly or slowly. A single exposure to 140 dBA of noise can cause hearing loss. Prolonged exposure to noise over 85 dBA can also produce the same results. Figure 10-4 shows an example of an audiogram from a person with noise-induced hearing loss.

Following are factors associated with noise hazards and potential hearing loss:

- Noise intensity
- Frequency of the sound
- Daily duration of noise exposure
- Total duration of noise exposure (e.g., number of years)

Recall that intensity is the loudness of a noise. For duration, both daily exposure and total exposure are factors. Other factors include the distance between the worker and the noise and the orientation of the ears to the noise (i.e., one ear might be turned toward the noise while the other is away from it).

Some individuals are more susceptible to noise, showing more sensitivity to sounds than others. Age and the health of the worker are also factors. Hearing loss can occur naturally the older a person becomes (age-induced hearing loss is called presbycusis). Hearing loss from old age is a relatively even loss over all frequencies.

FIGURE 10-4 Audiogram of a Person with Noise-Induced Hearing Loss

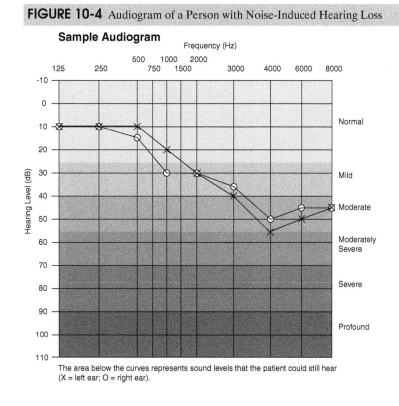

Sample Audiogram

The area below the curves represents sound levels that the patient could still hear (X = left ear; O = right ear).

Government regulations on noise hazards address noise level and duration. A general rule is if you must raise your voice to speak with someone less than a few feet away, then the noise in the environment could be hazardous. Also, if your ears ring or sounds seem dull after leaving a noisy environment, then the noise is probably hazardous.

Hearing Conservation and Protection

Noise levels at process facilities must be monitored on a regular basis. When changes occur in the work environment (such as when a new process is added or new equipment is installed), noise levels must be re-checked. Equipment that needs maintenance or is starting to fail can also create a change in the noise level. A special tool, such as a sound-level meter, can be used to measure noise levels. Employees can share with sound-level surveyors their knowledge of the work environment, the equipment, the process and other factors that can affect sound-level measurements.

Companies with high noise levels in one or more work areas must implement a hearing protection program to conserve the hearing of employees exposed to the noise. Exposure to noise levels of 80 dBA or less is generally considered safe. Prolonged exposure to any noise levels over 80 dBA requires hearing protection. The process industries generally require hearing protection in areas that exceed the 80-dB level. Exposure to levels over 115 dBA can be dangerous. A one-time exposure to a noise 140 dBA or higher can cause permanent hearing damage.

The Occupational Safety and Health Administration (OSHA) and the American National Standards Institute (ANSI) are the two main sources of noise hazard regulations and standards. OSHA defines a hazardous noise as "any sound for which any combination of frequency, intensity, or duration is capable of causing permanent hearing loss in a specified population." OSHA hearing protection regulations require companies to implement a hearing conservation program and provide hearing protection when employees are exposed to an eight-hour Time Weighted Average (TWA) of 85 dBA or above, based on a 40-hour work week. TWA is an average determined by dividing the total sample weight by the total sampling time.

A noise dose, or exposure, is expressed as a percentage of the allowable daily exposure. OSHA also has set a Permissible Exposure Limit, or PEL, of 90 dBA for an eight-hour TWA.

TABLE 10-2 OSHA Permissible Exposure Limits for Noise	
Number of Hours	*Decibel (dBA) Level*
8	90
6.2	92
4	95
3	97
2	100
1.5	102
1	105
0.5 (half an hour)	110
0.25 (fifteen minutes)	115

OSHA uses a rate relationship between the intensity and a dose of 5 dB. Thus, if the intensity of an exposure increases by 5 dB, then the dose is considered to be doubled. For example, for a dBA of 95, the duration must not exceed four hours.

The OSHA regulations cover the following basic requirements:

- Noise-level monitoring—checked on a regular basis or when changes occur in the work environment (e.g., a new process is added, an existing process is changed, or equipment is added).
- Medical monitoring—employees exposed to high noise levels must have their hearing tested. Employees are given audiometric hearing tests for both ears (the results are called an audiogram) when they are hired to establish a baseline and at least on an annual basis thereafter. OSHA requires records to be kept relating to hearing tests. Individual employee exposure to sound can be tracked, using a device called a dosimeter. A dosimeter is an instrument that measures and stores sound levels over a specified interval. This device allows a company to monitor an employee's exposure to sound and track historical data.
- Noise controls—used to address the source of noise. Noise control is required in settings where the noise level exceeds 90 dBA. Administrative programs, such as training and limited exposure, can be used unless the noise exceeds 100 dBA. In those cases, engineering controls (such as sound dampening or a barrier to keep workers away from the noise) must be used to control the noise hazard.
- Training and education programs—designed to address the effects of noise on hearing, the purpose of hearing testing, the purpose of hearing protection devices, and how to properly select and fit such devices.
- Personal Protective Equipment (PPE)—provided to workers for use around high noise levels; includes ear plugs, ear muffs, and similar devices. PPE is required when engineering and administrative program noise controls do not reduce the levels below the legal limits.

ANSI has published a set of standards, called S12.13-991 "Evaluation of Hearing Conservation Programs" to help employers determine the effectiveness of their programs.

TYPES OF HEARING PROTECTORS

Hearing protection is selected based on the following criteria, in order of importance:

1. Intensity of noise
2. Personal comfort
3. Availability

The main types of hearing protectors are earplugs or earmuffs. (Figure 10-5) Earplugs are made of materials such as foam, rubber, or plastic, and are inserted into

FIGURE 10-5 The Main Types of Hearing Protectors are Earplugs or Earmuffs

the ear canal. Earmuffs cover the entire ear with a cushioned cup held in place with a headband. Earplugs and earmuffs can reduce noise levels by up to 20-30 dBA. Worn together, they can reduce the noise level up to an additional 5 dBA (double hearing protection is required if the worker is exposed to noise levels of 110 dBA or higher).

One drawback to hearing protection equipment is the reduced ability to hear important sounds, such as alarms, announcements, vehicles, and equipment problems.

FIT, USE, AND MAINTENANCE OF HEARING PROTECTORS

Employers are required to provide hearing protection and train workers on its use and care. The following are some general safety procedures related to hearing protectors:

- Select the proper hearing protector based on the noise level and personal comfort. Cotton balls or fingers are never a substitute for proper hearing protection.
- Make sure your ear canals are free of ear wax, or that you do not have an ear infection. These can complicate the wearing of hearing protection.
- Before use, inspect hearing protectors for wear or damage. Replace any damaged hearing protectors.
- Make sure hearing protectors fit properly (e.g., adjust earmuffs or place plugs securely in the ear canal).
- Clean hearing protectors before use, based on manufacturer's recommendations.
- Do not remove hearing protection while exposed to the noise hazard, even for a short period. This significantly reduces the hearing protection's effectiveness.
- Wear hearing protection properly and consistently.
- Always wear your dosimeter (if one is provided).
- Maintain an accurate record of potential hearing loss; keep medical records update, including any non-occupational exposure to noise and any exposure to chemicals.

Government Regulations on Noise Hazards

OSHA regulation 29 CFR 1910.95 regulates occupational noise exposure, describing permissible noise exposures. It also outlines requirements for engineering and administrative controls, along with a hearing conservation program.

Following are other federal agencies with noise standards that may affect the process industries:

- Department of Transportation (DOT)
- Environmental Protection Agency (EPA)
- Coast Guard
- Mine Safety and Health Administration (MSHA)
- Federal Aviation Administration (FAA)

Summary

The process industries can be very noisy, with much of the noise being produced by equipment and the flow of materials through pipes. These sounds, which are measured in decibels (dB), can vary in frequency (pitch) and intensity (loudness).

The decibel scale does not follow a linear route; it is more of a rising curve. According to this scale, noise intensity doubles with each 3-dB increase. This means that a sound that is 88 dB is twice as loud as a sound that is 85 dB, and a 30-dBA sound is more than 1,000 times greater than one that is 1 dBA.

Sounds below 85 dB are considered safe and do not require hearing protection. Sounds over 85 dB do require hearing protection. Discomfort can occur with sounds in the range of 85-95 dB. Sounds over 120 dB can cause intense pain.

Hearing loss is the primary hazard associated with noise. Hearing loss can occur quickly or slowly. A single exposure to 140 dB of noise can cause permanent hearing loss. Prolonged exposure to noise over 85 dBA can produce the same results. Because of this, companies with high noise levels in one or more work areas must implement a hearing conservation program to conserve the hearing of employees exposed to the noise.

Process technicians must recognize the hazards that noise presents, understand how to reduce the risk of noise-induced hearing loss, and be familiar with the government regulations that address noise and hearing conservation in the workplace.

Checking Your Knowledge

1. Define the following terms:
   a. Amplitude
   b. Decibel (dB)
   c. Frequency
   d. Hertz (Hz)
   e. Intensity
   f. Noise
   g. Permissible Exposure Limit (PEL)
   h. Sound
   i. Sound wave
   j. Time-Weighted Average (TWA)
   k. Vibration
2. *(True or False)* Sound is caused by a vibration.
3. Frequency is the number of sound _____ per second.
4. Intensity is:
   a. A decibel measurement.
   b. A high-frequency sound that is beyond the range of the human ear to hear.
   c. Unwanted sound.
   d. The loudness of a sound.
5. A decibel is:
   a. A weighted frequency average.
   b. A sound measurement.
   c. Part of the inner ear.
   d. Another term for noise.
6. _____ is unwanted sound.
7. Name five potential effects that noise hazards can cause.
8. *(True or False)* Vibrations do not pose a hazard.
9. Which of the following are factors associated with the impact of noise hazards? (select all that apply)
   a. Intensity
   b. Daily duration
   c. Sound wave refractionation
   d. Long-term duration
   e. Vibratory covalence equation
10. What is a TWA and how does it relate to noise hazards?
11. OSHA 29 CFR 1910.95 regulates occupational _____ exposure.

Student Activities

1. Write your definition of sound and noise, and then survey your classmates to get their definitions. Make a list of all the definitions and compare them.
2. Brainstorm activities outside the workplace that can cause hearing loss and use the Internet to research their decibel levels.
3. Visit the OSHA Web site and find 1910.95, addressing occupational noise exposure. Write a one-page paper on how this regulation protects process technicians.

11

Construction, Maintenance, and Tool Hazards

Objectives

This chapter provides an overview of construction, maintenance, and tool hazards in the process industries, which can affect the safety and health of workers.

After completing this chapter, you will be able to do the following:

- Name specific hazards associated with construction and maintenance tasks in a process industry environment.
- Describe how hand and power tools can be hazards.
- Describe government regulations and industry guidelines that address construction, maintenance, and tool hazards.

Key Terms

Arc welding—a welding process that uses an electrical arc produced between two electrodes to generate heat.

Electric tool—a tool operated by electrical means (either AC or DC).

Hand tool—a tool that is manually powered.

Hydraulic tool—a tool that is powered using hydraulic (liquid) pressure.

Oxyacetylene welding—a welding process that burns a blend of oxygen and acetylene to generate heat.

Pneumatic tool—a tool that is powered using pneumatic (air or gas) pressure.

Powder-activated tool—a tool that is powered using a small explosive charge (e.g., nail gun).

Power tool—a tool that is powered by a source such as electricity, pneumatics, hydraulics, or powder-activation.

Resistance welding—a welding process that uses electricity generated through the material to be welded combined with pressure at the weld point, to create the weld.

Introduction

There are many tool and construction hazards in the process industries. Even though process technicians are not likely to be involved in construction-related tasks, they might encounter construction areas around the facility. Because of this, technicians must be aware of the safety hazards that construction areas present.

Process technician duties include many maintenance-related tasks. They must understand all the hazards that maintenance tasks pose, as well as some general safety recommendations.

Hand and power tools are an integral part of the process technician's daily job. Knowing how to use tools properly and safely is critical.

This chapter provides an overview of hazards that the process technician can encounter relating to construction, maintenance, and tool usage.

Construction

In most instances process technicians will not be involved in construction tasks at work. However, they could be required to work around construction or repairs at their facility. It is outside the scope of this textbook to cover all construction-related government regulations and safety precautions. However, there are some general guidelines that should be followed. Specifically, process technicians should follow these general guidelines:

- Obey all construction signs.
- Never cross a construction barricade without approval.
- Wear appropriate Personal Protective Equipment (PPE) when passing through a construction area (e.g., hard hat, safety glasses, or goggles).
- Be aware that conditions around the facility can change and that new hazards can appear during construction (e.g., vehicles, construction debris, pits, and utility line damage).
- Watch out for construction vehicles (e.g., cranes, hoists, backhoes, bulldozers, dump trucks, cement trucks, and powered lifts).
- Exercise caution near trenches and excavations.
- Monitor noise levels and increase hearing protection as needed.

OSHA guidelines for construction are similar to some of the general industry regulations (e.g., fall protection, ladders, and PPE). However, some construction regulations cover issues that OSHA does not address in general industry regulations, such as lighting requirements.

Maintenance

Process technicians may perform a variety of maintenance and repair tasks as part of their jobs, since making sure equipment is maintained regularly and working properly is key to efficient process operations. However, maintenance operations expose process technicians to a variety of risks:

- Chemical hazards or biological hazards, when opening equipment that had product moving through it, or product that was not properly isolated from the process
- Electrical hazards from energized equipment
- Pressure hazards from pressurized equipment
- Heat hazards from fired equipment and hot work environments
- Radiation hazards from testing equipment
- Fire or explosion hazards when working in flammable or hazardous atmospheres
- Hazardous atmospheres when working in confined spaces or doing emergency repairs for spills and releases
- Injuries from repetitive tasks or working in unnatural or ergonomically incorrect positions
- Confined spaces
- Exposure to high levels of noise

When performing maintenance tasks, process technicians should always follow these procedures:

- Use tools properly and follow safety procedures.
- Wear appropriate Personal Protective Equipment (PPE).
- Work safely with ladders; use fall protection if working at heights of six feet or more.
- Follow proper permitting processes for hot work, confined-space entry, hazardous-energy control, and other similar requirements.
- Follow proper processes and procedures as outlined by the company.
- Use monitoring and testing equipment properly.
- Be aware of conditions that could change while performing maintenance procedures and create a hazard.
- If working near an area where maintenance is being performed, be alert for hazards that might not have previously existed (e.g., open hatches, spills, loose parts, and tools in walkways). Obey all signs and instructions around these areas.

Tools

Tools are common to every process technician's job, whether they are hand tools or power tools. Tools can help workers perform their daily tasks effectively and efficiently. However, even the simplest tools can present hazards if not used properly and safely.

The types of tools that process technicians use vary based on the process industries in which they work, along with job requirements. While it is beyond the scope of this textbook to discuss every type of tool a process technician might use, this section does cover some general use and safety tips.

Following are some of the more common tools:

- Screwdriver
- Hammer
- Pliers
- Wrench
- Socket wrench
- Drill
- Vise™ grips
- Knife
- Saw
- Grinders

OSHA states that employers shall be responsible for the safe condition of tools and equipment used by employees. It is up to the employee to ensure that tools are used properly and safely.

The following are some general tool usage and safety tips:

- Never "horseplay" in the workplace or when working with tools.
- Always obtain required permits for dangerous or hazardous work (e.g., a hot-work permit for grinding operations).
- Use the right tool for the job.
- Understand how to use and maintain the tool.
- Review safety procedures to learn hazards associated with a specific tool.
- Select the proper handgrip, either left or right-handed, or both (ambidextrous).
- Inspect the condition of the tool before use.
- Check that edged tools are sharp.
- Look over metal tools for slivers, cracks, or rough spots.
- Check for wear or damage on any plastic or rubber parts or coatings.
- Inspect wood handles for splinters, chips, or weathering.
- Do not use damaged tools (make sure they are labeled "Do Not Use" and report the tool condition to your supervisor).
- Wear the appropriate Personal Protective Equipment (PPE).
- Make sure the work area is clear and obstacles are removed.
- Keep a firm grip on the tool and maintain proper balance and footing.
- Do not leave tools in walkways or high traffic areas.
- Make sure the floor is clean and dry to prevent slips or falls with or around tools.
- Make sure the work area is properly lit.
- Maintain a safe distance from other workers.
- Do not distract others while they are using a tool.
- Clean the tool when done and return it to the proper place.
- Be aware of other work that may affect the usage of the tool in that area.

Following are some general hazards that tools can cause:

- Sparks (especially from grinding tools) in flammable or hazardous environments, resulting in a fire or explosion
- Impact from flying parts or fragments (from the tool or material being worked on)
- Cuts, scrapes, bruises, broken bones, or similar injuries
- Harmful dusts, fumes, mists, vapors, or gases released by the material being worked on
- Impalement on sharp edges
- Impact if the tool is dropped from a height
- Ergonomic hazards such as carpal tunnel syndrome or tendonitis from improper design or use, or repetitive motion
- Noise hazards, such as a power tool, or a hand tool striking a surface
- Falls while trying to use the tool at heights

FIGURE 11-1 Common Hand Tools

HAND TOOL USE AND SAFETY

Hand tools are non-powered and come in a wide range of types and designs. (Figure 11-1) Misuse and improper maintenance pose the greatest hazard. The following are some general tips specific to the proper use and safety of hand tools:

- Check wooden-handled tools for splinters, cracks, chips, or weathering. Make sure the handle is securely attached (such as hammer heads).
- Do not use screwdrivers as chisels.
- Check impact tools (e.g., chisels and wedges) for heads that have been flattened from repeated use (called "mushroom" heads) as these can shatter.
- Inspect wrenches to make sure that jaws are not sprung to the point that slippage occurs.
- Use spark-resistant tools around flammable substances (e.g., tools made from non-sparking materials such as brass, plastic, aluminum, wood, titanium, bronze, and Monel™).
- Be aware when working around electricity; make sure tools are grounded and insulated.
- Inspect spark-resistant tools for wear or damage, since they are made of materials that are softer than other tools and can wear down quickly.
- Keep a proper grip, holding the handle firmly across the fleshy part of your hand.
- Do not overexert yourself. If you feel your hand or arm strength weakening, take a break and resume the task later.
- In moist environments, make sure your hand remains dry and your vision is not blurred by sweat.

POWER TOOL USE AND SAFETY

There are various types of **power tools**: **electric** (powered by an AC or DC current), **pneumatic** (powered by air or gas pressure), fuel-operated, **hydraulic** (powered with liquid pressure), or **powder-activated** (powered by a small gunpowder charge). Power tools can enhance performance, but they can also pose the hazards described previously for all tools, along with the following:

- Power-source related injuries such as electrocution (electrical) or line whip (pneumatic)
- High-speed impact from projectiles hurled by broken tools or materials
- Injuries from contact with moving parts or entanglement
- Falls or trips from cords or hoses
- Vibration and noise
- Slipping on leaking fluids

In addition to the general safety procedures outlined previously, the following are some safety procedures specific to power tools:

- Follow all manufacturers' specifications for proper and safe use and read all warning labels.
- Understand the tool's capabilities, limitations, and hazards.
- Check that all safety features work properly.
- Never point the tool in the direction of another person.
- Check that all guards and shields are in place.
- Do not use tools that are too heavy or too difficult for you to control.
- Inspect all cords or hoses for wear, fraying, or damage.
- Make sure the tool is kept clean and properly maintained, including lubrication, filter changes, and other manufacturer-recommended practices.
- Avoid using tools in dangerous environments (e.g. wet, flammable, extreme heat or cold).
- Use only intrinsically safe or explosion-proof tools in flammable environments.
- Do not carry the tool by its cord or hose.
- Do not yank the tool cord or hose to disconnect it.

- Keep cords and hoses away from heat, oil, and sharp edges.
- Do not let cords or hoses get knotted or tangled with other cords/hoses.
- Make sure cords or hoses are not lying across a walkway presenting a fall hazard.
- Keep cords or hoses away from rotating or moving parts.
- Choose the correct accessories and use them properly.
- Check all handles to make sure they are secure and any grips are on tight.
- Keep your finger off the switch when carrying the tool or before it is in position to avoid accidental startup.
- Hold or brace the tool securely.
- Disconnect the tool before servicing it or when changing accessories (e.g., blades and bits).

Following are some safety practices related to electric power tools:

- Make sure all electric tools are properly grounded or double insulated.
- Always use a ground fault circuit interrupter (GFCI) when using electric power tools.
- Know the hazards of electricity, including electrocution, shocks, and burns.
- Be aware of potential secondary hazards from electricity (e.g., a mild shock from an electric tool that startles a worker off a ladder).
- Never remove the grounding plug from a cord.
- Understand the mechanical hazards of the tool.
- Avoid using the tool in wet or damp environments.
- Do not use a tool if it becomes wet.

Pneumatic tools can be powered by compressed air or gas. (Figure 11-2) Safety practices related to pneumatic tools are as follows:

- Do not use compressed air for cleaning purposes, unless it is reduced to less than 30 psi and then only with effective chip guarding and PPE.
- Make sure attachments are properly secured with a tool retainer/safety clip.
- Check the hose and hose connection.
- Do not exceed the manufacturer's safe operating pressure for hoses, pipes, valves, filters, and fittings.
- For hoses with a ½ inch inside diameter, a safety device must be used at the source of supply or branch line to reduce pressure in case of hose failure to prevent line whip.
- Never kink the hose to cut off the air supply; turn off the air using the valve.
- Check for proper pressure before using the tool.
- Secure hose connections with a lanyard.

Many of the safety precautions for pneumatic tools relate to hydraulically powered tools as well, along with the added precaution about exposure to hydraulic fluids. Hydraulic fluids can be very hot and/or harmful if you are exposed to them.

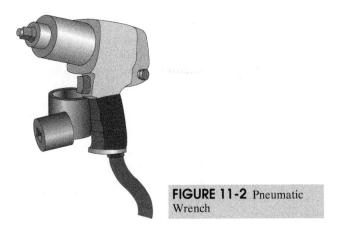

FIGURE 11-2 Pneumatic Wrench

For fuel-operated tools, use caution when filling. Remember that liquids can cause a static electricity buildup when filling the tool. Fill the tool in open air to avoid buildup of fumes.

For powder-activated tools, handle them carefully. Never point the tool at another person. Remember you are dealing with an explosive charge, even if it is a small one. Most of these types of tools will not activate unless the tip is pressed against a work surface. Never press the tip to anything other than the work surface.

WELDING

Welding, cutting, and brazing operations are widely used tasks in the process industries. Welding covers processes such as gas welding, electric **arc welding** (which uses an electrical arc produced between two electrodes to produce heat), **resistance welding** (which uses electricity generated through to the material being welded combined with pressure at the weld point, to create the weld) and similar processes such as brazing and soldering. This section describes some general welding hazards and safety precautions. Anyone doing welding work should be properly trained and understand the inherent dangers.

Welding hazards include heat, sparks, intense light (which can cause blindness), fumes, flammable gases (gas welding), and electricity (arc welding).

Welding is also one of the principal causes of industrial fires. OSHA welding requirements cover fire protection, personal protection, and ventilation. OSHA classifies, welding work under hot-work permits (discussed in more detail in Chapter 21, *Permitting Systems*). A hot-work permit must be obtained before any welding work is started. Welding also produces hazardous gases and fumes, so proper ventilation is required. Some gases can combine with water to form acids. Proper PPE must always be worn, including vision protection and clothing to protect from burns.

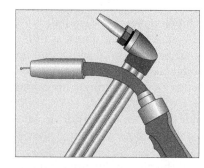

FIGURE 11-3 Welding is One of the Principal Causes of Industrial Fires

Acetylene gas, used as the fuel for the gas welding process (also known as **oxyacetylene welding**), is extremely unstable. It can be mixed with a suitable solvent, such as acetone, so it can be pressurized to around 200 psi and stored in a portable cylinder. The tradeoff for achieving this pressure is that the acetone solvent is highly flammable. Alternative gases can be used, but often do not produce the same high-temperature burns that acetylene does.

Acetylene containers must be stored and used in an upright position. Containers can leak near the cylinder valve stem, resulting in a "stem fire." Also, if the valve stem is broken off, the pressure release can turn the cylinder into a missile.

Oxygen cylinders must be handled carefully. The valve protection cap must remain on while the cylinder is in storage, being transported, whenever it's not in use. Do not use the slots on the valve protection cap to grasp or hold the cylinder. These vents are designed to safely channel the gas out if the valve is damaged with the cap on. Also, remember that oxygen is one of the prime elements in the fire triangle. Make sure your hands or gloves do not have grease or flammable materials on them when handling the cylinder.

Since acetylene is flammable and oxygen is part of the fire triangle, extreme caution should be exercised when oxygen and acetylene cylinders are stored in the same location. A barrier at least five feet high, made of non-combustible material, must be placed between the cylinders or the cylinders must be separated by at least 20 feet when being stored. Hoses must be inspected for wear and damage. Hoses must not become tangled with other hoses. The welding torch can experience a dangerous condition called flashback, in which the flame travels back up the mixture stream (oxygen and acetylene) and inside the torch. Flashback produces distinct popping, snapping, and humming sounds. Immediately turn off both the oxygen and gas valves if this occurs.

Acetylene is very explosive. Gas regulators on acetylene cylinders are designed to limit the gas pressure to 15 psig maximum. An explosion could occur if the gas pressure is above 15 psig. Acetylene is also very shock sensitive. Acetylene cylinders should never be placed on their side, dropped, or handled in a rough manner.

FIGURE 11-4 Acetylene is Very Explosive and Shock Sensitive; the Cylinders should Never be Placed on their Side, Dropped, or Handled in a Rough Manner

Arc welding can pose more hazards than gas welding. Arc welding requires large amounts of low-voltage, high-amperage power. The frame of the welding machine must be properly grounded. Cables can become overheated, damaging the insulation and causing a fire or electrocution hazard. Coiled cable must be spread out before welding starts. The welder must ensure that only the proper objects are made part of the welding circuit.

Resistance welding can pose shock hazards such as arc welding, along with mechanical hazards (e.g., pinch points).

Welding in confined spaces can be hazardous because of the sparks and potentially hazardous atmospheres. In such cases, a confined-space permit must be obtained along with a hot-work permit.

Government Regulations

There are a variety of government regulations relating either directly or indirectly to construction and tool use:

- OSHA 1926 Construction Standards
- OSHA General Industry Standards
  - 1910 Subpart P: Hand and Portable Power Tools and Other Hand-Held Equipment
  - 1910.241 Definitions
  - 1910.242 Hand and portable power tools and equipment, general
  - 1910.243 Guarding of portable power tools
  - 1910.244 Other portable tools and equipment
  - 1910 Subpart Q: Welding, Cutting, and Brazing

Summary

There are many tool and construction hazards in the process industries. While process technicians are not likely to be involved in construction-related tasks, they may encounter construction areas around their facility. Because of this, technicians must be aware of the safety hazards that construction areas and tools present.

Proper care, use, and maintenance of tools are integral for a safe work environment. Before working with tools, technicians should always select the proper tool for the job, and then inspect the tool for wear, damage, or other defects that can make a tool unsafe. In addition, technicians should always use proper techniques and safety practices when using tools. This includes wearing proper PPE, clearing obstacles or hazards from the work site, and ensuring that other individuals are a safe distance away before work begins.

Checking Your Knowledge

1. Define the following terms:
   a. Arc welding
   b. Electric tool
   c. Hydraulic tool
   d. Oxyacetylene welding
   e. Pneumatic tool
   f. Powder-activated tool
   g. Power tool
   h. Resistance welding

2. *(True or False)* When performing maintenance, workers are NOT required to use fall protection if working at a height six feet or higher.

3. A(n) _____ _____ permit and confined-space permit are examples of permits that might be required during maintenance tasks.

4. It is the _____'s responsibility that tools are used properly.

5. Which of the following are good safety practices for tools? (select all that apply)
   a. Inspect the tool before use.
   b. It is permissible to distract someone while they are using a tool.
   c. Return the tool to its proper place when done.
   d. Don't worry about obstacles in the work area before using a tool.
   e. Keep a firm grip on the tool.

6. *(True or False)* Tool use can pose an ergonomic hazard.

7. What types of tools are acceptable for use in flammable environments?
   a. Metal
   b. Intrinsically safe or explosion-proof
   c. Sparking
   d. Rubber coated

8. Name five materials that can be used to make spark-resistant tools.

9. _____ is one of the principal causes of industrial fires.

10. List five safety practices related to electric tools.

11. What is the name of the OSHA regulation covering tools?

Student Activities

1. What are the potential hazards associated with maintaining a piece of equipment, such as a pump or a furnace? Make a list.

2. Based on your personal experience, discuss some hazards you have encountered working with tools, along with any injuries you or others suffered. How could these hazards have been prevented? Discuss this with your fellow students.

3. Research tool safety on the Internet. Write a three-page paper, focusing on some specific tools and their safety procedures.

Vehicle and Transportation Hazards

Objectives

This chapter provides an overview of vehicle and transportation hazards in the process industries, which can affect the safety and health of workers, equipment, the facility, or the surrounding community, as well as the environment.

After completing this chapter, you will be able to do the following:

■ Name specific hazards associated with vehicles and transportation used in the process industry environment:

Forklifts

Powered platforms

Cranes

Trucks

Trains

Watercraft

Pipeline

Helicopters

Bicycles and carts

■ Describe government regulations and industry guidelines that address vehicle and transportation hazards.

Key Terms

Bonding—a system that connects conductive equipment together, keeping all bonded objects at the same electrical potential to eliminate static sparking.

Grounding—connecting an object to earth using metal in order to provide a path for electricity to travel and dissipate harmlessly into the ground.

Powered industrial truck—the American Society of Mechanical Engineers (ASME) defines a powered industrial truck as a mobile, power-propelled truck used to carry, push, pull, lift, stack, or tier materials. Forklifts and other similar vehicles are considered to be powered industrial trucks.

Powered platform—equipment designed to lift personnel on a platform to work at heights.

Working load limit—the maximum weight that can be lifted and should not be exceeded when working with a load.

Introduction

The process industries are extremely busy, mobile work environments. Many different types of vehicles and equipment are used to move materials in, around, and out of a facility, including forklifts, cranes, trucks, rail cars, tankers, barges, and pipelines. Process technicians must be aware of the hazards presented by these vehicles and equipment, along with safety procedures.

This chapter discusses the various hazards presented by the vehicles and other transportation methods associated with the process industries.

Vehicle and Transportation Hazards Overview

Moving vehicles can pose many hazards. Two hazards from vehicles are rollovers involving the driver and any passengers, and run-overs involving pedestrians. To avoid rollover, vehicle operators must be aware of vehicle handling characteristics, proper loading procedures, driving conditions, and potential obstacles and hazards. Two factors can contribute to run-overs: visibility issues (driver) and awareness (driver and pedestrian).

To maintain visibility, use headlights, wipers, and heaters as necessary. Remain alert; do not get distracted while driving. To minimize distractions, power off cellular telephones while driving. Slow down as you approach blind spots. As for awareness, drivers should familiarize themselves with high pedestrian-traffic areas in the facility. Pedestrians should keep an eye out for vehicles while walking, and slow down or stop if approaching a blind spot. Look and listen for approaching vehicles. Some facilities use mounted mirrors to help reduce blind spots.

Some companies require a vehicle entry permit to be issued if a vehicle is going to operate in certain areas of the facility. Check with your company's safety official about the permitting process, what vehicles are covered, and what areas are involved.

Forklifts and Powered Industrial Trucks

The American Society of Mechanical Engineers (ASME) defines a **powered industrial truck** as a mobile, power-propelled truck used to carry, push, pull, lift, stack, or tier materials. Following are some of the most commonly used powered industrial trucks:

- Forklifts
- Pallet trucks
- Rider trucks
- Fork trucks
- Lift trucks
- Aerial lifts

FIGURE 12-1 Forklifts are the Most Commonly Used Powered Industrial Truck in the Process Industries

Powered industrial trucks are used in the process industries to accomplish a wide range of materials-handling tasks. Forklifts are the most commonly used powered industrial truck in the process industries. (Figure 12-1) Forklifts unload and load materials (e.g., raw materials or finished products) to and from cargo trucks or railcars, store the materials in warehouses, and move the materials to where they are needed.

Forklift operations can be hazardous in a variety of ways. If electric forklifts are in use, battery charging can produce hazardous, flammable fumes. Fuel-powered forklifts can generate high levels of carbon monoxide gas, since they are typically operated indoors. Workers near the forklift can get crushed by a shifted load or pinned between the load and another object. Never stand under a forklift load or between the forklift and a fixed object. Always be aware of forklifts around your work area. If you cannot see the driver, he or she cannot see you.

The practice of using a forklift as a personnel hoist must be avoided unless a specially constructed safety platform is used. Workers should never ride as passengers on the forklift or on the forks. If a forklift is working nearby on an uneven surface, be aware that the forklift itself can shift and slide or fall sideways.

Forklift operations can also build up static electricity, which can prove dangerous in areas containing flammable or combustible materials. Some types of powered industrial trucks are not permitted in certain hazardous areas. Powered industrial trucks are categorized by a variety of factors, including fuel source, exhaust, fuel, electrical safeguards, and so on. Using a powered industrial truck in an area that it is not rated for it could result in a fire or explosion.

OSHA requires that forklift operators be formally trained and certified. OSHA also requires re-certification on a periodic basis, or after any accident or observation of unsafe driving. Training programs also address site-specific issues, such as facility hazards (e.g. ramps, narrow aisles, pedestrian traffic areas).

Forklift operators often assume that if they can drive a car, they can operate a forklift. However, different skills are required. Picking up a load, moving it, and placing it in a new location require skill and experience. Instability is a major hazard. Forklifts can easily become unstable and flip over. First, they have a shorter wheelbase and wheel size than a car. Second, the center of gravity of the forklift can shift when a load is lifted or the forklift is moving. Sufficient counterbalance weights must be added to make sure the load is properly balanced. Some forklifts have overhead guards to protect the operator from falling objects, or rollover protection systems. Forklifts use rear-wheel steering; this gives them greater maneuverability but decreases their stability at elevated speeds.

Visibility is often restricted for the operator. Many times, the forklift must be operated in aisles where there is heavy pedestrian traffic. Warehouses and storage areas often have blind corners where visibility is limited. Poor lighting can also affect visibility. The noise from a forklift can be loud, making it difficult for the operator to hear important sounds in the area.

Forklifts can also cause damage to pipes, cables, sprinklers, walls, machinery, and other equipment if the operator is not paying attention to potential obstacles when moving the fork.

One of the most dangerous hazards is the transition between the dock and a cargo vehicle or rail car, over a dockboard. The dockboard must be of adequate strength, properly placed, and anchored or locked into position. The forklift operator must drive slowly when moving over the dockboard.

The following are some forklift safety recommendations:

- Select the proper forklift for the job, based on load capacity, work area, operating conditions, and size.
- Handle only stable or safely arranged loads.
- Check to be sure the forklift has been inspected before use. OSHA requires daily safety inspections. Worn tires, damaged pressure lines, faulty brakes, or bent forks can all cause major hazards. If the forklift does not meet the required inspection, remove it from service.
- Only authorized personnel can make forklift repairs.
- Keep the forklift in a clean condition. Follow the manufacturer's cleaning recommendations.
- Before each use, check the horn, brakes, lights, play in the steering wheel, and controls.
- Make sure you wear the proper Personal Protective Equipment (PPE) when using a forklift.
- Know the material storage and handling guidelines for your facility.
- Always be alert for potential hazards.
- Do not rush or take shortcuts.
- Never exceed the lift capacity and overload the truck.
- If the forklift must be left unattended, shut off the motor and set the brakes.
- If the forklift must be left unattended, lower the fork and shift the controls to neutral.
- When ascending or descending grades in excess of 10 percent drive loaded forklifts with the load upgrade.
- Never charge an electric forklift in a non-designated area.

Powered Platforms

Process technicians may use **powered platforms**, manlifts, or vehicle-mounted work platforms to work at heights. (Figure 12-2) This type of equipment includes aerial baskets, aerial ladders, boom platforms, and platform elevation towers.

Employers must ensure that all such equipment is properly maintained and inspected every 30 days for mechanical operation and integrity. Employers must also train employees on the proper and safe use of powered platforms and associated equipment before allowing employees to use the equipment.

Emergency procedures must include action plans for how workers can escape the platforms in the event of a power failure, equipment failure, or other emergency. Employees must be trained on these procedures.

While on the platform, employees must wear personal fall protection (e.g., harnesses, lifelines, lanyards, or other approved devices). The fall protection must not

FIGURE 12-2 Process Technicians may Use Powered Platforms, Manlifts, or Vehicle-Mounted Work Platforms to Work at Heights
Courtesy of Eyewire/Getty Images

hamper the employee's work and must be securely anchored to prevent injury. Stabilizer ties must be used to allow an employee to move along the full length of the platform without becoming entangled.

In addition to the height hazard, other hazards include:

- Electrocution if the platform contacts a power line
- Injury if the platform strikes an object
- Impact from falling objects, such as tools accidentally dropped from the platform

Improper use of platforms causes more injuries than equipment failures. The following are some general safety procedures for powered platforms:

- Always wear your fall protection device.
- Wear appropriate PPE.
- Keep clear of power lines, using a safe distance as specified in company procedures.
- Be alert for obstacles and other hazards as the platform raises or lowers.
- Do not exceed the platform's weight limits.
- Keep the platform clear of tools, materials, and debris not related to the task being performed.
- Do not lean too far over the basket to perform a task; instead, reposition it.
- Never ride in the basket when a vehicle-mounted platform is in motion.
- If required by the operating unit, obtain a vehicle entry permit.

Cranes

Cranes are used in process facilities to handle various lifting needs, from small (50 lbs.) to large (50 tons or more). (Figure 12-3) Cranes use the basic principals of a block and tackle, a system of cables and pulleys, to lift heavy loads.

Crane designs come in all different types. They can be as simple as I-beams fitted with chains (usually called a hoist) for lifting light loads, or as complex as a pneumatically operated crane that lifts several tons. Cranes can be fixed or mobile. Fixed cranes are permanently attached to a structure, such as the ceiling of a warehouse, and move on track systems. Mobile cranes can be relocated to different areas as necessary, moving about on giant wheels or treads.

Operating cranes requires extensive training and certification due to all the complexities associated with them, such as load balancing, rigging, and safety. Although most process technicians will not operate cranes as part of their job duties, they may be

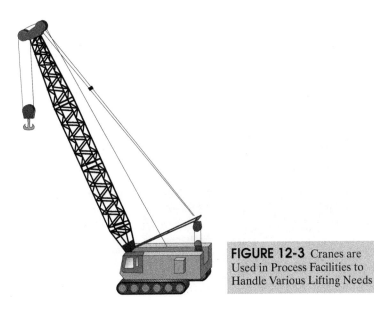

FIGURE 12-3 Cranes are Used in Process Facilities to Handle Various Lifting Needs

required to assist with crane loading and unloading. Two important safety concepts to understand are load limitations and how to properly rig a load. Cranes should never exceed their lifting-load capacity. The crane could collapse and drop its load, or the load could shift while moving. Snapped rigging can whip around violently. Workers can be injured or even killed by the falling load or rigging.

A vital task with crane operations is rigging the load. It is a time-consuming process and represents the biggest potential hazard for crane operation. When assisting with crane loads, know the **working load limit** (the maximum weight that can be lifted) and make sure the load does not exceed this limit. A stable load is critical to proper rigging. The load's center of gravity should be directly below the main hook and below the lowest point of sling attachment. Load stability is a combination of balance determined by center of mass, weight distribution, and rigging tightness. Finding the center mass for the load may require several attempts at rigging, thus ensuring the appropriate balance point.

Crane operators must follow these rules of operation:

- Be certified based on type of crane and weight of load.
- Know the load rating of the crane.
- Avoid hoisting or swinging loads over people.
- Test the brakes each time a load approaching the rated load is handled.
- Never leave the controls when a load is suspended.
- Check the upper-limit switch of each hoist under no-load conditions prior to the start of each shift.

The following are some general safety guidelines regarding crane loads:

- Know proper crane hand signals before working with cranes.
- Make sure each load is secure and properly balanced.
- Use the correct sling rating based on load weight.
- Check that the slings are not damaged by sharp edges on the load or excessive loading.
- Inspect all rigging; verify that there are no kinks in the slings.
- For multiple slings, check that each leg is the same length.
- Check that loose items are either removed from the load or secured.

Cranes can present an electrical hazard if the boom comes into contact with power lines or electrical equipment. Wind can make cranes unstable. Power failures also present an additional hazard, especially if a load is being lifted.

Loading and Unloading Liquids

Process technicians are often required to load or unload liquids from rail tank cars, tank trucks, tankers, barges, or other vehicles. (Figure 12-4) Bonding or grounding systems can reduce the potential for static electricity discharge, which could result in a fire or explosion.

FIGURE 12-4 Process Technicians are Often Required to Load or Unload Liquids from Rail Tank Cars, Tank Trucks, Tankers, Barges, or Other Vehicles

Courtesy of Eyewire/Getty RF Images

A **bonding** system connects conductive equipment, keeping all bonded objects at the same electrical potential to eliminate static sparking. **Grounding**, a type of bonding, connects a conductive object to a grounding electrode or building ground system.

Follow these safety procedures to avoid a potential static electricity discharge, or stray electrical currents.

- Know the liquid's flashpoint.
- Follow all bonding and grounding procedures.
- Ensure grounding or bonding connections are in good condition and properly placed at appropriate points.
- Remove all dirt, paint, rust, or corrosion from contact points before connections are made.
- Inspect containers (e.g., drums, tote bins, tank cars) for foreign objects, which can cause sparks.
- If loading a new liquid into a receptacle, ensure that the receptacle has been thoroughly cleaned.
- Avoid mixing air or steam with the liquid being handled.
- Use a low flow rate until the loading spout is submerged. Splashing and spraying increases static potential and can cause air to mix with flammable liquids, resulting in a fire or explosion hazard.
- Avoid moving through steam clouds if going into a hazardous area, because you can become charged with static electricity.

Trucks

Trucks haul a variety of materials to and from process facilities. Trucks come in a wide variety of designs, based on the type of material being hauled (e.g., cargo trucks, tank trucks). Collisions, fire, explosions, and spills are all potential hazards associated with trucking. Process technicians can be required to load or unload solid or fluid materials from trucks. Some materials may require cold or heat during transport. Be aware of the hazards that extreme temperatures might cause (e.g., burns) and follow appropriate safety procedures, such as wearing the proper PPE.

FIGURE 12-5 Trucks Haul a Variety of Materials to and from Process Facilities

Materials can also be hazardous. Know what you are handling; read all labels, warnings, and any documentation (e.g., Material Safety Data Sheet, or MSDS). Wear appropriate PPE and follow proper procedures when handling these materials.

If materials are flammable or combustible, understand the potential fire and explosion hazards associated with these materials.

Trains

Trains are used to transport materials to numerous companies in the process industries. (Figure 12-6) Process technicians can be required to load and unload materials from trains, either solid products in hoppers or boxcars, or fluid products from tank cars. Train-related safety hazards include fire, explosion, derailments, and spills or releases.

Loading and unloading of solid materials can present hazards. These hazards can lead to injuries such as pinches, broken bones, cuts, bruises, or fatalities from falling cargo. Workers should realize that materials could shift during transport. Exercise caution when opening or accessing materials. Straps can become damaged in transit, and

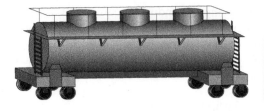

FIGURE 12-6 Trains are Used to Transport Materials to Numerous Companies in the Process Industries

any stacked materials can fall. Some materials might be hazardous, so wear appropriate PPE (e.g., respirators). Also, be aware of potentially flammable materials and fire and explosion hazards. Materials may give off dust that can ignite with just a spark.

Loading and unloading operations of tank cars are potentially dangerous, chiefly due to fire and explosion hazards. Proper grounding procedures must be followed. Rail cars are typically moved into a loading or unloading facility where product is loaded or unloaded one car at a time. Two basic types of loading and unloading systems are used:

- Closed, when products have toxic or flammable vapors (for loading operations, some facilities use a vapor recovery system).
- Simple (open), when products have low volatility and no vapor hazard.

Rail tank cars look similar to each other on the outside, but they can differ greatly on their internal design, based on the type of products they carry. Like tanker trucks, tanker cars are designed in a variety of sizes and can be modified to hold a variety of different products. A general-purpose tank car can carry about 20,000 gallons of liquid, while an LPG tank car can hold about 127 cubic meters of gas (about 33,500 gallons). Tank cars are monitored through their operating life and inspected regularly by the railroads that operate the cars. Along with a general safety inspection, the tank pressure rating and valves are checked. Tank car safety has been improved by the addition of couplers less likely to disengage during derailments, head shields on each end of the car to add protection against punctures, and thermal protection from special steel jackets or coatings.

The following are some general safety procedures to follow when working around trains:

- Be careful on multiple tracks, and aware that the sound of one train can hide the sounds of another train on different tracks.
- Watch out for uncoupled rail cars; they can unexpectedly roll.
- Cross tracks carefully; they can present a fall hazard.
- Remember that a train can extend out to three feet on either side of the tracks. Allow clearance between yourself and a moving train.
- Be aware that a train can appear to be traveling slower than it actually is.
- Tank cars can contain hazardous fumes and gases; exercise care when working with them.
- Do not cross between cars, as the train can move or shift while you are between the cars.

Watercraft

In many cases, process technicians do not work around watercraft. The type of industry, the location of the facility, and the job description determine whether a process technician will be involved with watercraft as part of their duties. Some common watercraft in the process industries include the following:

- Tankers are ships designed to carry petroleum products across large bodies of water. There are many different types of tankers, based on the products they carry (e.g., chemical tankers, liquefied-gas tankers).
- Barges are flat-bottom boats designed to transport cargo on inland waterways. Barges are typically connected in a long string.
- Supply boats are vessels used to carry materials to offshore platforms.

FIGURE 12-7 Process Technicians Dealing with Watercraft will Primarily be Involved with Loading and Unloading Operations

Process technicians dealing with watercraft will be involved primarily with loading and unloading operations. (Figure 12-7) Falls, drowning, dangerous materials, confined spaces, fire, explosions, and spills are all potential hazards. For watercraft workers, marine safety training includes water survival along with general safety training.

Pipelines

Pipelines have different types of uses, moving either liquids or gases through onshore or offshore systems. (Figure 12-8) Pipelines are a feasible, cost-effective way to transport large quantities of fluids (e.g., oil and gas). Compared to other transportation methods, pipelines are efficient and create less pollution. Pipelines are also one of the safest methods of transportation. However, potential pipeline hazards include ruptures, explosions, fires, and leaks. Pipeline accidents can result in injuries, loss of life, and property damage.

The most common threats to pipelines include corrosion and third-party incidents (digging onshore or marine accidents offshore). Pipeline materials are selected based on safety and integrity, as well as the type of product being transported. Corrosion is still a major problem for pipelines; most pipeline leaks occur due to corrosion. Periodic monitoring, inspection, maintenance, and leak detection are crucial safety-related tasks.

Another major concern for underground pipelines is damage from third parties, such as construction or excavation crews digging and causing accidental ruptures. Although pipelines are usually well marked and the industry constantly attempts to educate about "call before you dig," some older lines are not properly marked. Marine vessel hulls, fishing nets, anchors, and other gear have caused serious damage to offshore pipelines.

Safety, health and environmental requirements are similar to other petrochemical-related jobs and tasks:

- Personal Protective Equipment
- Fall protection
- Walking and working surfaces
- Hearing protection
- Equipment hazards (e.g., rotating equipment such as pumps and compressors).
- Grounding
- Radiation (from inspection equipment).

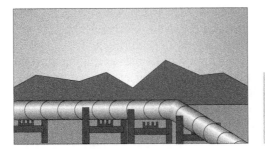

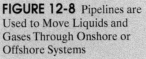

FIGURE 12-8 Pipelines are Used to Move Liquids and Gases Through Onshore or Offshore Systems

- Hot work
- Lockout/tagout
- Electrical hazards
- Fire safety
- Crane safety (during construction or maintenance).

Helicopters

Most process technicians will not encounter any type of aircraft as part of their jobs, but if they do it will most likely be a helicopter. Any other types of aircraft and safety-related procedures are outside the scope of this book. If the facility is located offshore or in an isolated region, workers might be shuttled to and from the site on helicopters.

FIGURE 12-9 Helicopters are Frequently Used to Carry Process Technicians to Offshore Platforms

The greatest threats from helicopters are the turning rotor blades (main rotor and tail rotor). Coming into contact with one of the blades can result in injury or death. Turbulence, mechanical failures, and contact with trees or power lines can cause hazardous conditions or crashes.

The following are some safety procedures to follow around helicopters:

- Enter or exit the helicopter in a crouched position from the front or side of the helicopter; make sure the pilot can see you.
- Do not walk toward the rear of the helicopter unless unloading a baggage compartment and only if the pilot is aware of what you are doing. Never cross underneath the tail boom when the engine is operating.
- Watch out for flying debris.
- Listen carefully to all safety briefings and obey all instructions from the pilot.
- Never distract the pilot or touch the flight controls.
- If a safety belt is available, make sure it is adjusted properly and securely fastened. Keep it on until the pilot indicates that you can remove it.
- Wear hearing protection.
- Never smoke near a helicopter or in one.
- Some flying situations require that you wear appropriate gear (e.g., cold-weather survival gear for flights over the North Sea).
- If flying over water, make sure you know the location of floatation devices and how to use them.
- In some situations, you may be asked to know how to open the emergency exit, use a fire extinguisher, locate the first-aid kit, use the headset and microphone, or other similar tasks.

Company Vehicle Safety

Process technicians might be required to drive a company vehicle. Company vehicles can be used to transport workers (e.g., during shift changes, or shuttling between the gate and work areas), haul materials, or move equipment. (Figure 12-10)

Some companies might require their employees to earn a "company driving permit" by completing a short course and passing a quiz.

Inside facility grounds, watch for posted speed or traffic signs and obey them. Know your company's driving policies. For example, policies might state where you can and cannot drive the vehicle, or what to do during an emergency or alarm (i.e., stop the

FIGURE 12-10 Company Vehicles can be Used to Transport Workers, Haul Materials, or Move Equipment

vehicle, turn off the engine due to a potential spark hazard, and leave the keys in the ignition).

The following are some general safety regulations for company vehicles:

- Do not park in fire zones.
- Always wear your safety belt, and have passengers wear them also.
- Watch for potential hazards, including pedestrian traffic, bicycles and utility cart traffic, forklifts, and cranes.
- Be aware of the changing work environment, where hazards can appear quickly.
- Sight lines and visibility can be reduced at a facility. Remain alert; do not get distracted while driving.
- Most facilities have a lot of pedestrian traffic, so always be aware of the people around you. Remember that pedestrians always have the right-of-way.
- Some facilities require that you leave the keys in the vehicle, in case it must be moved during an emergency.

Personal Vehicle Safety

Each process facility sets its own driving regulations and personal vehicle safety requirements. Most facilities require drivers to park their vehicle outside the plant, and then walk in.

FIGURE 12-11 Process Technicians Driving Within a Process Facility should Always Obey Traffic Signs and Posted Speed Limits

In cases where personal vehicles are allowed into a plant, obey all traffic signs and speed limits. Follow the safety recommendations under the company vehicle safety section.

Bicycles and Carts

Some facilities also use bicycles or utility carts for workers to get around, especially at large, spread-out complexes. (Figure 12-12) Carts can be electrical or fuel driven; most often, carts will have four wheels, although some could have three.

Workers using bicycles or carts to get around a facility must be aware of the dangers presented by other vehicles such as forklifts, cranes, and trucks. Bicycles or carts are no different than cars; pedestrians always have the right of way.

The following are some safety procedures for operating bikes and carts:

- Follow the same regulations and signs that are posted for automobiles, along with all other facility-specific rules about bike and cart use.

FIGURE 12-12 Some Facilities Use Bicycles or Utility Carts for Workers to Get Around, Especially in Large, Spread-out Complexes

- Always be alert for hazards; you do not have the same protection as when you drive an automobile.
- Only ride or drive in approved driving areas.
- Make sure the bike or cart is properly maintained before use.
- For a cart, make sure all safety equipment works (lights, horn, backup signal, flashing lights).
- Familiarize yourself with the cart manufacturer's recommended use and safety practices.
- Only charge electric carts in designated areas.
- Do not overload a cart with cargo or people.
- Do not let toolboxes or cargo obscure your vision in a cart.
- Only one person should be on a bike; never give rides to others.
- Be aware of hazardous riding or driving conditions, such as wet or slick roads or uneven surfaces, and adjust your speed accordingly.
- Slow down when approaching areas with heavy traffic, either vehicular or pedestrian, or blind spots.
- Do not engage in other activities while riding or driving, such as using a walkie-talkie or eating.
- Never park a bicycle or cart in an area that blocks emergency exits, fire lanes, fire hydrants, stairs, doors, or sidewalks.
- Keep your arms and legs inside a cart, unless using an arm to signal a turn.
- Do not act recklessly or engage in horseplay.

Government Regulations

The following are various government regulations relating to vehicle and transportation-related hazards:

- OSHA General Industry Standards
  - Subpart F: Powered Platforms
    - 1910.66 Powered Platforms
    - 1910.67 Vehicle-Mounted Elevating and Rotating Work Platforms
  - Subpart H: Hazardous Materials
  - Subpart N: Materials Handling and Storage
    - 1910.176 Handling Materials - General
    - 1910.178 Powered Industrial Trucks
    - 1910.179 Overhead and Gantry Cranes
    - 1910.183 Helicopters

Vehicles are regulated by different government agencies, based on type:

- The Department of Transportation (DOT) regulates all types of vehicles, including trucks and trains.

- The Federal Aviation Administration (FAA) regulates aircraft, including helicopters. Coast Guard regulations can also apply if the helicopter flies over bodies of water.
- The Coast Guard regulates watercraft.

A range of federal government agencies and their regulations impact pipelines: OSHA, EPA, Department of Transportation, Coast Guard, and the Minerals Management Service.

Summary

The process industries are extremely busy, mobile work environments. Many different types of vehicles and equipment are used to move materials into, around, and out of a facility, including forklifts, cranes, trucks, rail cars, tankers, barges, and pipelines. Each of these vehicles poses a unique set of safety hazards. It is important for process technicians to be aware of these hazards, and know how to prevent them.

For example, forklifts are one of the most commonly used industrial vehicles in the process industries. Forklifts can produce fumes that could be hazardous if operated indoors. They can also pose safety issues if used improperly, such as to lift a load that is too heavy.

Powered platforms are used to lift employees to work areas above ground level. The height of these work areas can range from near to the ground to high in the air. When working at elevations, technicians must always be aware of possible hazards (e.g., power lines overhead), wear proper protective equipment, and be familiar with operational and safety procedures.

Cranes are used for lifting heavy loads. Operating cranes requires extensive training and should not be performed by untrained personnel. Technicians operating cranes must always secure the load properly and must never exceed the lift limit.

Process technicians loading and unloading flammable liquids from trucks, ships, trains, pipelines, or other vessels must be familiar with the substances they are working with (e.g., know the flash point, toxicity, and reactivity levels). They must also make sure that equipment is properly grounded to prevent static electricity from igniting flammable vapors.

Process technicians who work around helicopters (e.g., those on offshore oil rigs) should be extremely cautious around the turning rotor blades. Technicians should always obey the pilot and follow all safety rules associated with the air craft. When flying over water, technicians must also know how to locate and use survival gear and flotation devices in the event of an emergency.

Individuals required to drive company vehicles must always obey posted speed limits, follow proper safety precautions, and watch for pedestrians. Likewise, pedestrians and bicycle riders must watch for other vehicles and hazards they might encounter.

Checking Your Knowledge

1. Define the following terms:
   a. Bonding
   b. Grounding
   c. Powered industrial truck
   d. Powered platform
2. How does the ASME define a powered industrial truck?
3. *(True or False)* Workers are required to wear a fall-protection device when using a powered platform.
4. A vital task with crane operations is _____ the load.
5. To prevent stray currents from igniting flammable or combustible liquids during loading or unloading operations:
   a. Do not smoke.
   b. Use grounding or bonding cables.
   c. Turn off all electricity to the area.
   d. Never perform these tasks in bad weather.

6. What are two basic types of tank car loading and unloading systems?
   a. Closed and complex
   b. Open and shut
   c. Closed and simple
   d. Simple and complex
7. Name and describe three types of watercraft that process technicians could encounter on the job.
8. The greatest threats from helicopters are the turning _____ _____.
9. *(True or False)* Personal vehicles are banned from every process facility.
10. Which of the following are incorrect statements about cart safety? (select all that apply)
    a. Carts may be banned from certain areas.
    b. Honk repeatedly to get a pedestrian to move out of the way.
    c. It is permissible to use the walkie-talkie while driving.
    d. Do not act recklessly or engage in horseplay.
    e. Never stick your arms or legs outside a cart unless signaling a turn.
11. What is a major concern for pipeline safety?
    a. Fugitive emissions
    b. Accidental damage from third parties
    c. Fall protection
    d. PPE

Student Activities

1. Select a type of vehicle that could be used in the process industries. Research potential hazards, any relevant government regulations, and ways to address the hazards. Write a two-page paper on your findings.
2. Use the Internet to research grounding and bonding procedures, including the hazards of static electricity and stray electrical current and how the bonding or grounding system works. Discuss the findings with your fellow students.
3. Walk around your campus, work area, or neighborhood and notice all potential hazards relating to vehicles, bicycles, and pedestrians. Make a list of the hazards, and then think about how you would react to each hazard.
4. Use the Internet to research regulatory citations that have been issued for powered industrial trucks and cranes. Discuss the research findings with your fellow students.
5. Identify the location of pipelines in your community and verify that the "Call Before You Dig" hotline telephone contact information is posted.

13

Natural Disasters and Inclement Weather

Objectives

This chapter provides an overview of natural disasters and inclement weather in the process industries, which can affect people, facilities, and resources.

Upon completion of this chapter you will be able to do the following:

■ Name specific hazards associated with natural disasters that could impact the process industries:

Hurricanes

Tornados

Floods

Lightning and rain storms

Earthquakes

Extreme temperatures

■ Describe how emergency preparedness plans address natural disasters.

Key Terms

Earthquake—a shaking and moving of the earth resulting from a sudden shift of rock beneath the surface (geologic stress).

Flood—the rising of water to cover normally dry land.

Hurricane—an intense, low-pressure tropical (warm area) weather system with sustained winds of 74 miles per hour or more. Hurricanes can rotate clockwise or counter-clockwise, depending on their location of origin.

Storm surge—danger created by hurricane force winds which push water toward the shoreline. As a storm surge advances on a shoreline, it can create a wall of water 15 feet or more above normal water levels.

Tornado—a destructive, localized windstorm. Tornadoes are produced by severe thunderstorms. A funnel cloud may or may not occur during a tornado.

Warning—a weather advisory issued when certain weather conditions (e.g., thunderstorm, hurricane, flash flood) are expected in the specified area.

Watch—a weather advisory issued when certain weather conditions (e.g., thunderstorm, hurricane, flash flood) are possible in the specified area.

Introduction

Process facilities operate around the clock all year long, through all types of weather. Process technicians must work in good or bad weather. They are expected to work despite rain, wind, ice, heat, storms, or other fairly common weather conditions. Process facilities have plans to deal with various weather conditions, including extreme weather and natural disasters.

Overview

The process industries provide vital products and services to meet a variety of critical needs for everyday life:

- Electrical generation and nuclear power
- Food manufacturing
- Oil and gas production
- Chemical manufacturing
- Water treatment
- Pharmaceuticals

Because of the critical nature of these products and services, facilities operate 24 hours a day, 7 days a week, 365 days a year. This means process technicians must work in all weather conditions to ensure the production of these products and services. Working in different weather conditions can range from being somewhat unpleasant to dangerous and life-threatening. Process technicians must also assist their companies during natural disasters, as emergency plans are put into action for continuing, shutting down, stabilizing, or resuming operations. Natural disasters can take many forms:

- Hurricanes
- Tornadoes
- Floods
- Extreme temperatures
- Earthquakes

Additional natural disasters can include wildfires, volcanic activity, and solar storms.

Natural disasters and inclement weather (e.g., extreme heat or cold and rain) affect process operations, as the different weather conditions may impact equipment, utilities, processes, or the entire facility. For example, humidity affects some equipment and processes (humid days can affect cooling tower capability). Hot weather can break down insulation on wiring and strain water cooling systems. Ice can coat power lines and the resulting extra weight can make them fall.

Freezing conditions can cause lines to freeze and crack, or the frozen material can plug the line. Flooding can cause water damage to equipment. It can also make parts of the plant difficult to access. Standing water from floods can lead to mosquito infestations. Lightning can damage electrical equipment or cause flammable liquids to ignite. High winds can buffet towers and columns.

The process technician must understand the hazards of natural disasters and inclement weather, and be prepared to work safely in these conditions. The following are some general tips relating to natural disasters and inclement weather:

- Familiarize yourself with your facility's emergency plans for specific natural disasters and inclement weather conditions. You might also need to refer to unit-specific plans in addition to facility plans.
- Understand that weather can prevent a shift change from occurring. This can cause fatigue in workers, which could result in errors that cause safety issues. If you work during a natural disaster or inclement weather, make sure you eat properly, drink plenty of water, and rest when possible.
- If you are not at work, check with the facility. You might be called in to work. Personnel deemed as critical can be required to work. Some facilities use an emergency notification plan (e.g., a notification pyramid, in which one person contacts two people, those people each contact two people, and so on).
- Regularly check and maintain emergency equipment, such as emergency power sources (e.g., generators, uninterrupted power supplies, sump pumps, and lighting).
- Verify that emergency radios are in place and working properly; control rooms and other areas should have weather radios that automatically come on when weather alerts are issued.
- Check communications equipment periodically (e.g., ensure radios are charged and operating).
- Practice good housekeeping in your work environment to ensure that the area is clean and secure.
- Make sure to use intrinsically safe (ones that are not powered by an ignition source) flashlights, in case of ruptured gas lines or the release of flammable materials.
- Government agencies, assistance organizations, and the media can provide further information about how to prepare.
- Process technicians should have their own emergency preparedness plans and be familiar with their company's emergency plans.

Hurricanes

Hurricanes are an extremely destructive force of nature. A **hurricane** is an intense, low-pressure tropical (warm area) weather system, featuring a well-defined counter-clockwise circulation (spiral) around a calm center (or eye) in the Northern Hemisphere and sustained winds of 74 miles per hour or more. (Figure 13-1)

Hurricanes can be accompanied by rain, lightning and thunder, and tornadoes. Areas around the United States that have experienced hurricanes include the Atlantic coast states and the Gulf of Mexico states. Hurricanes also occur in nearby regions off the west coast of Mexico and the Caribbean Sea. Hurricanes actually occur around the world, but are referred to in different regions as typhoons and cyclones.

Process industries in coastal areas prone to hurricanes can experience extensive devastation from their occurrence. Processes must be shut down before the hurricane arrives, and then started back up after the hurricanes passes. Water and wind damage from a hurricane can affect the startup of a facility after the storm.

FIGURE 13-1 Aerial View
of a Hurricane

Courtesy of the FEMA Photo
Library

The National Weather Service issues hurricane watches and warnings for the United States. A hurricane **watch** is issued when hurricane conditions are possible in the specified area, usually within the next 36 hours. A hurricane **warning** is issued when hurricane is expected to reach the specified area, usually within the next 24 hours.

Hurricanes are ranked using a scale (called the Saffir-Simpson scale), based on sustained wind speed.

A Category 4 storm could create up to 100 times the damage of a Category 1 storm. Massive evacuation of areas may be required for higher-category storms. However, lower-category storms can still cause extensive damage. Intensity of threat determines the level of planning.

TABLE 13-1   Saffir-Simpson Scale for Hurricane Wind Speed

Category	*Wind Speed*
Tropical Storm	39-73 mph
Category 1 Hurricane	74-95 mph; storm surge usually 4-5 feet above normal
Category 2 Hurricane	96-110 mph; storm surge usually 6-8 feet above normal
Category 3 Hurricane	111-130 mph; storm surge usually 9-12 feet above normal
Category 4 Hurricane	131-155 mph; storm surge usually 13-18 feet above normal
Category 5 Hurricane	156 mph and up; storm surge usually greater than 18 feet above normal

HURRICANE HAZARDS

Following are some of the hazards of hurricanes:

- High winds
- Storm surge
- Rain and flooding
- Tornadoes
- Lightning
- Loss of services

Winds can do the following:

- Hurl debris, turning objects into potentially harmful projectiles
- Knock over small buildings
- Damage roofs
- Break windows
- Blow down trees and poles
- Rip off parts of buildings and damage process equipment (sway)

FIGURE 13-2 Hurricanes can be Very Destructive, Hurling Debris and Turning Objects into Potentially Harmful Projectiles
Courtesy of the FEMA Photo Library

The type and amount of damage can vary with wind speed. People outside in high winds are at risk of being hit by falling objects and trees, or wind-driven debris. (Figure 13-2)

Rain can reduce visibility, create hazardous driving or working conditions, and cause flooding (coastal and inland). Water can damage electrical components and sensitive equipment.

Hurricanes produce storm surges and rain, which result in land erosion and flooding. Storm surges and flooding cause the most deaths of any weather-related disasters. They can cause water damage to equipment, sensitive electronic components, buildings, and more. As a **storm surge** approaches the coast, it can create a wall of water 15 feet or more above normal levels and pound the shoreline, destroying many structures in its path. Flooding also presents a hazard as it sweeps debris, which can potentially damage or destroy anything in its path. The section on flooding discusses this hazard in more detail.

Hurricanes can also spawn tornadoes. **Tornadoes** are localized windstorms, one of the most violent and destructive weather events, with winds reaching over 300 mph. Tornadoes can resemble a funnel-shaped cloud extending toward or touching the ground, but such a cloud may not necessarily appear. Some hurricanes produce no tornadoes, while others can produce multiple ones. Tornadoes produced by a hurricane are most likely to occur on the northeast (right front) side of the hurricane, and can occur anytime of the day or night. Tornadoes can occur elsewhere in the bands of rain hurricanes produce. Hail and lightning do not accompany hurricane-generated tornadoes to the same degree as in other tornadoes. Tornadoes can occur for several days after the hurricane passes over land. The next section discusses tornadoes and their hazards in more detail.

Lightning can be generated during a hurricane. Lightning is an electrical discharge in Earth's atmosphere, either from cloud-to-cloud or cloud-to-ground. Lightning results in a brilliant flash and heat, up to 50,000 degrees F (five times hotter than the Sun). As lightning heats the surrounding air, the associated sound of thunder is created. Lightning presents a variety of hazards, such as damaging electronic equipment, starting fires, and electrocuting people (resulting in severe injuries or death). The section on lightning and rainstorms discusses lightning in more detail.

The combined hazards of hurricanes (wind, rain, tornadoes, flooding, and lightning) can result in great devastation, including deaths, injuries, property damage, and financial loss. Hurricanes also can cause a general disruption of important services, including the following:

- Power generation
- Clean water supply
- Food delivery
- Telephone and communications
- Fuel distribution
- Emergency response
- Medical treatment
- Transportation and delivery (raw materials, parts, finished products).

As mentioned previously, the process industries are responsible for providing many products and services we depend on for daily life. The section on emergency planning describes how companies can prepare for business continuation and resumption during and after disasters.

HURRICANE PREPAREDNESS

Process technicians working in areas where a hurricane can potentially hit should have their own emergency preparation plans and be familiar with their company's plans. These plans include shutdown procedures before the storm and startup procedures after the storm (including damage inspection and safety review). The following are some important tips about hurricane preparedness:

- First and foremost, have a disaster plan that includes a primary evacuation route and an alternate route and communicate your plan with family and friends. (Figure 13-3)
- Follow the advice of local officials about preparations or evacuation plans.
- Check weather reports for updated storm information (TV, weather radio, radio, the Internet).
- Make sure any loose objects that could be picked up by the wind are secured.
- Make sure vehicle gas tanks are filled.
- Windows should be covered with storm shutters or plywood. Tape will not prevent a window from breaking.
- Make sure you have enough food supplies for everyone in your household for a minimum of three days. Store them in a waterproof container that can be easily transported.
- Battery-operated flashlights and radios are critical items to have, along with fresh batteries.
- Make sure you have plenty of clean drinking water for everyone in your household for a minimum of three days.
- If your area is not evacuated, stay inside and away from windows. Avoid travel.
- Take only essential items if you evacuate. Make sure to include eyeglasses, contact lenses, and any prescription medication you are taking.
- Be alert for tornadoes and flooding. An interior closet or bathroom without windows is a good place to remain if a tornado strikes. Get to high ground in case of flooding.
- Do not travel to damaged areas.

Process technicians might need to perform these tasks at the process facility:

- Secure any loose objects (e.g., bolts, tools) that could be picked up by the wind.
- Tie down any materials or objects that could be blown away.
- Make sure the rigging is secure on any tall objects that use guy wires.
- Certain areas might need to be sandbagged or reinforced from flooding.

FIGURE 13-3 Individuals on the Coast Should Always have a Disaster Plan that Includes Primary and Alternate Evacuation Routes

- Check that drains and grating are clean and unobstructed (to facilitate the drainage of rising waters).
- Shut down non-critical processes and systems as required. Shutdowns should be determined based on the proximity of the storm (i.e., more critical procedures are shut down the closer a storm gets).
- Process levels should be lowered and tanks and vessels might need to be partially drained. However, a certain amount of liquid must be left in tanks and vessels to prevent them from floating off their foundations in flood conditions.
- As required, make sure the unit is totally secure (which can include a maintenance-type shutdown and process block-in).

Tornadoes and High Winds

As mentioned previously, tornadoes are destructive, localized windstorms. (Figure 13-4) Tornadoes are produced by severe thunderstorms, and contain swirling winds that can reach over 300 mph and leave a devastating path of destruction. A funnel cloud may or may not be visible during a tornado.

Tornadoes can form anywhere in the world, but the United States has more tornadoes per year than any other country. A good number of those occur in the Midwest and South, a region that has earned the nickname Tornado Alley.

A tornado watch is issued when tornado conditions are possible in the specified area. A tornado warning is issued when a tornado has been spotted or appears on radar in a specified area.

Tornadoes generally occur on the trailing edge of thunderstorms. The following are some potential signs of a tornado:

- A freight train-like sound may be heard.
- Hail (small ice particles) and lightning may be present.
- A cloud of debris can show where a tornado is, even if no funnel cloud is visible.
- Winds may die down and the air become still before a tornado hits.

Tornadoes are measured using a scale called the Fujita (or F) scale. Damage is used to determine a tornado's wind speed.

TABLE 13-2   Fujita (F) Scale Descriptions

Fujita Scale	*Wind Speed*	*Damage*
F0 Gale Tornado	40	Damage to chimneys; tree branches broken off; shallow trees uprooted.
F1 Moderate Tornado	73	Roof surfaces peeled off. Mobile homes or small sheds and buildings overturned. Moving autos pushed off roads.
F2 Significant Tornado	113	Considerable damage. Roofs torn off frame houses. Large trees snapped or uprooted. Light-object missiles generated.
F3 Severe Tornado	158	Severe damage. Roofs and some walls torn off well-constructed homes. Trains overturned. Most trees in forests uprooted. Heavy cars lifted off ground.
F4 Devastating Tornado	207	Well-constructed houses leveled. Structures with weak foundations blown some distance. Cars thrown and large missiles generated.
F5 Incredible Tornado	261	Strong frame houses lifted off foundations and disintegrated. Automobile-sized missiles fly through the air at speeds in excess of 100 mph. Trees debarked.

Tornado hazards include the following:

- Extremely high winds
- Large debris flying through the air
- Hail
- Rain
- Lightning

FIGURE 13-4 Tornadoes are Extremely Destructive, Localized Windstorms with Swirling Winds that can Exceed 300 mph

Hail rarely kills people, but it can cause injury and property damage. Hail is a frozen raindrop produced by an intense thunderstorm. Most hail is small, less than two inches across, but it can still cause damage because of the speed it reaches as it falls.

TORNADO PREPAREDNESS

Tornado watches and warnings are issued on a county or parish basis. Tornadoes give far fewer warning signs than hurricanes. However, new advances in weather radar provide more indications of conditions that could produce tornadoes. When a tornado is coming, a person has only a short time to make critical decisions. Advance planning, preparation, drills, and rapid response are keys to safely surviving a tornado. (Figure 13-5)

The following are some general tips on preparedness:

- Follow the advice of local officials about preparations.
- Check local weather reports for updated storm information (e.g., TV, radio, the Internet).
- Make sure any loose objects that could be picked up by the wind are secured.
- Make sure vehicle gas tanks are filled.
- Be alert to current weather conditions. Watch for the potential signs listed in the previous section.
- Make sure emergency supplies are well stocked. Battery-operated flashlights and radios are critical items to have, along with fresh batteries.
- Make sure you have plenty of clean drinking water and enough food for several days.
- Avoid travel.
- An interior closet or bathroom without windows is a good place to remain if a tornado strikes.
- Do not travel to damaged areas.
- Keep blinds or shades drawn over windows to keep glass from flying in if a window gets broken due to wind.
- If you are outside or in a vehicle, hurry to the closest sturdy building. If one is not available, lie flat in a ditch or low area.
- Watch for fallen power lines.

Tornados are hard to predict and often little or no warning is available. For process technicians, it is vital that you know the location of the closest secure building is to which you can go. Also, follow your facility's emergency plans for tornados.

FIGURE 13-5 Tornadoes can Leave a Devastating Path of Destruction and Provide Far Less Warning than Hurricanes

Courtesy of the FEMA Photo Library

Flooding

Flooding is the rise of water on normally dry land. Flooding can occur when water overflows its natural or artificial boundaries, such as levees or dams, or a body of water (e.g., stream, river, lake, and ocean). Flooding can also occur when drainage systems cannot remove the water fast enough from roadways. (Figure 13-6) Flooding is responsible for the most weather-related deaths.

Different events can produce flooding:

- Heavy rains (most common)
- Melting snow
- Storm surge (such as from a hurricane)
- Underwater earthquake or volcanic activity (creating a tsunami)
- Barrier failure (e.g., dam break)

Every state in the United States can experience flood conditions. Factors that affect flooding are:

- Already-soaked ground (when the ground is already saturated with water, usually from recent heavy rainfalls)
- Type(s) of soil in surrounding land (e.g., sandy soil compared to clay)
- The geography of surrounding land (e.g., low-lying terrain)
- Proximity and size of bodies of water
- Frequent storms and heavier-than-normal rainfall

River flooding is natural, typically happening on a seasonal basis with spring or winter rains, or melting ice and snow. Coastal flooding occurs when winds from tropical storms, hurricanes, or low-pressure systems drive ocean water inland. Coastal flooding can also be caused by giant tidal waves (tsunamis), produced by earthquakes or volcanic activity. Coastal flooding occurs when land upstream has received a lot of water; it heads down the rivers and can eventually end up at the coast. Urban flooding occurs when water is moved from its natural state to roads, parking lots, and buildings, usually because the land loses its ability to absorb rainfall and drainage systems have not been adequately designed to compensate. Runoff from heavy rains can turn streets into rivers and can fill low-lying areas with water.

A flash flood is flooding that rises rapidly, but can fall quickly. Flash floods can occur with little or no warning. Most flood-related deaths are caused by flash floods. The main factors that contribute to flash floods are the rate (intensity) of rainfall intensity and how long it lasts (duration). Soil conditions, lay of the land (topography) and ground cover and vegetation also affect flash flooding. Flash floods usually happen within just a few minutes of an excessive rainfall, or because a barrier, either manmade or natural, fails (e.g., levee, dam, ice).

Flash floods can carry debris such as trees, rocks and other items in its path, causing much destruction. If debris builds up, it can create a barrier that holds back the water and allows it to build up. The resulting barrier can then break and unleash more flash flooding. Flash floods can also cause landslides, which are discussed later in this chapter.

FIGURE 13-6 Flooding can Occur When Water Overflows Natural or Artificial Boundaries

Courtesy of the FEMA Photo Library

Flash flooding is a relatively short event, while other flooding can last for several days or longer. Flood or flash flood watches are issued if flooding is possible in the specified area. Flood or flash flood warnings are issued if flooding has been reported or is imminent in the specified area. Urban or small stream advisories can be issued about flooding of streets, low lying areas, or streams.

The following are some signs that point to potential flooding:

- Thunder from approaching storms
- Weather reports of heavy rainfall or thunderstorms in the area, even upstream
- Reports of an approaching hurricane
- Heavy rainfall
- Saturated ground from previous rainfall
- Flood history of the area
- Rapidly rising water

FLOODING HAZARDS

Floods can produce the following hazards:

- Drowning
- Water damage to structures
- Swiftly moving water (even six inches of fast-moving water can knock people off their feet, and two feet of water can float a car)
- Damage from debris
- Contaminated water
- Landslides
- Animals and insects in the water (e.g., snakes, fire ants, and mosquitoes)

FLOOD PREPAREDNESS

Prepare for a flood by following these suggestions:

- Know your area's flood risk.
- Follow the advice of local officials about preparations and evacuations.
- Check local weather reports for updated storm information (e.g., TV, radio, the Internet).
- Take only essential items if you evacuate.
- If your area is not evacuated, stay inside. Avoid travel.
- Make sure vehicle gas tanks are filled.
- Be especially careful at night, since it is more difficult to recognize flood dangers.
- Do not drive on flooded roads. If your vehicle stalls, leave it immediately and seek higher ground. Even a couple of inches of water on the roadway may be enough to sweep your car away.
- Make sure emergency supplies are well stocked. Battery-operated flashlights and radios are critical items to have, along with fresh batteries.
- Make sure drains and storm sewer gratings are cleared.
- Go to higher ground if floodwaters are rising.
- Be careful of submerged dangers e.g., submerged power lines.
- Do not travel to damaged areas.
- Check electrical equipment before trying to use it.

Lightning and Rainstorms

Lightning is a naturally occurring, electric current created when atmospheric electricity is discharged, either from one cloud to another, or from a cloud to the ground. Lightning is an extremely destructive force of nature, generating high voltage and current along with heat up to 50,000 degrees F—temporarily hotter than the surface of the Sun. See Chapter 5, *Equipment and Energy Hazards* for more information on lightning.

The following clues indicate that lightning might be nearby:

- A rise in wind speed
- Darkening skies
- Towering clouds
- Static on AM radio stations
- Thunder
- Flashes of light in the distance

If you can hear thunder, you are potentially close enough to the storm to be struck by lightning. If you see lightning, you can estimate its distance from you. Since light travels faster than sound, you can see the flash of lightning before hearing the resulting thunder. Count the number of seconds between the flash of lightning and the next sound of thunder. Divide this number by five to get the distance in miles. Follow the procedures in the section on lightning and rainstorm preparedness for safety information.

Lighting occurs during thunderstorms, generally on the trailing edge of the storm. A thunderstorm is a potentially severe weather condition characterized by heavy rainfall, lightning, winds, and possible hail. Thunderstorms can form the basis of various severe weather conditions, including tornadoes, hail storms, flash flooding, and hurricanes. Because of this, weather forecasting prominently addresses thunderstorms.

A thunderstorm is classified as severe when winds go over 57 mph and/or when it produces hail the size of a dime or larger. Thunderstorms can last less than 30 minutes at times, or form long-lasting bands (called squall lines). Thunderstorms can spawn tornadoes. Some thunderstorms grow powerful, last for hours, and spawn numerous tornadoes. These are called supercells. Approximately 10% of the thunderstorms in the United States become severe, causing most of the damage, injuries, and death. Every state in the United States is prone to thunderstorms.

A severe thunderstorm watch is issued when conditions are favorable for developing into a severe storm. Watches are issued before actual storms are sighted or indicated by weather radar. Watches typically cover a large area and can last for several hours. A severe thunderstorm warning is issued when a thunderstorm has the characteristics to become severe. Warnings are issued for a county or parish, or parts of that area. A warning is issued when a sighted thunderstorm is about to enter or has entered a county. Warnings are usually in effect for a short period of time (the estimated time it will take the storm to move through the county).

Rainstorms occur when droplets of water fall from clouds. Rain can range from light to heavy, and last for varying periods of time. Rain is one type of precipitation, or water that comes from the sky as part of the weather. Other types include snow, sleet, freezing rain, hail, and dew. Rainfall is typically measured in inches, using a rain gauge. The primary difference between a rainstorm and a thunderstorm is whether lightning is present.

FIGURE 13-7 Landslides, Which Carry Mud and Debris from the Size of Small Rocks to Large Boulders and Trees, can Occur During Intense Rainfall or Fast-Melting Snow

Courtesy of the FEMA Photo Library

LIGHTNING AND RAINSTORM HAZARDS

Lightning can be extremely destructive, creating the following hazards:

- Electrocution, including severe injury or death
- Fires
- Explosions
- Equipment and property damage
- Damage to electronic components and electrical systems

Rainstorms can produce winds, lightning, hail, flooding, tornadoes, and landslides or mudslides. Along with the hazards associated with these weather events, rainstorms create other hazards such as the following:

- Reduced visibility and poor driving conditions
- Unsafe working conditions
- Water damage

Landslides can be a major hazard, taking place in almost every state in the United States. Landslides cause much destruction, injury, and death. Landslides can occur during intense rainfall or fast-melting snow. (Figure 13-7) They generally start on steep hillsides, and then accelerate to 10-35 mph. Depending on their force, landslides carry mud and debris ranging in size from small rocks to large boulders and trees. Sloped areas with burned or damaged vegetation can be prone to landslides, especially if a lot of water has soaked into the ground.

LIGHTNING AND RAINSTORM PREPAREDNESS

Process facilities take great care to protect against lightning strikes, installing special rods, masts, and other equipment to reduce the hazards.

Observe these instructions for your own personal safety:

- Follow the advice of local officials about preparations.
- Check local weather reports for updated storm information (e.g., TV, radio, the Internet).
- Be alert to current weather conditions. Watch for the potential signs mentioned in the lightning and rainstorms section.
- Do not use the phone.
- Stay away from windows.
- Check with the appropriate personnel about whether you should stop using electrical equipment and/or computers.
- Avoid taking a shower during a thunderstorm.
- If you are in a boat, return to land as quickly as possible.

If you are outdoors observe the following precautions:

- Find shelter in a building or vehicle.
- Stay low.
- Put down any metal objects you are carrying (e.g., umbrellas).
- Avoid trees, tall objects, power lines, metal (including fences), and water.
- Find a low-lying area, but watch out for flooding.
- If you are in a group, spread out.

If you feel your hair standing on end or a tingling sensation, lightning may be about to strike. Make yourself as small a target as possible, and minimize your contact with the ground. Drop to your knees, bend forward, and put your head between your knees. Place your hands on your legs or head. DO NOT ever lie flat.

If you are in a vehicle:

- Remain in the car until the storm passes.
- Do not touch any metal objects in the car, if possible.
- Keep the windows closed.

The following tips can protect you during rainstorms:

- Wear proper protective gear.
- Be careful walking, since surfaces may become slippery.
- Be aware of potential submerged hazards (such as power lines).
- Watch for lightning.
- Avoid working at heights.

Practice these tips to remain safe when driving during rainstorms (refer to Chapter 12, *Vehicle and Transportation Hazards*, for more details):

- Use extra caution and reduce your speed.
- If visibility is severely reduced, pull off the road to a safe distance, making sure you do not stop near tall objects such as trees or poles that can blow down or attract lightning.
- Avoid driving on roads covered by water. Water can be deeper than anticipated or hide potholes and other hazards.
- If spray from an oncoming vehicle blinds you, hold the wheel firmly and do not brake. Be prepared to brake when the windshield clears.
- Remember that the roads will be slick because of the rain.

Following are tasks that process technicians might need to perform at the process facility:

- Maintain lightning arresters.
- Be prepared for power outages.
- Check for water buildup in critical areas such as dikes or containment walls.
- Make sure there are no rainwater leaks in facility roofs and coverings.
- In sustained, heavy rains, you might need to place certain equipment controls in manual mode as per procedure; instrumentation can be affected and give false readings.
- Some facilities require that drums be stored on their side to prevent water accumulation during heavy rains.

Extreme Temperatures

Process technicians must often work in extreme temperatures, either very hot or very cold. Both pose dangers not only to people, but also equipment and facilities. (Figure 13-8)

HOT WEATHER

Hot weather is characterized by high temperatures (often over 100 degrees F). Some areas experience high humidity (moisture in the air) along with heat. Lack of rainfall can produce drought conditions and periods of low humidity. A heat wave is a prolonged period of extremely hot temperatures and high humidity. The heat index, given in degrees F, indicates how hot it feels when relative humidity is combined with the air temperature. If a person is exposed to full sunlight, the heat index can increase by as much as 15 degrees F.

The human body keeps cool by sweating. High humidity can reduce the amount of sweat evaporation. This makes working outdoors dangerous, because it is harder for the body to cool itself in humid conditions. Overheating can result in harmful conditions such as heat cramps, fainting, heat exhaustion, or heat stroke. Chapter 7, *Pressure, Temperature and Radiation Hazards*, covers these conditions in more detail.

HOT-WEATHER PREPAREDNESS

Several important tips, when practiced, can protect you during hot weather:

- Wear protective sunglasses.
- When working outdoors, pace yourself.
- Stay indoors if possible, or if working outdoors take frequent breaks inside.

FIGURE 13-8 Extreme Temperatures Pose Dangers to People, Equipment, and Facilities

- Use a fan to help your body remove sweat.
- Avoid becoming dehydrated.
- Drink plenty of water. Drink regularly and often.
- Drink even if you do not feel thirst.
- Water is best to drink. If you are sweating a lot, you can also drink a sports drink. Avoid caffeinated drinks.
- Eat small meals, and eat more often.
- Avoid foods high in protein.
- Do not use salt tablets unless directed to by medical personnel.
- Use sunscreen to protect against sunburn and skin cancer.
- Make sure equipment is properly maintained to prevent overheating. Check for wear from heat or sunlight.
- Be careful when driving (heat on the road can cause tire problems).

Processes and equipment also can be affected by hot weather:

- Equipment can easily reach its engineering limitations.
- Humidity and evaporation can affect equipment and processes.
- Cooling towers, fans, and other similar devices might not be as efficient.
- It can be difficult to maintain proper pressures.
- Metal parts can swell and equipment can expand.
- Blocked-in lines and vessels can rupture.

SNOWSTORMS AND COLD WEATHER

Snow, like rain, is a form of precipitation. Snow is water falling from clouds in the form of ice crystals joined together as snowflakes. A snowstorm combines snow and/or sleet with cold temperatures and winds. Cold weather can involve temperatures at or below freezing, ice, snow, winds, and freezing rain.

Wind chill is the combined effect of wind and temperature. Wind chill is important to know, especially its effect on exposed skin.

A blizzard is a dangerous snowstorm (Figure 13-9), with the following conditions:

- Low temperatures
- Winds blowing over 35 mph
- Visibility of one-quarter mile or less
- A duration of three hours or more
- Blowing snow
- Heavy snowfall

A winter storm watch is issued if a winter storm is possible in the specified area. A winter storm warning is issued when a winter storm is headed for the specified area. A blizzard warning is issued when strong winds, blinding wind-driven snow, and dangerous wind chill are headed for the specified area.

FIGURE 13-9 Blizzards can Produce Avalanches and Other Dangerous Conditions
Courtesy of the FEMA Photo Library

SNOWSTORM AND COLD-WEATHER HAZARDS

The following hazards can be produced by snowstorms and cold weather:

- Exposure, or hypothermia, in which the body fails to maintain heat
- Frostbite, an extreme condition in which flesh freezes
- Snow depth (drifts or banks of deep snow)
- Snow weight (piled up on objects, potentially causing damage)
- Wet conditions
- Dangerous driving conditions
- Ice
- Winds

Chapter 7, *Pressure, Temperature and Radiation Hazards*, covers hypothermia and frostbite in detail.

SNOWSTORM AND COLD-WEATHER PREPAREDNESS

Be aware of the following hazardous working conditions that may occur during freezing weather:

- Icing of platforms and stairways
- Icicles forming on equipment
- Skin contact with frozen metal
- Water and steam leaks due to damaged equipment
- Plugging of piping or equipment by frozen liquids
- Follow the advice of local officials about preparations
- Check local weather reports for updated storm information (e.g., TV, radio, the Internet)
- Make sure vehicle gas tanks are filled
- Be alert to current weather conditions
- Make sure emergency supplies are well stocked. A supply of blankets is important. Battery-operated flashlights and radios are critical items to have, along with fresh batteries
- Make sure you have plenty of clean drinking water and enough food for several days
- Avoid travel
- Do not travel to damaged areas

If working outdoors, observe the following precautions:

- Avoid overexertion.
- Wear appropriate clothing and minimize the exposure of skin. Several layers of lightweight clothing are better than one single heavy coat.

- Gloves and a hat are vital.
- Cover your mouth to protect your lungs.
- Move to stay warm.
- Keep clothing dry. Wet clothing can transfer heat away from the body 240 times faster than dry clothing.
- Walk carefully to avoid slipping and falling.

These tasks may be necessary at the process facility:

- Keep snow and ice from accumulating on equipment.
- Make sure equipment is properly maintained to prevent freezing.
- Check that process fluids (e.g., heating oil) do not solidify at lower temperatures.
- Block, drain, and blind unused water lines.
- Check steam traps and service lines.
- Drain and rod pumps and other equipment.
- Drain out-of-service steam lines and equipment, and remove condensation.
- Block off instrument leads for equipment that is to be steam-purged, then check for condensation.
- Use bypass lines or open drains for adequate flow of water lines in service (but even a flowing line can freeze).

EARTHQUAKE

An **earthquake** occurs when the earth trembles and moves, resulting from a sudden movement of rock beneath the surface (geologic stress). (Figure 13-10) Earthquakes can cause damage through ground motion (shaking), ground failure (landslides), fires, ruptured dams, damaged pipes, tsunamis, and so on. Ground motion is the most dominant force and causes widespread damage.

The point where an earthquake starts is called the epicenter. The intensity of the ground shaking diminishes as it radiates out from the epicenter through the earth. Earthquakes can be followed by smaller ones, called aftershocks, after the main quake. In a few cases, foreshocks occur before the main quake.

Earthquakes occur around the world on a daily basis, but most can be felt only by measurement instrumentation (called a seismometer or seismograph). In the United States, California and Alaska have experienced numerous earthquakes. The size of an earthquake is referred to as its magnitude, measured using the Richter scale.

TABLE 13-3 Richter Scale Values for Earthquakes	
Magnitude	*Potential Effect*
2.5 or less	May not be felt, but can be recorded by a seismometer
2.5 to 5.4	May be felt, but results in only minor damage
5.5 to 6.0	Can cause slight damage to property
6.1 to 6.9	Can cause damage in populated areas
7.0 to 7.9	Serious damage (major earthquake)
8.0 or greater	Major destruction (great earthquake)

For each whole-number increase on the scale, the amount of ground motion goes up by a factor of 10. A magnitude 5 earthquake results in 10 times the amount of ground-motion caused by a magnitude 4 earthquake.

EARTHQUAKE HAZARDS

Earthquakes create a variety of hazards:

- Damage to equipment and buildings
- Falling debris
- Cracks in the earth
- Landslides and mudslides
- Destruction of roadways and railways

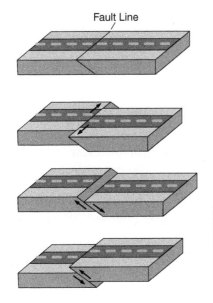

Fault Line

FIGURE 13-10 Earthquakes Occur When the Earth Trembles and Moves Because of Sudden Movements in the Rock Beneath the Surface

- Damaged pipes (e.g., gas, water)
- Downed power lines
- Fires and explosions
- Tidal waves (tsunamis)
- Ruptured dams

EARTHQUAKE PREPAREDNESS

Scientists study where earthquakes are likely to occur, but there is no reliable way to predict when one will occur at a specific location. The following tips will help you prepare for an earthquake:

- Follow the instructions of local officials about preparations.
- Check local reports for updated earthquake information (e.g., TV, radio, the Internet).
- Make sure emergency supplies are well stocked. Battery-operated flashlights and radios are critical items to have, along with fresh batteries.
- Make sure you have plenty of clean drinking water and enough food for several days.
- Identify a place in your various work areas you can use to protect yourself, such as under a sturdy table or desk, or against an inside wall. You should never be more than a few steps from cover.
- Remember "Drop, Cover, and Hold On." During an earthquake, drop under a sturdy desk or table, hold on, and protect your eyes by pressing your face against your arm. If there is no desk or table, sit on the floor against an interior wall away from windows, loose items, and tall objects that can fall on you.
- If you are outdoors, go to a spot away from buildings, trees, power lines, and tall objects. Drop to the ground.
- If you are in a vehicle, make sure to stop in a clear area (see above) and remain in the car until the shaking stops.
- Use the telephone only to report major emergencies.
- Use flashlights instead of candles or open flames, in case of ruptured gas lines.
- Do not travel to damaged areas.

The following processes and equipment can be affected by earthquakes:

- Process fluids can slosh and their flow can be interrupted during the quake.
- Incorrect instrumentation readings can occur (e.g., levels, gauges, vibration monitors) due to the seismic activity.
- Towers can sway.

- Equipment can topple.
- Materials can shift and fall.

Emergency Plans

Emergency plans analyze potential vulnerabilities combined with threats. Vulnerability is a weakness, or "hole" in a defense. A threat is an event that could harm people or property, such as a natural disaster or accident.

Companies create emergency plans to address vulnerabilities and threats. A company's emergency plans address what responses and actions will need to occur in a wide range of crisis situations. These plans help facilities deal with potential natural disasters, accidents, personnel safety, communications, evacuation, process shutdown and startup, and so on. Drills and training help process technicians understand what to do during an emergency.

Continuity and resumption plans define the steps a company should take to reduce the impact of a crisis or emergency, ensuring that the business continues normal operations, or resumes as quickly as possible.

Government Agencies and Aid Organizations

The OSHA Process Safety Management (PSM) 1910.119 regulation addresses emergency planning and response. For process facilities, emergency planning and response can include natural disaster plans.

A number of government agencies and aid organizations address natural disaster, inclement weather, and preparedness planning:

- National Weather Service (NWS)
- Federal Emergency Management Agency (FEMA)
- United States Geological Survey (USGS)
- OSHA
- American Red Cross
- State and local emergency management agencies

You can also find weather-related information from local news stations (e.g., TV and radio), publications, and the Internet.

Summary

Process facilities operate around the clock all year long, through all types of weather. Process technicians must work in good or bad weather. They are expected to work despite rain, wind, ice, heat, storms or other fairly common weather conditions.

Some of the more serious conditions a process technician may encounter include hurricanes, tornadoes and high winds, heavy rains, lightning and flooding, earthquakes, and extreme temperatures.

Hurricanes, which primarily affect coastal areas, are an extremely destructive force of nature. Hurricanes are usually accompanied by heavy rains, lightning, thunder, high winds, and tornadoes. They can hurl debris and turn normally harmless objects into deadly projectiles. Hurricanes can also produce storm surges that cause flooding that can damage equipment and electronics, and cause people to be stranded in one location for a prolonged period of time.

Whenever a hurricane warning is in effect, process technicians should monitor the storm carefully, stock adequate provisions (e.g., food, water, medicine, and batteries), create an evacuation plan, and prepare their homes and the process facilities for the impending storm.

Checking Your Knowledge

1. Define the following terms:
   a. Earthquake
   b. Flood
   c. Hurricane
   d. Storm surge
   e. Tornado
   f. Warning
   g. Watch

2. *(True or False)* A tropical storm is upgraded to a hurricane when its sustained winds reach 74 mph or higher.

3. Define hurricane watch and hurricane warning.

4. What part of the world experiences the most tornadoes?
   a. Asia
   b. California
   c. Canada
   d. The United States

5. What is the top wind speed listed for an F5 tornado?
   a. 74 mph
   b. 260 mph
   c. 318 mph
   d. There is not one.

6. A flash flood is flooding that _____ rapidly, but can fall quickly.

7. *(True or False)* Six inches of fast-moving water can knock people off their feet.

8. Lightning can generate heat up to _____ degrees F.

9. *(True or False)* If you hear thunder, there is no potential for being struck by lightning.

10. To determine the distance you are from lightning, count the number of seconds between the flash of lightning and when you hear the sound of thunder, then divide this number by ____ to get the distance in miles.
    a. 2
    b. 1
    c. 60
    d. 5

11. Frostbite is an extreme condition that occurs when _____ freezes.

12. Earthquakes are measured in magnitude, using the _____ scale.
    a. Fujita
    b. Richter
    c. Seismic
    d. Bernoulli

13. *(True or False)* Earthquakes can generate tidal waves (tsunamis).

14. Emergency plans take into account vulnerabilities and _____.

15. What is the name of the government agency that provides weather-related information and notifications?

Student Activities

1. Research several famous natural disasters, then select one for a report. Write at least two pages on what caused the disaster, its impact, and how you might have prepared for such a disaster.

2. Make a list of potential disasters and weather-related hazards in your area. Create a set of emergency preparedness plans for each one.

3. Role play with a group of fellow students, discussing how you would prepare emergency plans for the XYZ Corp. to either continue operations during, or resume operations after, a hurricane.

14

Physical Security and Cybersecurity

Objectives

This chapter provides an overview of physical security and cybersecurity in the process industries, which can affect people, facilities, resources, information, and computers.

Upon completion of this chapter you will be able to do the following:

- Identify special vulnerabilities and risks associated with the process industries:

  Terrorist organizations and hostile nation-states

  Insiders

  Criminal elements

- Recognize the nature of threats to physical security and cybersecurity:

  Terrorist threats and acts

  Workplace violence

  Criminal acts

  Industrial espionage

- Describe the activities involved with maintaining physical security in these areas:

  Access and perimeter

  Operations

  Communications

  Personnel

- Describe the tasks associated with protecting electronic information through sound cybersecurity practices, including the following:

  Password protection

  Malicious software, or malware

- Describe the government regulations that address physical security and cybersecurity.

Key Terms

Cyber—relating to computers and computing items, such as data, the Internet, and computer networks.

Cyber attack—an attack against information, computers, and communication systems, to cause harm, steal information, disrupt productivity, or take control of a computer system.

Hostile nation-state—a country that poses a threat to other countries.

Insider—a person inside a company who causes harm, either intentionally or unintentionally.

Malware—computer programs developed to cause intentional harm.

Network—two or more computers linked together for sharing data, programs, and resources such as printers and scanners.

Program—computer software (sometimes called an application).

Risk—a combination of vulnerabilities and threats.

Terrorist—a radical person who uses terror as a weapon to control others.

Threat—a perceived or implied feeling or communication; an individual or a group that will harm people or property.

Vulnerability—a weakness, or "hole" in a defense system.

Introduction

The process industries face external and internal threats to physical security (e.g., people, facilities) and cybersecurity (e.g., information, computers). Process technicians must understand the nature of these threats, recognize them, and understand the impact on themselves and the plant or facility for which they work.

Physical Security and Cybersecurity Overview

The process industries face a wide range of threats to security, many that have become prominent only within recent years (e.g., terrorism and computer hacking). These threats are called security hazards and include a hazard or threat from a person or group seeking to intentionally harm people, computer resources, or other vital assets. Threats can take many forms, but they typically fall into one of two categories:

- Physical security threats, such as those to people, equipment, products, and facilities (e.g., pipelines, control centers, and other vital areas)
- Cybersecurity threats, such as those to information, computer, and communication resources. **Cyber** is a term that relates to computing and computing-related items, such as data, computer networks, and the Internet.

The process industries provide vital products and services to meet a variety of critical needs for everyday life:

- Electricity generation and nuclear power
- Food manufacturing
- Oil and gas production
- Chemical manufacturing
- Water treatment
- Pharmaceuticals

If the production of these products or services was severely interrupted, such as in the event of a terrorist attack, a country could be critically debilitated.

Many companies in the process industries are viewed as high-profile organizations, meaning the general populace recognizes their name, image, and reputation. Thus, companies in the process industries are prime targets from threats such as terrorists, hackers, hijackers, blackmailers, and more.

Companies must take steps to protect themselves from threats, both external and internal, using physical security and cybersecurity measures to safeguard assets and workers.

Threats to Process Industries

Different individuals and groups of individuals can pose threats to the process industries. A lone person can bring a gun to work and start shooting. A hacker can access and steal sensitive information. A terrorist group can plant a bomb or release biological agents. An unknown person can extort a company by tampering with products.

The following are some general descriptions of individuals or groups that can threaten the physical security or cybersecurity of a company in the process industries:

• **Terrorists** and/or terrorist organizations: radicals that use terror as a weapon. They use force and violence, typically against civilians and property, to achieve political or ideological objectives. Terrorists aim to create fear and panic, trying to intimidate or threaten the public and seeking publicity for their cause. They attempt to convince the public that their government is powerless to prevent terrorism.

• **Hostile nation-states**: a nation-state is a sovereign territory in which most of the citizens are united by common language, ideology, or descent. A hostile nation-state is one that is belligerent (hostile) to other countries near it or around the world, posing a threat to the general peace.

• **Insiders**: persons working within a company whose actions cause harm to that company, its people, and its resources. These actions can be intentional or unintentional. Such actions can include spying, workplace violence, carrying out terrorist-related activities (e.g., gathering information for a terrorist group), leaking sensitive information, or even accidentally allowing outsiders access to a facility or resources.

• **Criminal elements:** a person or groups that break local, state, and/or federal laws for profit or other reasons. Extortionists and hackers can be considered criminal elements.

The Nature of Threats to Physical Security and Cybersecurity

Along with understanding which individuals or groups of individuals can pose threats, process technicians should also recognize the nature of these threats. A **threat** is a potential event that could result in harm to either people or property (e.g., equipment, computers, and information). For example, the threat can be from a terrorist attack, criminal act, or workplace violence. A threat is often meant to exploit a known vulnerability.

A threat can be deliberate or accidental. A deliberate threat has a purpose behind it, while an accidental threat is unintended. A kidnapping is a deliberate event, while someone inadvertently erasing vital information from a computer is accidental.

The typical goals behind deliberate threats include the following:

• Threats to the safety and security of people
• Destruction or damage of critical facilities, resources, or equipment
• Theft or damage of vital equipment, important materials, or sensitive information
• Demands (e.g., money, release of prisoners)
• Delivering a message or creating adverse publicity

Threats can take the form of a message from an individual or group, indicating intent to inflict harm on people or property. The threat can be real or false. For example, someone may phone in a bomb threat to a company, even though no bomb exists. Or a terrorist group can announce an intended attack through a TV broadcast. Such threats can also include demands to be met to avoid the threat being carried out.

In broad terms, even bad weather can be a threat to physical security and cybersecurity. For example, flooding can disrupt operations or lightning can damage computers with sensitive information. For more details on natural disasters, refer to Chapter 13, *Natural Disasters and Inclement Weather*.

The following sections briefly describe a wide range of potential threats, but by no means provide a complete list of all possible threats to the process industries.

BIOLOGICAL ATTACK

A release of a biological agent, such as bacteria, viruses, or toxins (see the *Biological Hazards* chapter for more details), can injure or kill people, animals (e.g., livestock), or crops. Anthrax and smallpox are two potential biological agents. Biological agents can be dispersed in the following ways:

- Aerosols, or spraying them in the air as a fine mist or releasing spores (that may look like a dust) that can potentially drift for miles. Inhaling the biological agent can cause disease in people, animals, or crops.
- Animals, by turning loose infected animals that can carry the disease to humans. Animals such as mice, fleas, and mosquitoes can be potential carriers of diseases.
- Person-to-person, if an infected person deliberately exposes himself or herself to other people. People can spread diseases such as smallpox and the plague.
- Contamination of food and/or water, by placing a biological agent in food or water supplies, causing harm to anyone ingesting them.
- Biological agents sent through the mail.

CHEMICAL ATTACK

A chemical attack can release poisonous materials in solid, liquid, or gas form, causing toxic effects on people, animals, or plants (see the *Chemical Hazards* chapter for more details). Hazardous chemicals can be dispersed in a variety of ways, including bombs, sprays from aircraft or other vehicles, or as released liquids. Chemical agents may have no taste or odor. Toxic effects can be immediate (within seconds) or delayed (over several days). Even though chemical agents are potentially deadly, they are hard to deliver in lethal concentrations and difficult to produce.

Chemical hazards come in different forms:

- Pulmonary agents (lung-damaging) such as phosgene
- Blistering agents, such as mustard gas
- Nerve agents, such as sarin or VX
- Riot-control agents, such as tear gas

Some chemical agents also cause incapacitation, induce vomiting, or affect the blood. Chemical agents can potentially be sent through the mail.

Did You Know?

Anthrax is a serious disease caused by Bacillus anthracis, a bacterium that forms tiny spores that can be ingested, inhaled, or injected.

After the attack on the World Trade Center on September 11, 2001, terrorist organizations waged a biological attack and began mailing these spores to key public officials with the intent of maiming or killing the recipient.

ANYCorp
Dylan Road
Elizabeth, TX
71005

Johnny Collins
32798 Joshua Lane
Megansville, TX 70400

FIGURE 14-1 Radioactive Symbol

Nuclear or Radiological Attack

With more countries building nuclear weapons than ever before, there is a possibility that a terrorist group or hostile nation-state can use a nuclear weapon. Nuclear explosions cause extensive devastation by generating an extensive blast, blinding light, extreme heat, radiation and fallout, shockwaves, and fires. Fallout can be carried on air currents for hundreds of miles. Terrorists would favor using a smaller-sized nuclear bomb, called a "suitcase" weapon, because of its compact size.

A more likely scenario than a nuclear explosion is the use of a radiological dispersion device (or "dirty bomb"), built by combining conventional explosives with radioactive materials. Such a device would be easier for a terrorist group to construct than obtaining a nuclear weapon. The explosion of such a device would scatter radioactive particles into the air currents, dispersing them over a large area. Factors that affect the success of such a device are the amount of radioactive materials used and wind conditions. A terrorist group could target radioactive materials, used in various forms throughout most process industries, for theft. (Figure 14-1)

Nuclear weapons detonated in the atmosphere also create an effect called an electromagnetic pulse (EMP) that is a high-density electrical field. EMP can damage electronic devices connected to power sources or antennas, such as computers, communication systems, electrical appliances, vehicle ignition systems, and other similar devices.

Terrorist Attack

Terrorism aims to maximize the fear produced by an attack. A terrorist attack could include multiple coordinated threats, including gunfire; chemical, biological, or nuclear attack; bombings, and so on.

GUNFIRE

One or more individuals can shoot at people or property with guns (e.g., pistols, rifles, shotguns, or automatic weapons). The most likely cause of a shooting in a process industries setting is workplace violence. (Figure 14-2) A laid-off or distraught worker can bring a gun to work to cause damage to property or to an individual. A terrorist attack on a facility, although not likely, could result in gunfire.

Companies can use metal detectors, security guards, controlled entries, and other such methods for reducing the gunfire hazard. Companies can also implement policies that prohibit weapons from being brought onto company property.

EXPLOSIVES AND BOMBINGS

A person or a group can plant explosives to harm people, buildings, and equipment. A bomb is an explosive device; it typically consists of a container filled with an explosive material and designed to cause damage. Bombs are typically triggered by a clock,

FIGURE 14-2 The Most Likely Cause of a Shooting in a Process Industries Setting is Workplace Violence

remote control, or sensor (e.g., contact, pressure). Bombs can be any of the following types:

- Conventional—uses chemical explosives to cause destruction
- Dispersive—uses explosives to spread materials such as chemicals or radioactive substances, or shrapnel
- Nuclear—creates an atomic explosion and radioactive fallout

Bombs are a central part of a terrorist's arsenal. Terrorists usually build custom-made bombs, using different designs and explosives to maximize the terror caused.

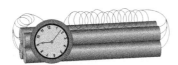

FIGURE 14-3 A Bomb is an Explosive Device (Typically Triggered by a Clock, Remote Control, or Sensor) Designed to Cause Damage

The most likely event to occur in a process industries setting is a bomb threat. All bomb threats must be treated seriously and appropriate actions taken. Bombs can potentially be sent through the mail.

KIDNAPPING

Kidnapping involves an individual or group taking a person against that person's will. The person is held captive, either for ransom (money or other demands) or to make a political or ideological statement. A person could also be forced to reveal sensitive information, such as trade secrets, for use in committing another crime.

HIJACKINGS

Hijacking involves an individual or group stopping a vehicle in transit, and taking control of it by force. All types of vehicles (e.g., trucks, cars, boats, and aircraft) have been hijacking targets. The hijacking can be for criminal or terrorist purposes. For example, a truck can be hijacked so that its cargo can be stolen. Or, a helicopter could be hijacked and forced to fly terrorists to an offshore oil facility.

FIGURE 14-4 Hijacking is When an Individual or Group Takes Control of a Vehicle by Force

CYBER ATTACK

A **cyber attack** is when an individual or group targets computer or communication systems to gain sensitive information or cause damage. Targets of cyber attacks can include critical infrastructures for power systems, financial systems, process control systems, or confidential information.

WORKPLACE VIOLENCE

Workplace violence is a rapidly growing crime. Workplace violence can take many forms, including assault with a weapon (typically a gun) or physical violence (i.e., no weapon is involved). Workplace violence, often due to despair or depression, can be triggered by downsizing, termination, conflict with a boss or co-worker, or personal problems.

INDUSTRIAL ESPIONAGE

Industrial espionage is spying, done for commercial reasons instead of national security purposes. An individual, group, or even a government can be seeking proprietary information from a company, such as trade secrets, inventions, stock-related information, or business processes. Information can fall into one of three types: public, private, and confidential. Industrial espionage typically targets confidential information. A variety of methods are used to obtain information: computer hacking, coercing a person trusted with the information, or breaking into secured areas and stealing it. Companies often implement policies that require their employees to sign a non-disclosure agreement.

EXTORTION AND BLACKMAIL

An individual or a group can threaten to inflict harm on a person or company, unless a demand is met (e.g., for money, goods, services, behavior, or action). This is called extortion or blackmail. The threatened harm could be against a person or people, property, or even reputation. For example, a person could threaten to tamper with a company's products unless given money. Or a group could threaten to detonate a bomb unless prisoners are released.

SABOTAGE

Sabotage involves the deliberate damage, or even destruction, of equipment or systems, in order to limit or halt operations of a facility. The facility's operations usually perform a vital task, such as provide power or water, create chemicals or oil products, supply food, and so on. For the most effective results, sabotage usually involves explosives, but not always. A facility can be the target of an attack; even something as simple as a threat can at least temporarily disrupt operations.

ASSASSINATION

Assassination involves killing an important person, which is most often a political figure. The assassination can be related to ideological or political reasons, or for money.

FIGURE 14-5 Assassination Involves the Killing of an Important Person

Countering Threats

Security, whether it is physical or cyber, is about managing risk. **Risk** is a combination of vulnerability plus threat. **Vulnerability** is a weakness, or "hole" in a defense. A threat is an individual or a group that could harm people or property. A threat could also be a natural threat, such as a disaster or accident. For example, a vulnerability might be a computer system without adequate security and the threat could be a hacker looking to steal confidential information. Or another type of vulnerability is a facility with an unsecured perimeter that allows unrestricted access, for which the threat could be a terrorist group planning an attack.

Companies address security issues by minimizing risk, which means reducing vulnerabilities and countering threats. Companies analyze their vulnerabilities and potential threats, then typically develop threat response plans to counter any physical or cybersecurity issues. Threat plans determine the level of risk and describe the processes and procedures for reducing or eliminating threat.

A company's emergency plans address what responses and actions will need to occur in a wide range of crisis situations. Continuity and resumption plans define the steps a company should take to reduce the impact of a crisis or emergency, ensuring that the business resumes normal operations as soon as possible. Your company will notify you of these plans, including emergency communication information, and train you on vital elements of the plans as necessary.

In the United States, the Homeland Security Advisory System provides a way to notify the general public and government agencies about the risk of terrorist acts. This system uses warnings, called "Threat Conditions," that increase as the risk of a threat increases. Process technicians should pay close attention to the threat condition level. This level can determine what additional security measures a company enacts. These conditions are as follows:

- Low Condition (Green)—declared when there is a low risk of terrorist attack
- Guarded Condition (Blue)—declared when there is a general risk of terrorist attack
- Elevated Condition (Yellow)—declared when there is a significant risk of terrorist attack
- High Condition (Orange)—declared when there is a high risk of terrorist attack
- Severe Condition (Red)—declared when there is a severe risk of terrorist attack

Signs of Potential Terrorist Activity

Following are the seven signs of potential terrorist activity:

- Surveillance—monitoring or recording activities at potential targets
- Elicitation—gathering information, usually from people "in the know," about operations, capabilities, and people
- Tests of security—attempting to measure reaction times to security breaches, and assessing strengths and weaknesses
- Acquiring supplies—buying or stealing explosives, weapons, uniforms, badges, access cards and other necessary materials
- Suspicious persons—people that do not seem to belong ("out of place") in the environment they are in
- Trial run—holding a "run-through" of the plan without actually committing the attack
- Deploying assets—placing people and supplies in place to commit the attack; this is the last chance to alert authorities before the attack happens

Physical Security

Physical security involves putting measures in place that will safeguard personnel and others (e.g., the surrounding community). Physical security also aims to protect a

FIGURE 14-6 Physical Security is Used to Protect a Company's Critical Resources

company's critical resources, such as equipment and materials, from unauthorized access, to prevent these resources from being harmed or stolen.

Physical security monitors and protects resources through the following means:

- Access barriers (gates, fences, doors)
- Alarm and monitoring devices (sensors, cameras, motion detectors)
- Security forces (guard posts, patrols)
- Pass-protected door entries and gates

Companies analyze vulnerabilities and threats to determine what must be protected, how it must be protected, and to what level. Based on this analysis, physical security policies and procedures are implemented to address the following five areas:

- Deterrence—discouraging or preventing access to a facility and certain areas within it. Examples include fences, lighting, signs, and patrols.
- Detection—identifying threats to security. Examples include alarms, motion detectors, and camera systems.
- Assessment—determining the nature of a threat and the potential threat level. Security personnel usually review the threat and assess whether it is minor or major.
- Communications—informing appropriate personnel that a threat exists. Examples include communications such as a process technician reporting a suspicious person to security, an alarm going off, or a security guard monitoring a camera system.
- Response—acting on the threat. Examples include security guards responding, local police being called, or evacuating a facility.

Physical security applies these four procedures to the following areas:

- Access and perimeter control
- Operations and procedures
- Communications
- Personnel

ACCESS AND PERIMETER CONTROL

The aim of access and perimeter control is to deny access to unauthorized people while monitoring those authorized for access. The perimeter (surrounding property) of a facility is typically secured. Fencing and gates can be used to restrict access to the facility, with a limited number of access points (entry or exit). Barriers can be placed across roadways to slow or stop incoming vehicles. Warning signs can be placed around the perimeter and the facility.

Adequate lighting is important for good security and perimeter control. Alarms can be used to monitor unauthorized access. Cameras, recorders, and other devices also can be used to monitor a facility. Some critical areas might also have independent alarm systems to monitor entry, including silent alarms.

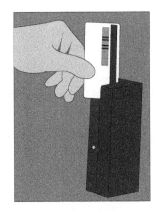

FIGURE 14-7 Access and Perimeter Controls are Used to Deny Access to Unauthorized People

Access points can be staffed with security personnel. Security personnel can monitor access, making sure authorized people can enter or exit the facility while unauthorized people are denied access. A facility could have measures such as photos IDs, access cards, or passes to process and identify personnel, contractors, and visitors while on the property. Additionally, workers might need keys or key cards to gain access to the facility or specific areas in it.

Security personnel, using cameras, patrols and guard stations, can also conduct surveillance of the perimeter and facility to make sure no unauthorized people have entered the facility.

OPERATIONS AND PROCEDURES

Physical security and cybersecurity affect a company's operations. Monitoring is a key component of security that affects operations. It is vital that all employees be alert for any signs of threats or suspicious behavior, and immediately report incidents to the proper personnel.

Physical security and cybersecurity measures can also affect operations by hampering productivity. Lax security exposes vulnerabilities to threats and increases risk, but overly tight security can impact operations through restricted access, excessive security policies, and procedures and uneasy working conditions.

Threat response plans and emergency plans are also part of a company's operations. Any threats or potential disasters trigger the appropriate plan, and employees will be instructed on specific actions to take. Certain employees may be asked to participate in threat response planning and emergency planning as part of their job. Operations can also include drills and training.

COMMUNICATIONS

Communications are a vital element to both physical security and cybersecurity. Up-to-date, accurate information about threat situations must be communicated between different personnel. Process technicians must help security personnel by watching for potential threats and reporting them through proper channels immediately. Their work

FIGURE 14-8 Process Technicians Must Help Security Personnel by Watching for Potential Threats and Reporting them Through Proper Channels Immediately

can take them all over a facility, letting them provide additional "eyes and ears" to identify potential threats. Every process technician must diligently watch for threats, take any threat seriously, and report threats immediately.

Decision-makers must analyze and determine identified threats. Management must communicate threat responses and emergency plans to personnel. Communication systems such as alarms, radios, phones, and other devices can keep all involved personnel updated on threats and how to respond.

PERSONNEL

People applying for jobs in the process industries may be subjected to a thorough background check before being employed. This background screening can involve a variety of checks (drug testing, criminal history, references, past employment, and credit history) to make sure that only trustworthy and reliable people are hired.

Pre-employment background checks are a vital element of physical security and cybersecurity measures. Such a screening can mitigate the threat from insiders. Insiders can pose a high threat, since most physical and cybersecurity measures aim to prevent unauthorized access from outsiders, not authorized personnel.

PRACTICING GOOD PHYSICAL SECURITY

Process technicians must be constantly aware of their surroundings. The very nature of any kind of attack depends upon surprise. There may be little or no warning before an attack. All threats should be presumed real and reported immediately.

Process technicians must also be watchful for individuals behaving suspiciously around the facility. A process technician should immediately report anyone observed acting in a strange way. If the person seems to be out of place, it is better to report the individual than run the risk of failing to stop a potential attack.

Anyone that is conducting surveillance must be reported. The person will appear to be recording or monitoring activity around the facility. The person could be doing any of the following:

- Taking notes
- Taking pictures
- Using binoculars, telescopes, or other similar objects
- Shooting video
- Drawing diagrams
- Marking a map, or counting off steps
- Loitering near the facility
- Observing people and vehicles come in and out of the facility

Someone might try to get vital information from the process technician that could be used to plan an attack. This person could contact the process technician in person, over the phone, via e-mail, and so on. The process technician must also report any stolen or missing items, such as badges, uniforms, access cards, or parking passes. Someone could try to use these to gain access to a facility.

Process technicians must know how to report potential threats and suspicious persons. Immediate reporting is critical. Process technicians should be aware of the location of the nearest communication device (e.g., radio and phone).

Process technicians should also know where emergency exits are located in relation to where they are working. If working in an unfamiliar area, exits should be noted before starting work.

During a threat situation or emergency, process technicians should follow all instructions immediately, and remain calm.

Cybersecurity

Process technicians must understand that information, computing systems, and communications systems can be targets for attack as well as people and facilities. These attacks are called cyber attacks. Cyber attacks can take different forms. Individuals or

FIGURE 14-9 Cyber Attacks can Destroy Important Electronic Equipment and Data

groups can hack in and steal sensitive information, unleash a dangerous computer virus, take control of a computer system, or affect the performance of a computer system. Criminal elements can seek to profit from cyber attacks by selling or using confidential information. Terrorists can use cyber attacks to damage critical resources such as power generation and oil and gas production. Disgruntled employees can cause harm to important information by altering or deleting it.

Cyber attacks can take different forms, such as the following:

- A direct attack from an outside source through a network ("hacking")
- A physical attack against a computing or communication system
- An attack from inside, from a person (or persons) trusted with access to the computing or communication system (either intentional or accidental)
- Many analysts believe that cyber attacks will become a new form of terrorist attack and a new type of warfare that hostile nation-states can unleash

Cyber attacks target the following aspects of process facilities:

- Information
- Productivity and capability
- Control

INFORMATION

Information is a valuable asset for a company. Information can take many forms:

- Trade secrets
- Proprietary information
- Inventions and patent applications
- Employee records
- Financial data
- Forecasts
- Customer lists
- E-mail messages

In a company's computer system, computers are usually **networked**, or linked together. The computer system can also be linked to the Internet (and thus, the outside world). Networked computers and Internet access create an environment that allows external threats (e.g., hackers) to potentially access and misuse this information. Even insiders can cause harm, whether maliciously or unintentionally.

It is critical that a company protect proprietary or sensitive information from unauthorized access. A company must ensure that those with authorized access are trustworthy. Computer access and usage must also be monitored to ensure that information and the computer system are not compromised.

PRODUCTIVITY AND CAPABILITY

Cyber attacks can impact productivity by unleashing malicious software that causes computers to behave erratically, important data to be erased, and computer networks to perform slowly. This can significantly impact a company's productivity, and affect its capability to do business.

CONTROL

A cyber attack can impact physical security as well. Although most computer-based control systems are isolated from access by the outside world (i.e., through the Internet), some are connected to an internal computer network that is connected to the outside world through the Internet. Potentially, a hacker could gain access to a facility's critical control systems, then shut down or modify process operations.

CYBERSECURITY PRACTICES

Cybersecurity involves protecting information, computing systems, and communications systems from unauthorized access, modification, or destruction.

Cybersecurity involves the following:

- Preventing unauthorized computer access (e.g., through passwords, computer "fire-walls")
- Making sure those with access are trustworthy
- Monitoring computer access and usage
- Preventing physical threats (e.g., direct access to sensitive information and computer networks)
- Safeguarding communication (e.g., cell phone and PDA protection)
- Keeping groups or individuals from using sensitive information and computer networks to carry out physical threats (e.g., hacking into a pipeline monitoring network and taking control of the system)

Cybersecurity seeks to protect valuable information (e.g., trade secrets, financial data) and capabilities (e.g., Internet access, e-mail) from unauthorized access, either from external or internal sources. Cybersecurity also aims to protect information from unauthorized changes or deletions. Cybersecurity policies and procedures allow a company to protect its valuable information, computer systems, and communications systems.

Each company establishes its own policies and procedures for cybersecurity. However, everyone in the company, including process technicians, are responsible for cybersecurity. The actions of one person not following the company's cybersecurity practices can jeopardize the entire computer system. For example, the system can be compromised when a user does not log off a computer or gives a password to a co-worker.

Cybersecurity practices typically fall into one of four categories:

- Confidentiality—preventing unauthorized access to information
- Integrity—preventing unauthorized modification of information
- Availability—allowing authorized individuals to access information
- Accountability—establishing a record or trail of who accessed or modified information

Hackers can use spyware and other tools to intercept sensitive data sent over the Internet (e.g., through e-mail). Another hacker tool allows them to crack and steal passwords.

A variety of practices can be used to improve cybersecurity:

- Encryption—using special software that creates a code for disguising information
- Account lockout—locking out a user after a certain number of incorrect attempts to log in to a network
- Password practices—creating strong, complicated passwords
- Network monitoring—checking user activity on their computers (e.g. visiting Web sites, using email, accessing files).
- Access devices—requiring someone to use a special device (called an access token) to log in to a computer

- Biometrics—using fingerprints or retinal scans to identify a user (although not widely in use by the process industries, companies can use them to control access to sensitive areas)
- Information classification—applying labels to information, such as "confidential" and "sensitive"
- Backups—storing crucial data on removable media that can be placed in a safe location

Cybersecurity threats can come from inside a company as well. Even co-workers should not be trusted with your password. A disgruntled co-worker who has access to your password can cause harm but make it look like your doing. A malicious insider can also be looking to steal information or cause damage to vital data. An insider can attempt to sell information, steal financial or customer records, or publicize confidential data. Some insiders cause harm accidentally, meaning they did not intend to do anything wrong. For example, they can accidentally delete or damage files, misuse equipment, spill drinks on computer hardware, and so on.

THE INTERNET

The Internet is not anonymous, and a user's activity can be traced. Also, remember that e-mails and sensitive information can be intercepted. Malicious software can be hidden on files that you download from the Internet. Also, a company can monitor and log Internet activity as well as e-mail usage. An individual's right to privacy does not apply if company resources (e.g., computers, Internet access, e-mail system, network) are used to send or receive information.

PASSWORDS

Many process technicians are required to use a computer as part of the job. You will most likely need a user login and password to access the network. Protect your login and password. If a hacker or co-worker uses your password to access the network, he or she could harm information, view inappropriate material over the Internet, or send e-mails; computer usage logs or records would make it appear that you did those things.

Following are some tips for protecting your password:

- Do not give your password to anyone.
- Do not write your password down.
- Do not use personal information as your password.
- Change your password often.
- Do not reply to e-mails that ask you to send your user name and password.

If you are allowed to set your own password, create a strong, complicated password (see the recommendations below).

Hackers can use a tool called a password cracker to figure out passwords. Computer users can make it easy on hackers by creating short, simple passwords with names, words from the dictionary, no numbers, and so on. The password cracker can easily figure out such passwords. A hacker who figures out a password can spy on the computer network, steal other passwords, monitor activities, download important data, or erase files.

To create a strong, complicated password (and make it harder for a password cracker to figure out) follow these tips:

- Use a combination of alphanumeric characters (letters and numbers).
- Do not use words from a dictionary, in any language.
- Use upper and lower case letters.
- Use at least 8 characters.
- Use special characters and punctuation, such as # . !
- Substitute actual numbers for words, such as 2 for to or too, or 4 for four
- Do not create a password consisting of only numbers.

Check with your company's IT area for rules about passwords.

Follow your company's procedures for passwords. If allowed to create your own password, the best way to create a password is to use the "Phrase" method. Come up with a phrase that has letters and numbers in it, along with upper or lower case, punctuation, and special characters. Then, take the first letter of each word along with the numbers or special characters or punctuation to create the password.

For example:

The phrase "Texas is one really really great state!" becomes the password Ti1rgs!

The phrase "Kailey Alex is one password for me." becomes the password KAi1pfm.

The phrase "A strong password helps me to be secure." becomes the password Asphm2bs.

Once you have created a password phrase, do not remember the password, just the phrase.

VIRUSES, WORMS, AND OTHER MALWARE

Computer viruses, worms, and other types of **malware** (software developed to cause harm) are a huge problem. Millions of dollars have been lost due to the destruction and data loss cause by malware. Malware can damage files, e-mail or transmit itself to other computers, make computers shut down or work improperly, install spyware, steal passwords, and cause computer networks to not perform properly. The following are the different types of malware:

• Virus—a computer program that can cause harm to other files or **programs** (computer software, sometimes called applications). Viruses can be hidden in all different types of seemingly innocent files, but like real viruses can cause harm. Viruses commonly spread through e-mail, and require user action to spread (e.g., by opening an infected e-mail).

• Worm—a computer program that can change or damage files and other programs. Worms are like viruses, but more dangerous. A worm can spread without user interaction, can spread quickly, and can send itself to other computers automatically.

• Trojan horse—a computer program that can collect, destroy, falsify, or exploit data. Trojan horses trick users into installing a harmful program onto a computer by hiding inside interesting-looking software (like animation or games).

• Spyware—software that monitors computer usage, allowing hackers to steal passwords or even take control of the computer.

FIGURE 14-10 Computer Viruses, Worms, and Other Types of Malware Cost Companies Millions of Dollars Each Year

Process technicians must be careful about bringing computer files and media from home for use at work. Many companies do not allow this practice, nor do they allow workers to take electronic media home from work.

PRACTICING GOOD CYBERSECURITY

Following are some tips for practicing good cybersecurity:

• Understand your company's security and privacy policies about computers and the network.

- Never give your password to anyone else, even if they claim to be a technical support person (unless you are talking with a trusted source).
- Use your work computer only for approved tasks and work-related purposes.
- Report suspicious behavior to the proper personnel at your company.
- Always log off your computer anytime you leave it unattended.
- Do not open e-mails from unknown senders.
- Only use legal software (properly licensed); do not make copies of work software for use at home.
- Do not open e-mails from known senders if the subject line seems suspicious and there is an attachment.
- Use up-to-date virus software and a firewall at home (most companies will have anti-virus programs and firewalls installed). A firewall is a computer program or hardware that protects the resources of a computer or computer network from access by outsiders.
- Scan all files with anti-virus software before copying them to your computer.
- Download files over the Internet only from trusted sources.
- Shred any documents containing sensitive or confidential information.
- Never give out information to anyone except a trusted authority.
- Never discuss sensitive or confidential information over a cell phone, or view such information on a laptop or similar device in public.
- Make sure that any devices such as laptops, cell phones, computer storage devices, pagers and PDAs (Personal Digital Assistants) are properly secured and cannot be easily stolen.
- Wipe data off any media (e.g., diskettes or hard drives) or destroy CDs before throwing them away.
- Never divulge cybersecurity procedures to anyone outside your company.
- Remember that radio communications may not be secure. Someone can electronically eavesdrop.
- Most companies do not allow cameras, recording devices, or cell phones with cameras to be brought into a facility. Some do not allow cell phones at all. Know what your company's policy is and follow it.

PHYSICAL SECURITY ASPECTS OF CYBERSECURITY

Critical computing resources can be protected using physical security as well as cybersecurity. Some computers and related components can be secured in locked rooms with limited access. Portable devices, such as laptops and PDAs, can be locked up when not in use. Media with critical data should be secured in areas safe from theft or damage from disasters.

Government Regulations Affecting Physical Security and Cybersecurity

The following are some government regulations and information that relates to physical security and cybersecurity:

- Patriot Act, to deter and punish terrorist acts
- Homeland Security Advisory System, to provide threat condition levels
- Maritime Security (MARSEC), to provide maritime threat condition levels
- National Strategy to Secure Cyberspace, to protect information and computing resources from attack
- Computer Fraud and Abuse Act, to protect against unauthorized computer access for the purposes of fraud or damage
- Electronic Communications Privacy Act, to protect against interceptions of wire, spoken, or electronic communications during transmission or storage
- Copyright Act, to protect intellectual property

Some government agencies involved in physical security and/or cybersecurity include the following:

- Department of Homeland Security
- National Security Agency
- FBI
- Nuclear Regulatory Commission
- Department of Transportation
- OSHA

Summary

The process industries face a wide range of security threats. These threats include hazards or threats from individuals or groups seeking to intentionally harm people, computer resources, or other vital assets. Threats can take many forms, but they typically fall into one of two categories: physical security or cybersecurity.

Physical security threats are threats to people, equipment, products, and facilities (e.g., pipelines, control centers, and other vital areas). Cybersecurity threats are directed toward data and information, computer resources, and communication resources.

Threats can come from a variety of sources, including terrorist groups, hostile nation-states, company insiders, and outside criminals. Threats can take many forms.

Some of the most serious threats a person may face include biological, chemical, nuclear, or terrorist attacks. It is important for process technicians to understand the nature of these threats, be able to recognize them, understand the impact on themselves and the plant or facility where they work, and be familiar with ways to prevent them.

Checking Your Knowledge

1. Define the following terms:
   a. Cyber
   b. Cyber attack
   c. Cybersecurity
   d. Hostile nation-state
   e. Insider
   f. Malware
   g. Network
   h. Physical security
   i. Program
   j. Risk
   k. Security hazard
   l. Terrorist
   m. Threat
   n. Vulnerability
2. *(True or False)* Cybersecurity involves protecting people and facilities from harm.
3. _____ aim to create fear and panic in order to achieve political or ideological objectives.
4. *(Choose the best answer)* Threats can be accidental or _____.
   a. Vulnerable
   b. Deliberate
   c. Inadvertent
   d. False
5. A(n) _____ attack involves the release of bacteria, viruses, or toxins.
6. *(True or False)* Sarin and mustard gas are chemical hazards.
7. *(Choose the best answer)* Security is about managing:
   a. Insiders
   b. Sabotage
   c. Bomb threats
   d. Risk
8. *(Choose the best answer)* A vulnerability is:
   a. A weakness
   b. A response
   c. A threat
   d. Risk

9. List the seven signs of potential terrorist activity.
10. Which of the following is NOT one of the five areas of physical security?
    a. Deterrence
    b. Password protection
    c. Detection
    d. Risk
11. *(True or False)* Process technicians do not need to watch for potential threats; watching is the job of security personnel.
12. *(Choose the best answer)* Cyber attacks target:
    a. Information
    b. Power generation only
    c. Key personnel
    d. Government agencies only
13. Name the four categories of cybersecurity practices.
14. Which of the following is the best password?
    a. billyh
    b. JD3i1gp!
    c. 5823145
    d. rocketman
15. *(True or False)* Computer viruses can spread without user interaction.
16. Which of the following is a good cybersecurity practice?
    a. Give your password to a co-worker, in case you forget it.
    b. Always log off your computer anytime you walk away from it.
    c. Open e-mails from unknown senders.
    d. Write down your password.

Student Activities

1. Go to the Department of Homeland Security Web site and research what you can personally do to prepare for a potential terrorist attack. Write up a two-page action plan.
2. Based on the cybersecurity discussion in this text, evaluate the degree to which your home computer is secure. Make a list of ways you can improve its security. If you do not have a home computer, learn what your institution's computer security policies and procedures are for computers available for student use. Compare them to the information in this text and make a list of how they are similar or different.
3. Research the topic of workplace violence. Find several sources that discuss what it is and how to protect against it. Write a three-page paper on the topic.

15

Recognizing Ergonomic Hazards

Objectives

This chapter provides an overview of ergonomic hazards found in the process industries.

After completing this chapter, you will be able to do the following:

■ Name certain activities performed in the process industries and discuss the potential ergonomic hazards posed by these activities:

Lifting and handling materials

Working at heights

Working in confined spaces

Using repetitive motions

■ Demonstrate proper lifting techniques.

■ Demonstrate proper ergonomics for repetitive motions.

■ Describe government regulations and industry guidelines that address ergonomic hazards.

Key Terms

Ergonomics—the study of how people interact with their work environment.

Musculoskeletal Disorder (MSD)—a health condition characterized by damage to muscles, nerves, tendons, ligaments, joints, etc.

Repetitive Motion Injury (RMI)—an injury caused by repeating the same motion.

Vertigo—a sensation or illusion of movement in which a person feels as if revolving in space (called subjective vertigo) or senses the surrounding environment to be spinning (called objective vertigo).

Introduction

Ergonomics is the study of how people interact with their work environment. The term *ergonomics* was coined by a group of scientists in the 1950s to describe their efforts to design equipment and work tasks to "fit the workers." So, ergonomics is the science of designing the job to fit the worker, rather than physically forcing the worker's body to fit the job.

The goal of ergonomics is to minimize accidents and illnesses due to chronic physical and psychological stresses while maximizing productivity, quality, and efficiency.

Ergonomics and Stress

The National Science Council, based on work produced by the Swedish Work Environment Fund, describes pairs of factors that can affect physical stress that occurs on the job (which leads to ergonomic hazards). See Table 15-1 on the following page for a comparison of some physical stress factors.

Even hazards such as confined spaces, heights, noise, and vibration can contribute to ergonomic stress. In light of all the types of tasks that process technicians must perform on a daily basis that produce physical stress, it is vital that technicians understand the importance of ergonomics. Your company can help you with information and training on how to reduce ergonomic hazards.

TABLE 15-1 Comparison of Physical Stress Factors

Factor #1	*Factor #2*
Sitting Sitting is better than standing, but sitting can cause stress (incorrect posture, improperly adjusted chair, non-ergonomic chair design).	**Standing** Standing, especially for extended periods in one position, can produce stress on the back, legs, and feet.
Stationary Tasks that require a worker to move infrequently create less stress than tasks that require constant motion (e.g., hauling, climbing, and turning).	**Mobile** Tasks that require constant motion cause stresses on a variety of body parts.
Small demand for strength and power Tasks that require little strength and power cause less stress than tasks that require more strength and power.	**Large demand for strength and power** Tasks that require strength and power, whether lifting heavy objects occasionally or lighter objects frequently, cause stress on the back, arms, hands, legs, and feet.
Good horizontal work areas Properly designed horizontal work areas (such as desks or workstations) require little side-to-side motion (e.g., twisting) by the worker.	**Bad horizontal work areas** Improperly designed horizontal work areas require constant side-to-side motion by the worker, causing stress on the neck, back, arms, hands, legs, and feet.
Good vertical work areas Properly designed vertical work areas require little up or down (e.g., bending, reaching over the head) motion by the worker.	**Bad vertical work areas** Improperly designed vertical work areas require constant up or down motion by the worker, causing stress on the neck, back, arms, hands, legs, and feet.

(Continued)

TABLE 15-1 (Continued)	
Factor #1	*Factor #2*
Non-repetitive motions Tasks that do not require frequent repetitive motions cause less stress than tasks that require ongoing repetitive motions (e.g., turning valves, climbing ladders).	**Repetitive motions** Tasks that require frequent repetitive motions cause stress on the wrists, arms, back, legs, and feet.
Low contact Tasks that do not require the worker to make contact with an object's surface (e.g., tools, equipment) cause less stress than tasks that require more contact.	**High surface contact** Tasks that require more contact between a worker and an object's surface, such as using a tool, cause stress to a variety of body parts.
Few negative environmental factors Work areas with few negative environmental factors (e.g., hazardous environments, heat or cold, humidity, pollution) cause less stress than areas with more negative factors.	**More negative environmental factors** Work areas with a number of negative environmental factors means that workers must wear many different types of Personal Protective Equipment (PPE). Although PPE forms a protective barrier against chemical, physical, and/or biological hazards, it can contribute to physical stress.

Ergonomic Hazards

Ergonomic hazards are produced when workstations, equipment, or tools are not engineered to match the needs of the workers who will interface with them. A **Musculoskeletal Disorder (MSD)** is a condition caused by physical stress that damages muscles, nerves, tendons, ligaments, and/or joints. A common MSD is carpal tunnel syndrome, a painful and potentially disabling condition affecting the wrists.

The process technician can encounter several types of ergonomic hazards within the workplace:

- Lifting and material handling—moving heavy or bulky objects, adding quantities of product to a container
- Repetitive motions—using tools such as screwdrivers, turning valves, frequently climbing stairs or ladders
- Poor posture—placing the body in awkward positions or moving incorrectly
- Improper use of tools—using the wrong tool for a particular task, or incorrectly using the right tool for a task
- Extended workstation or computer use—sitting at a Distributed Control System (DCS) workstation or computer and frequently using the keyboard and mouse
- Working at heights—performing tasks at elevations while on ladders, platforms, tanks, tall equipment, and other similar locations
- Working in confined spaces—performing permit work in a confined space

---

Did You Know?

The term *ergonomics* comes from two Greek words: "ergon," meaning human work and strength, and "nomos," meaning law or rule.

Ergon-Nomos

Impact of Physical Stress

Physical stress can lead to short and long-term health problems. It can also affect a person emotionally and mentally. The following are some situations that can result from physical stress:

- Accidents
- Unsafe working conditions
- Poor product quality
- Absenteeism
- High worker turnover
- Worker complaints

Lessening Physical Stress

The following are some general tips that can help reduce physical stress:

- Get proper exercise.
- Get sufficient sleep.
- Eat properly and drink enough water.
- Take breaks.
- Learn how to stretch and limber up.
- Start strenuous tasks slowly, and then build up the pace.
- Avoid using wrist splints, if possible (they can cause muscle decline).

Lifting and Material Handling

Common sources of ergonomic stress are activities involving lifting and material handling. Back injuries are one of the most frequently occurring work-related injuries. Process technicians are required to do a fair amount of lifting, carrying, pushing or shoving, and pulling to fulfill their job requirements. For example, a process technician may need to move a 55-gallon drum from one area to another, or may need to lift, carry, and dump a bag of a chemical into a process vessel.

CAUSES AND SYMPTOMS OF ERGONOMIC STRESS
FROM IMPROPER LIFTING OR MATERIALS HANDLING

Ergonomic stress from improper lifting or material handling can be caused by any of the following activities:

- Lifting from the floor
- Lifting while twisting
- Lifting objects that are too heavy
- Lifting objects of odd shapes
- Repetitive lifting

Improper Heavy Lift

FIGURE 15-1 Process Technicians Should Always Use Proper Lifting Technique When Moving Heavy Objects

- Lifting from shoulder height
- Lifting while seated
- Pushing or pulling loads without assistance
- Bending while moving

Symptoms of ergonomic stress resulting from improper lifting or materials handling include the following:

- Back pain
- Knee pain
- Shoulder pain

PREVENTING ERGONOMIC STRESS WHEN LIFTING OR HANDLING MATERIALS

To prevent exposure to the ergonomic stress associated with improper lifting or improper material handling, a process technician should remember the following rules:

- Employ correct lifting, pushing, and pulling techniques.
- Use the proper techniques no matter the weight or size of the material.
- Use tools and mechanical aids (e.g., carts, dollies, lift tables, rollers) when necessary.
- Make sure the area where you will be walking is clean and free of obstacles.
- Check that the walking surface is dry and not slippery. Wear non-skid footwear.
- Reduce the load when necessary.
- Ask for help when necessary.
- Store materials at a proper height when possible, so when the material is moved again less bending or reaching will be required.

PROPER LIFTING TECHNIQUES

When lifting items, use the following lifting techniques:

1. Plan for your lift to help avoid making awkward movements while you are holding something heavy. Have a path already cleared before you begin.
2. Get help if needed, or if you think you might need it. If an object is too heavy or shaped awkwardly, ask someone else to help you lift. If you are lifting with someone else, make sure you both agree on the lifting plan before you begin.
3. Stand with your feet shoulder-width apart to help give you a more solid base.
4. Place your feet close to the object and center your body over it.
5. Bend your knees and keep your back straight.
6. Tighten your abdominal muscles to hold your back in a good lifting position and prevent excessive force on the spine.
7. Lift with your legs rather than your back. Your legs are many times stronger than your back muscles.
8. Lift the object in a smooth motion; avoid jerking upward.
9. Lift close to your body. This will give you a stronger, more stable lift. Make sure you have a firm hold on the object and keep it close to your body.
10. Take short steps if walking with the object.
11. Keep your torso facing forward; avoid twisting.
12. Set the object down in its new position slowly and smoothly, then release it. Keep your back straight and knees bent (if necessary) while setting the object down.
13. Wear a back belt or back support while lifting heavy items, as required by your company's policy. Figure 15-2 illustrates proper lifting technique.

MOVING OBJECTS

When moving objects, remember that pushing is better than pulling. Pulling places more strain on your back than pushing. Use a tool or mechanical aid, such as a hand cart or rollers, if possible.

FIGURE 15-2 Proper Lifting Technique

Repetitive Motions

Repetitive Motion Injuries (RMI) are the result of repeated overuse of body joints. The connective tissues become sore and can become unusable if they are exposed to repeated trauma. The symptoms often occur over a long period of time, sometimes causing people to ignore the condition until the symptoms are more pronounced and/or a permanent injury occurs.

Not all joint pain results in a long-term condition. Sometimes, muscle pain develops after overexertion; however, muscles can heal quickly if properly managed. Early detection of RMI helps ensure that the problem can be corrected.

PROPER ERGONOMICS FOR REPETITIVE MOTIONS

Proper ergonomics for repetitive motion can help prevent injury. The best method for preventing RMI is to limit the amount of time spent performing repetitive motions. Process technicians tend to develop RMI from performing tasks that require extreme and awkward postures and excessive force to operate tools. Work practices to help reduce RMI include the following:

- Take rest breaks as needed to relieve fatigue.
- Get assistance; two or three workers performing a repetitive task reduces the stress on individual workers.

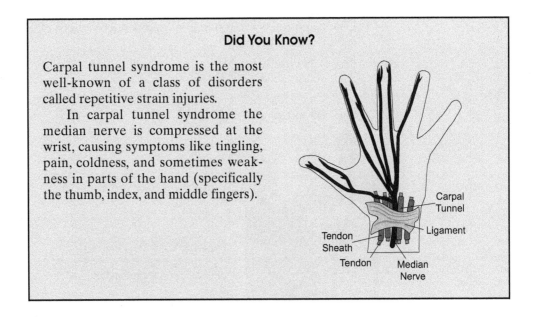

Did You Know?

Carpal tunnel syndrome is the most well-known of a class of disorders called repetitive strain injuries.

In carpal tunnel syndrome the median nerve is compressed at the wrist, causing symptoms like tingling, pain, coldness, and sometimes weakness in parts of the hand (specifically the thumb, index, and middle fingers).

Carpal Tunnel

Ligament

Tendon Sheath

Tendon

Median Nerve

- Rotate the tasks performed to involve different movements.
- Stretch every few hours to relieve tension and body aches.
- Lift using the legs and arms rather than the back.
- Avoid twisting while lifting.
- Use the proper tools for the task being performed.

Poor Posture

Process technicians frequently must perform tasks that involve an awkward or unnatural posture or motions that can result in physical stress. (Figure 15-3) Following are some examples of tasks that, process technicians often perform:

- Reach over their heads
- Perform repetitive motions (e.g., turning a valve or torquing a bolt)
- Twist their wrists repeatedly
- Use tools
- Bend down
- Twist their torso (trunk of the body)
- Climb ladders
- Lean on equipment or surfaces
- Exert force with their hands, arms, and legs
- Maintain a static (stationary) position

Following are some tips to help alleviate physical stress in these situations:

- Always use proper ergonomic techniques.
- Wear proper footwear.
- Get assistance.
- Maintain proper balance.
- Use the proper tool.
- Keep a firm grip on tools.
- Reduce any wasted motions (actions not directly affecting the task).
- Do not overextend your reach; limit reaching over your head or far from your body (use a ladder or reposition your body).
- Keep your elbows close to your body.
- Keep your arms straight or slightly inclined upward.
- Minimize forearm rotation and wrist-twisting motion.
- Avoid twisting your body. Reposition yourself as necessary.
- Squat down instead of bending over.
- Take regular breaks.
- Change positions when standing for extended periods of time. Try to prop one leg, then the other, up.

Improper Posture

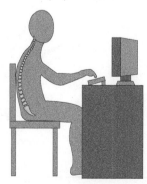

FIGURE 15-3 Poor Posture Puts Unnecessary Stress on the Joints and Bones in the Body

Improper Use of Tools

One of the most important ways to reduce ergonomic stress is to select the proper tool for the task. Use an ergonomically designed tool if available. The following are some recommendations on tool selection. (Note: This list does not consider specific tools or tasks, just the ergonomic factors of tool use). Select the following:

- Well-maintained tools
- Power tools when possible
- Tools with firm grips (avoid sharp, hard, or slippery grips)
- An ambidextrous (left or right hand) tool, so you can switch hands if you start to lose hand strength. If one is not available, select a tool designed for your dominant hand
- A tool with a handle or grip that extends across the fleshy part of your hand (this distributes the pressure of gripping the tool)
- The proper size tool at the lightest weight
- For tools with triggers, a trigger that depresses easily using two or more fingers
- Power or pneumatic tools with the least vibration

Manufacturers often provide instructions on the proper use of the tool, or your company might provide information and training.

Extended Workstation or Computer Use

Process technicians can spend a significant amount of their workday in front of a process control workstation (e.g., Distributed Control System, or DCS) or computers. Therefore, office-related ergonomics are important. They include the following factors:

- Chair adjustment, height, and positioning
- Workstation or desk height
- Keyboard position and placement
- Mouse placement and grip
- Monitor placement (some DCS monitors are fixed, however)
- Distance between the worker's eyes and the monitor
- A light touch while typing on the keyboard or using a mouse

Refer to company information and manufacturer recommendations for correct positioning and use of office furniture and computer/DCS equipment. Table 15-2 provides some general recommendations.

To achieve the proper positions, you might be required to adjust one or more of the following:

- Chair
- Desk or work surface

TABLE 15-2   Recommended Body Positions for Extended Workstation or Computer Use

Body Part	Position
Head	Positioned directly over your shoulders, without tilting forward
Eyes	About an arm's length from the monitor
Shoulders and neck	Relaxed; do not slouch
Back	Straight or leaning slightly forward (keep your back in the same position as when standing)
Elbows	Relaxed and at an upward angle
Wrist	Relaxed and straight, not angled
Knees	Lower than your hips
Feet	Placed firmly on the floor

- Keyboard
- Mouse
- Monitor

Following are some additional recommendations:

- Make sure the monitor screen is clean.
- Minimize glare on the monitor.
- Reduce the lighting around the monitor, if possible, to increase the contrast.
- Do not cradle the phone between your shoulder and cheek.
- Take a break during extensive typing or mouse use.

Using a computer or DCS can result in repetitive motions, also. Review the recommendations under the Repetitive Motions section.

Working at Heights

Another fairly common activity that can lead to ergonomic stress is working at heights. Process technicians may be required to do the following:

- Climb scaffolding
- Climb tall equipment (e.g., distillation towers, furnaces)
- Work on top of large storage vessels
- Load or unload rail cars, trailer trucks, tanker trucks, or barges
- Use powered platforms

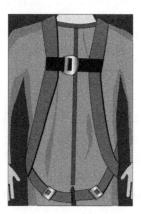

FIGURE 15-4 Process Technicians Should Use Fall Protection if Working at Heights Without Adequate Guarding

CAUSES AND SYMPTOMS OF ERGONOMIC STRESS FROM WORKING AT HEIGHTS

Ergonomic stress from working at heights can be caused by any of the following:

- Repetitive movement
- Falls
- Sore muscles and joints
- Strains

Symptoms of ergonomic stress resulting from the activities listed above include the following:

- Back pain or injury
- Knee pain or injury
- Shoulder pain or injury
- Foot or arch pain
- Wrist problems
- Bruising or swelling
- Anxiety

- Sweating
- Nausea
- **Vertigo** (a sensation or illusion of movement)

PREVENTING ERGONOMIC STRESS WHEN WORKING AT HEIGHTS

To prevent exposure to the ergonomic stress associated with working at heights, a process technician should follow these tips:

- Stay in good physical condition.
- Use fall protection if working at heights without adequate guarding.
- Inform a supervisor about any fear or concerns of working at heights.

Working in Confined Spaces

Many process technicians are required to occasionally work in confined spaces. Working in confined spaces can include entry into a voided storage tank, entry into a drained rail car, or entry into an empty reaction vessel to perform different types of tasks.

CAUSES AND SYMPTOMS OF ERGONOMIC STRESS FROM WORKING IN CONFINED SPACES

Ergonomic stress from working in confined spaces can be caused by any of the following activities:

- Repetitive movement
- Exposure to oxygen-deficient atmospheres
- Exposure to toxic atmospheres
- Fear of small, tight spaces

Symptoms of ergonomic stress resulting from the activities listed above include:

- Muscle or joint pain
- Loss of consciousness
- Inability to communicate
- Inability to think logically
- Dizziness
- Anxiety
- Sweating
- Nausea
- Vertigo

FIGURE 15-5 Process Technicians Should be Aware of Ergonomic Hazards When Working in Confined Spaces

PREVENTING ERGONOMIC STRESS WHILE WORKING IN CONFINED SPACES

In order to prevent exposure to the ergonomic stress associated with working in confined spaces, a process technician should follow these tips:

- Stay in good physical condition.
- Use proper Personal Protective Equipment (PPE) for the job and the possible hazards in the space.
- Inform a supervisor of any fear or concerns of small, tight spaces.

Vibrations

Vibration occurs when an object (usually something rigid) moves back in forth in a periodic motion. It is inaudible and can only be felt, which can result in ergonomic hazards. Vibrations can be experienced all over a body (such as from operating heavy equipment or driving or riding in certain types of vehicles). It can cause effects such as motion sickness and even spinal injury.

Vibration experienced through the hands and arms can result in a condition called Hand-Arm Vibration Syndrome (HAVS). This condition, which causes damage to nerves and blood vessels, can result from using power tools that generate vibrations (e.g., pneumatic tools). Some manufacturers now produce lower-vibration power tools to reduce such hazards. Using power tools in cold temperatures or with too-tight of a grip can worsen the effects of vibration-related hazards.

Following are some tips on vibration hazards:

- Take breaks from tools or areas where vibration is present.
- Use vibration-absorbing devices, such as floor mats or vehicle seat covers.
- Maintain a firm yet loose grip when operating vibrating tools.
- Use thick gloves while operating vibration tools if your company policy or manufacturer's recommendations permit them.
- Keep warm when using vibrating tools in cold environments.

Ergonomic Design

Many manufacturers consider ergonomics when designing new products (e.g., tools, equipment, furniture, work spaces). Even in a process industry environment, plant designers consider factors such as valve placement, flange or bolt arrangement, equipment spacing, pinch points, and other similar considerations.

Many process facilities use behavioral safety studies, where workers perform observations on each other's work habits and report anonymous data. These data are used to improve safety by addressing hazards, including ergonomic ones (e.g., were tools being used correctly, did the worker lift a heavy object correctly). Some ergonomic hazards can be addressed with engineering controls or administrative controls, so be sure to report such situations.

Ergonomic Products

A wide range of ergonomic products are available, such as the following:

- Tools
- Chairs, desks, and workstations
- Computer keyboards and mice
- Shoe inserts
- Floor mats
- Foot rails

Government Regulations and Industry Guidelines

OSHA developed guidelines for ergonomic design in the workplace in 2002. While these guidelines do not directly apply to the process industries, they are a first step toward addressing workplace ergonomics. Various organizations publish ergonomic guidelines.

In the process industries, industrial hygienists study work conditions and tasks to make suggested improvements that can limit and/or prevent ergonomic injuries.

Summary

Ergonomics is the study of how people interact with their work environment. The goal of ergonomics is to minimize accidents and illnesses due to chronic physical and psychological stresses while maximizing productivity, quality, and efficiency.

There are many different types of ergonomic hazards (e.g., improper lifting, repetitive motions, poor posture improper use of tools, extended workstation or computer use, working at heights, and working in confined spaces). Many of these ergonomic hazards can cause physical stress or injury. The impacts of physical stress include accidents, unsafe working conditions, poor product quality, absenteeism, high worker turnover, and worker complaints.

There are techniques process technicians can use to reduce physical stress. These include: get proper exercise, get sufficient sleep, eat properly and drink enough water, take breaks, learn how to stretch and limber up, and start strenuous tasks slowly and then build up the pace.

Many manufacturers consider ergonomics when designing new products (e.g., tools, equipment, furniture, workspaces). Even in a process industry environment, plant designers consider factors such as valve placement, flange or bolt arrangement, equipment spacing, pinch points, and more.

OSHA developed guidelines for ergonomic design in the workplace in 2002. While these guidelines do not directly apply to the process industries, they are a first step toward addressing workplace ergonomics. Various organizations publish ergonomic guidelines.

In the process industries, industrial hygienists study work conditions and tasks and make suggested improvements that can limit and/or prevent ergonomic injuries. Process technicians should always be aware of their work environment and report ergonomic issues whenever possible so they can be addressed.

Checking Your Knowledge

1. Define the following terms:
    a. Ergonomic hazard
    b. Ergonomics
    c. Musculoskeletal Disorder (MSD)
    d. Repetitive Motion Injury (RMI)
    e. Vertigo
2. *(True or False)* The goal of ergonomics is to minimize accidents and illnesses due to chronic illnesses caused by chemical exposure.
3. *(True or False)* High-contact tasks (e.g., using a tool or leaning on equipment) are less stressful than low-contact tasks.
4. _____ injuries are one of the most frequent work-related injuries.
5. List five activities related to lifting and handling materials that can cause ergonomic stress.
6. Which of the following precautions should be taken to prevent stress when lifting or handling materials?
    a. Reduce the load
    b. Ask for help
    c. Use tools or mechanical aids
    d. Use correct lifting techniques
    e. All of the above
7. Lift with your _____ rather than your back.

8. Name three possible symptoms of ergonomic stress from working at heights.
9. What is the condition resulting from repeated overuse of body joints?
    a. Repetitive Motion Injuries
    b. Repetitious Movement Conditions
    c. Muscular Degenerative Syndrome
    d. Spinal and Musculature Deterioration

Student Activities

1. Use the Internet or other resources to identify at least three medical conditions caused by ergonomic stress. Describe each condition, its cause, and treatment.
2. Observe people at your work, school, or home performing tasks that could cause potential ergonomic stresses. Then, make a list of the stresses you identified and describe possible solutions.
3. Demonstrate the proper technique for lifting a heavy object such as a box or bag. Have a fellow student or instructor observe your technique and make comments.
4. If available, practice opening and closing a manual valve using correct body posture.

16

Recognizing Environmental Hazards

Objectives

This chapter provides an overview of environmental hazards found in the process industries.

After completing this chapter, you will be able to do the following:

■ Identify specific categories of hazardous chemicals used in the process industries.

■ Explain the EPA regulations that impact the process industries.

■ Identify the various factors that can lead to leaks, spills, and releases.

■ Describe the potential dangers of leaks, spills, and releases in the environment and the community.

Key Terms

Corrosion—the eating away of materials by a chemical process (e.g., iron rusting).

Erosion—the wearing away (abrading) of materials by a physical process (e.g., sand-blasting).

Leak—a condition that occurs when a container or equipment is compromised, allowing a material to escape.

Release—a controlled or uncontrolled discharge of process materials into the environment.

Spill—an uncontrolled discharge of a liquid that usually involves more volume than a leak.

Superfund—a monetary fund that comes from tax dollars paid by the chemical industry to pay for the cleanup of abandoned waste sites in the event no responsible party can be found.

Introduction

Hazardous agents do not present a threat to the environment or to the average citizen in the community unless they are released into the environment. This chapter reviews the types of hazardous chemical categories, along with environmental regulations that impact the process industries. Also, this chapter discusses the various factors and dangers associated with leaks, spills, and releases.

The process technician must be familiar with the following:

- Pollutants and their sources
- Hazardous chemicals and their effects on the environment
- Factors behind leaks, spills, and releases
- Environmental controls
- Government regulations

Pollutants

The different types of pollution are covered in Chapter 2, *Types of Hazards and Their Effects*. This section reviews some sources of pollution:

- Burning of fossil fuels (e.g., coal, oil, natural gas)
  - SO_x
  - NO_x

- Particulates
- CO_2
- Unburned fuels and the evaporation of chemicals
  - Volatile Organic Compounds (VOCs)
- Pollutants formed in the air by combining VOCs, NO_x, and sunlight
  - Ozone

Following are some effects of pollution:

- The "Greenhouse Effect" from CO_2
- Acid rain from SO_x and NO_x
- Ozone-layer depletion from VOCs
- Ground-level health effects from ozone (e.g., shortness of breath and burning of the eyes and nose)

Hazardous Chemical Categories

CHEMICAL COMPOSITION AND CLASSIFICATION

Hazardous chemicals are often identified and grouped into categories on the basis of their hazardous characteristics. Chemicals are identified as flammable, corrosive, and/or toxic because their chemical composition causes them to exhibit these hazardous characteristics.

It is important for process technicians to understand the concept of categorization because categorization provides a foundation for understanding other fundamentals of material handling and storage, such as how materials should be labeled and used, and what materials can or cannot be safely stored alongside others.

OSHA, DOT, and EPA may use similar terminology when discussing chemical hazards, but the terms are not always interchangeable. OSHA's terminology relates to employee health and safety; DOT's terminology relates to the safe transportation of materials or wastes; and EPA's terminology relates to the protection of the environment.

Combustible and Flammable Materials

Combustible and flammable materials (gases, liquids, and solids) present fire hazards. These materials are classified as follows:

- Combustible liquid is any liquid that has a flashpoint at or above 100 degrees F but below 200 degrees F.

- Flammable liquid is any liquid that has a flashpoint below 100 degrees F.

- Flammable gas is any gas that at ambient temperature and pressure forms a flammable mixture with air at a concentration of 10% or less by volume.

- Flammable solid is any solid, other than a blasting agent or explosive, that is liable to cause fire through friction, absorption of moisture, spontaneous chemical change, or

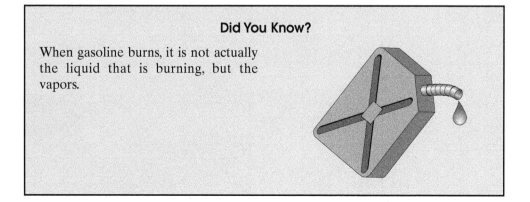

Did You Know?

When gasoline burns, it is not actually the liquid that is burning, but the vapors.

retained heat from manufacturing or processing. It can also be ignited readily, and when ignited burns so vigorously and persistently as to create a serious hazard.

Water-reactive, Pyrophoric, and Explosive Materials

• Water-reactive materials are chemicals that react with water to release a gas that is either flammable or that presents a health hazard (e.g., sodium reacts with water to produce hydrogen, which is an extremely flammable gas).

• A pyrophoric material is a chemical that will ignite spontaneously in air at a temperature of 130 degrees F or below (e.g., iron sulfites react with air to produce heat).

• An explosive is defined as a chemical that causes a sudden, almost instantaneous release of pressure, gas, and heat when subjected to sudden shock, pressure, or high temperature (e.g., ammonium nitrate and blasting powder).

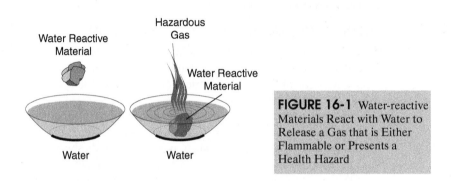

FIGURE 16-1 Water-reactive Materials React with Water to Release a Gas that is Either Flammable or Presents a Health Hazard

Peroxides and Oxidizers

• A peroxide (e.g., hydrogen peroxide) is a compound that contains the bivalent –O-O- structure and that may be considered to be a structural derivative of hydrogen peroxide where one or both of the hydrogen atoms has been replaced by an organic radical.

• An oxidizer (e.g., oxygen and chlorine) is a chemical other than a blasting agent or explosive that initiates or promotes combustion in other materials, thereby causing fire either of itself or through the release of oxygen or other gases.

Peroxide and oxidizers produce "loose" oxygen in the atmosphere. This can create an oxygen-rich environment that creates health and fire hazards. They can make objects burn that would not normally burn. When exposed to heat, they can decompose with explosive force.

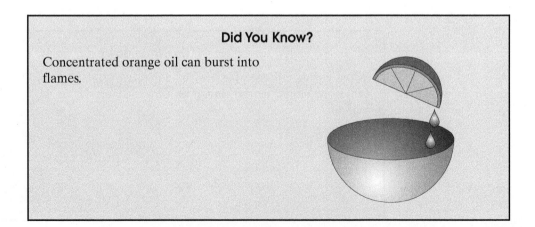

Did You Know?

Concentrated orange oil can burst into flames.

Leaks, Spills, and Releases

When process fluids are not handled properly, they can create dangerous situations for the process technician. When toxic process fluids manage to escape the confines of their containers or equipment, other dangerous situations are created. Process technicians need to understand the hazardous characteristics associated with process fluids and how process fluids become even more dangerous when they leak or spill or are released into the environment.

Leaks, spills, and releases all involve the uncontrolled discharge of hazardous process fluids. A **leak** occurs when a container is compromised and allows a small amount of liquid to escape. A **spill** is similar to a leak in that it involves the uncontrolled discharge of a liquid, but usually involves more volume than a leak. A **release** refers to the uncontrolled discharge of a gas into the atmosphere.

FIGURE 16-2 A Spill Occurs When a Container is Compromised and the Compromised Area Allows a Large Quantity of Liquid to Escape

FACTORS LEADING TO LEAKS, SPILLS, AND RELEASES

There are a number of factors that can lead to leaks, spills, and releases:

- Human error
- Exceeding operating limits
- Corrosion and erosion
- Improperly designed equipment
- Equipment failure
- Process changes
- Weather conditions

FIGURE 16-3 Statistics Indicate that Many Problems are Created by Human Error

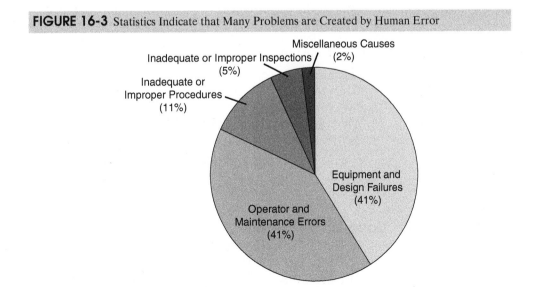

Human Error

Statistics indicate that many problems are created by human error (Figure 16-3). Though we all make mistakes from time to time, it is very important to be aware of your actions at all times because one minor mistake may injure or kill you and/or others near you. The following is a partial list of factors that frequently contribute to human error:

• Skill level—Some equipment seems very simple to use and, as a result, sometimes semi-skilled or unskilled operators are allowed to operate it. An experienced operator will have a better chance of recognizing equipment malfunctions and how to properly operate the equipment in a safe manner than a less-experienced operator.

• Attitude, judgment, and ability to concentrate—The psychological and physical status of operators can dictate how well they can perform on the job. If an operator is facing too many distractions, whether from personal or work-related issues or physical injuries, their ability to concentrate on the job is compromised. Distractions of this sort can lead to accidents or injuries.

• Operator fatigue—If operators become too tired to perform their job, then concentration (as well as physical response) is compromised. Some situations that can lead to fatigue include excessive pressure from multiple job assignments, boredom from monotonous operations, awkward work motions or positions, and excessive overtime.

• Failure to follow procedures—Each facility is required to maintain procedures on how to perform tasks correctly and safely. Unfortunately, not everyone follows the procedures correctly. Failure to follow documented procedures can have dramatic repercussions.

Human Error Analysis (HEA) is a method used to identify and correct any potential human errors that may arise on the job. HEAs are usually conducted in one of two ways:

• By observing employees at work and noting any hazards or hazardous behavior. This is often referred to as a Job Safety Analysis (JSA).

• By having analysts perform the desired job task to gain first-hand knowledge of potential hazards. The analyst can then suggest realistic ways of performing the job that will increase the probability of doing the job safely and correctly.

Exceeding Operating Limits

All equipment has a specific procedure that must be followed to operate it within proper operating limits. To understand the consequences of deviating from the equipment's operating limits and how to take corrective steps when deviation occurs, process technicians need to consult the written operating procedures established for the equipment.

The OSHA Process Safety Management of Highly Hazardous Materials (PSM 1910.119) regulation recommends that the following be documented for equipment being used in a process covered by the regulation:

• Pressure limits
• Temperature ranges
• Flow rates
• What to do when an upset condition occurs
• Alarms and instruments pertinent to an upset

Process technicians must familiarize themselves with the documented operating procedures and follow these to the letter in order to operate within specified safety limits.

Corrosion and Erosion

• **Corrosion** occurs when materials are eaten away by a chemical process (e.g., iron rusting). Corrosion can occur for a variety of reasons, whether it is from direct contact with chemicals (e.g., acids, bases and salts), temperature extremes, oxidation, or water exposures. Corrosion can have a catastrophic effect upon the integrity of pipelines, vessels, and equipment.

• **Erosion** occurs when materials are worn (abraded) away by a physical process (e.g., sandblasting). Erosion can be a problem in pipelines that transport fluids containing particulate matter, such as sand and salt. Erosion can also occur when fluids are moved through pipes at high velocities. Usually, erosion involves the degradation of the pipe at the elbows and flanges.

Normal Pipe   Corroded Pipe   Eroded Pipe

FIGURE 16-4 Corrosion and Erosion can Seriously Impact the Integrity of Pipelines, Vessels and Equipment

Corrosion and erosion may not always be visible. For example, piping and vessels covered with thermal insulation may have corrosion that is shielded from view, buried pipelines are totally concealed from view, and erosion occurs on the inside of the equipment. These factors have led to active inspection and maintenance programs as an industry practice. When conducting an active inspection, technicians use non-destructive testing methods (e.g., sonic and X-ray) to determine pipe and vessel wall thickness.

Improperly Designed Equipment

A major goal in equipment design is to reduce the probability that the equipment can be used or configured incorrectly. However, various factors can result in improperly used equipment, such as installation, the nature of the process, the work environment, and the workers.

Equipment Failure

Equipment failure can be described as the unintended shutdown of a piece of equipment due to a mechanical malfunction or process upset. There are several factors that can lead to equipment failure. Following are some of the most common:

• Corrosion and/or erosion of vessels and piping
• Operating outside the equipment safety limits
• Improper operation of electrical equipment
• Using equipment for unintended purposes
• Overloading electrical equipment and/or circuits
• Using incorrect equipment in hazardous areas
• Interruption of process utilities (e.g., electrical power, steam, and cooling)
• Not following equipment or process procedures

Problems such as those listed above, coupled with the lack of an effective inspection and preventative maintenance program, can allow these factors to result in damage to equipment or the creation of hazardous situations. Examples of these situations include the rupture or explosion of vessels or equipment and electric motor burnout.

Process Changes

It is common for a plant's process or processes to be changed throughout its operating life. Sometimes the process changes can be minor, such as introducing a new piece of equipment or finding a new way to make the process slightly more efficient. On other occasions, the process change can be a major redesign or retooling of the process. Whether the change is minor or extensive in nature, the process changes must be thoroughly documented according to the OSHA PSM standard (1910.119 section I) unless it is an exact replacement ("replacement in kind," according to OSHA).

Weather Conditions

Severe weather is always a potential hazard. In the process industries, severe weather such as hurricanes, floods, tornadoes, lightning, high winds, and rain can cause extremely hazardous situations. When water freezes, it expands and exerts pressure that can burst pipes or cause other problems. Flooding can result in wastewater problems if dikes and other containment vessels are compromised. High winds can blow out the flare flame, resulting in gas or vapor release.

Many chemicals that are normally liquid can suddenly become solids when the temperature drops (e.g., diesel fuel). Excess moisture causes critical instrument failures, such as when moisture condenses in the instrument air supply.

POTENTIAL DANGERS TO THE ENVIRONMENT

The environment and the surrounding community can be harmed as a result of leaks, spills, or releases in the following ways:

- Shock waves from explosions
- Hazardous emissions and heat from fire
- Other toxic vapor releases
- Inhalation, absorption, and ingestion hazards (e.g., chemicals or radiation)
- Leaks or spills seeping into offsite soil
- Leaks or spills into groundwater, rivers, and lakes

Environmental Controls

A variety of equipment and systems are used to control environmental hazards. Process technicians should be familiar with the functions, operations, and maintenance of environmental control equipment and systems. (Further information is outside the scope of this textbook, and is typically covered in Equipment and Systems textbooks and courses.)

- Clarifiers for primary water treatment
- Trickling filters for secondary water treatment
- Activated sludge digesters, dewater, dryers, and incinerators for secondary water treatment
- Baghouses
- Cyclone separators
- Electrostatic precipitators
- Wet scrubbers
- Carbon absorbers
- Packed towers
- Dikes and containment walls

Process Technician Responsibilities

Process technician duties that are directly or indirectly related to the environment include the following:

- Familiarizing themselves with the hazardous chemicals and materials they work with, their potential hazards to the environment, along with worker safety and health
- Monitoring and sampling air, water, and soil
- Promptly responding to potential leaks, spills, and releases
- Operating processes at optimum performance
- Maintaining equipment
- Troubleshooting process problems
- Performing general housekeeping
- Reporting any potential issues
- Complying with federal, state, local, and company regulations

Environmental Regulations: Air

CLEAN AIR ACT

The Clean Air Act was established to regulate air emissions from a variety of sources (area, stationary, and mobile). The act authorized the EPA to establish National Ambient Air Quality Standards (NAAQS) to protect public health and the environment by requiring every state to comply with NAAQS by 1975. As with many of the EPA programs, federal efforts to control air pollution have gone through several phases. The first federal legislation to address air pollution, the Air Pollution Control Act, was passed in 1955. Since 1955, air pollution control legislation has been amended many times to address new or changing air pollution issues.

The Air Pollution Control Act of 1955 was the first federal legislation covering air pollution. Federal funds were appropriated to conduct air pollution research. In 1966, the passage of the Clean Air Act placed more emphasis on air pollution control. The program established by the Clean Air Act placed emphasis on identifying techniques for monitoring and controlling air pollution.

In 1967, the Air Quality Act expanded the federal government's role from air pollution research, monitoring, and control to include enforcement. However, the government's enforcement activities were limited to air issues dealing only with interstate pollution.

The enactment of the Clean Air Act of 1970 shifted the federal government's role again. This act authorized the development of federal and state regulations that would place limits on stationary (industrial) and mobile sources of pollutant emissions (e.g., cars, trucks, and trains). The Clean Air Act of 1970 brought about the establishment of four key regulatory programs:

- National Ambient Air Quality Standards (NAAQS)
- State Implementation Plans (SIPs)
- New Source Performance Standards (NSPS)
- National Emission Standards for Hazardous Air Pollutants (NESHAPs)

Clean Air Act Amendments of 1977

Significant changes were made to the federal air pollution standards with the enactment of the Clean Air Act Amendments of 1977. One of these amendments provided for the Prevention of Significant Deterioration (PSD) of air quality in attainment areas. The other significant amendment dealt with air quality issues in non-attainment areas. Attainment areas are contiguous geographic locations that have met the NAAQS goals, whereas non-attainment areas have not met their NAAQS goals. Key permit review requirements were established by both of these regulations to further strengthen the nation's objective to reach and maintain attainment of the NAAQS.

Clean Air Act Amendments of 1990

In 1990, a new group of amendments substantially increased the power and responsibility of the EPA. Following were the most important issues addressed by these amendments:

- Control of acid deposition (acid rain)
- Issuance of stationary source operating permits
- Incorporation of NESHAPs into an expanded program for controlling toxic air pollutants
- Control of stratospheric ozone
- Expansion of enforcement authority
- Growth of research programs

TITLES I, III, IV, AND V

The 1990 amendments organized the Clear Air Act into nine separate "Titles." The following titles directly affect the process industries.

Title I – Provisions for Attainment and Maintenance of National Ambient Air Quality Standards (NAAQS)

It wasn't until the enactment of the Clean Air Act of 1970 that Congress delegated to the EPA the establishment of uniform National Ambient Air Quality Standards (NAAQS). Two separate levels of compliance were established: primary standards were set to protect health and secondary standards were set (at slightly higher levels) to protect crops and other community property.

The 1990 amendments identified six criteria pollutants: sulfur dioxide, nitrogen oxides, carbon monoxide, ozone, lead, and particulate matter.

Title I – State Implementation Plans (SIPs)

In 1970, each state became responsible for developing plans to ensure that the NAAQS were implemented, maintained, and enforced. These plans were referred to as State Implementation Plans (SIPs). Each plan had to contain the following components:

- Detailed pollutant-source emission inventories
- Pollutant modeling and calculating data
- Pollutant-monitoring programs
- Strategies to reach the NAAQS within the designated timeframe
- A summary of the state's legal authority to administer and enforce the plan

The amendments of 1990 also had a significant impact on the SIPs. Plans had to be modified in order to address the increasing threat of ozone, particulate matter, and carbon monoxide. These three pollutants have proven the most difficult to control due to the variety of sources from which they are generated.

Title I – New Source Review (NSR)

The 1977 amendments created the New Source Review (NSR) program. The NSR program requires permits for all major new sources, or substantially modified sources, in both non-attainment and attainment areas.

Acquiring a permit for major new sources in non-attainment areas is much more difficult than obtaining a permit in an attainment area. Non-attainment areas have not yet reached their NAAQS goals, whereas attainment areas have. It is difficult to secure permission to add another source of air pollution to an area where emissions have not been brought into compliance. It is also difficult to obtain these permits under the NSRP due to the requirement that facilities must install control equipment to ensure the Lowest Achievable Emission Rate (LAER) and must demonstrate offsets of their current pollutant emissions.

Title III – Hazardous Air Pollutants (HAPs)

The control of HAPs was based on the NESHAPs prior to 1990. NESHAPs regulated "noncriteria" pollutants while the NAAQS regulated the localized emission sources. Only seven HAPs were regulated under the NESHAPs program between 1970 and 1990. These seven consisted of beryllium, asbestos, mercury, vinyl chloride, benzene, radionuclides, and inorganic arsenic.

The 1990 amendments greatly expanded the number of HAPS regulated under NESHAPs. Under Title III, 189 compounds or groups of compounds were classified as hazardous, including the original seven. The EPA adds and deletes compounds from the NESHAP list as new data from research becomes available.

Title IV – Acid Deposition Control

The control of acid deposition, or acid rain, was also addressed by the 1990 amendments. Research has proven that sulfur dioxide and nitrogen oxides are the primary precursors responsible for acid rain. Title IV was added to the Clean Air Act in order to minimize the adverse effects of acid rain by reducing air emissions of these two precursor compounds. The act limited the total amount of sulfur dioxide emissions to 10 million tons per year and nitrogen dioxide emissions to two million tons per year.

In order to accomplish this goal, owners and operators of stationary sources that generate these emissions must install continuous emission monitoring systems to track their emission rates. These data must be reported to the EPA regularly. The EPA reviews the data to confirm that the emissions are within the limits set forth for each facility.

Title V – The Operating Permit Program

Title V is a federal program that controls and restricts the amount of emissions from a process facility. The details of this program are extensive and outside the scope of this textbook.

Environmental Regulations: Water

CLEAN WATER ACT

The Clean Water Act and its amendments also have an enormous impact on the operations of most process industry facilities. Some of the tasks that may be assigned to the process technician, such as wastewater sampling and treatment, are performed in order for facilities to remain in compliance with the requirements of this regulation. As with air pollution control, water pollution control legislation has gone through several phases. The first legislation to address water pollution, the Federal Water Pollution Control Act, was passed in 1948. Since then, water pollution control legislation has been reauthorized and amended many times to address new or changing water quality issues.

Today's Clean Water Act consists of two major parts. One part, located in Titles II and IV, authorizes federal financial assistance for the construction of municipal sewage treatment facilities. Another part, found throughout the act, deals with the regulatory requirements that apply to industrial and municipal dischargers.

This act has been termed a technology-forcing statute because of the rigorous demands placed on those who are governed by it to achieve higher and higher levels of pollution abatement. Industrial facilities were given until July 1, 1977, to install "best practicable control technology" (BPT) to clean up their wastewater streams. Municipal treatment plants were required to satisfy a similar goal, termed "secondary treatment" by that same date. The primary focus of BPT was controlling discharge of conventional (biodegradable) pollutants – substances that can be broken down by bacteria. Such substances include suspended solids, biochemical oxygen-demanding materials, fecal coliform and bacteria, and pH.

The act stipulated even tighter effluent standards to be achieved by 1989. These standards required industry to install the "best-available technology" (BAT) that was economically achievable. BAT generally focuses on toxic substances, whereas BPT focuses on conventional pollutants.

The act utilizes both water-quality standards and technology-based effluent limitations to protect water quality. Technology-based effluent limitations are established by the EPA, placed on certain pollutants from certain sources and applied to these sources through the generators' discharge permits. Water-quality standards are for the overall water quality. They designate how bodies of water can be used, such as for recreation, water supply, or industrial purposes. Water-quality standards also identify the maximum concentrations of various pollutants for all bodies of water in the state.

Water Quality Act of 1987

Prior to the 1987 amendments, programs in the Clean Water Act were primarily directed at point-source pollution – wastes discharged from discrete and identifiable sources, such as pipes and other outfalls. In contrast, little attention had been given to non-point source pollution, such as storm water runoff, despite estimates that runoff represents more than 50% of the nation's remaining water pollution problems.

The 1987 amendments directed states to develop and implement non-point source pollution management programs. States were encouraged to pursue groundwater protection activities as part of their overall non-point pollution control efforts.

Under the Act, federal jurisdiction is broad. The EPA issues regulations containing BPT and BAT effluent standards particular to categories of industrial sources, such as petroleum refining and organic chemicals.

As with other environmental laws, certain responsibilities are delegated to the states. Qualified states issue discharge permits and enforce the requirements of these permits. The EPA issues discharge permits in states that have not been qualified to do so. This results in a federal-state partnership philosophy, with the federal government setting the standards and state governments carrying out implementation and enforcement.

Titles II and IV – Municipal Wastewater Treatment Construction

Federal law has authorized grants for the planning, design, and construction of municipal sewage treatment facilities since 1956. Congress greatly expanded this grant program in 1972. Since that time, Congress has authorized $65 billion and appropriated $69 billion to aid wastewater treatment plant construction. Grants are allocated among the states according to population and an estimate of treatment needs identified through surveys.

Permits

Under the Clean Water Act, all discharges into the nation's waters are unlawful unless specifically authorized. Industrial and municipal dischargers must obtain permits from the EPA or the municipal government before allowing any effluent to leave their premises. Penalties for permit violations can result in fines and/or criminal charges.

National Pretreatment Standards

Federal pretreatment standards apply to those who discharge water from any industrial source into a publicly owned water treatment facility. The standards' objectives are to protect local groundwater, improve the quality of effluents and sludge so they can be used for beneficial purposes, and protect the treatment plants from any threat posed by untreated industrial wastewater. In order to maintain pretreatment standards, the industrial facility must perform the following tasks:

- Report discharges
- Sample discharge materials for standards compliance
- Provide upset provisions
- Allow for bypassing when sampling indicates inappropriate pollutant levels

Environmental Regulations: General

RESOURCE CONSERVATION AND RECOVERY ACT (RCRA)

The Resource Conservation and Recovery Act (RCRA) of 1976 controls the handling of solid and hazardous waste. It actually amends earlier legislation and is very comprehensive in its coverage. The act defines solid and hazardous waste, authorizes the EPA to set standards for facilities that generate or manage hazardous wastes, and establishes a permit program for hazardous waste treatment, storage, and disposal facilities.

RCRA was last reauthorized by the Hazardous and Solid Waste Amendments (HSWA) of 1984. The amendments set deadlines for permit issuance, prohibited the land disposal of many types of hazardous waste without prior treatment, required the use of specific technologies at land disposal facilities, and established a new program regulating underground storage tanks.

Subtitle C of RCRA created the hazardous waste management program. A waste is considered hazardous if it is ignitable, corrosive, reactive, toxic, or appears on a list of about 100 industrial process waste streams or more than 500 discarded commercial products and chemicals.

The 1976 law expanded the previous definition of solid waste to include "sludge . . . and other discarded material, including solid, liquid, semi-solid or contained gaseous materials." This broadened definition is particularly important with respect to hazardous wastes, at least 95% of which are liquids or sludges. Some wastes are specifically excluded, however, including irrigation return flows, industrial point source discharges regulated under the Clean Water Act, and nuclear material covered by the Atomic Energy Act.

Under RCRA, hazardous waste generators must comply with regulations concerning recordkeeping and reporting, the labeling of wastes, the use of appropriate containers, the provision of information to transporters, treaters, and disposers on the waste's general chemical composition, and the use of a manifest system.

RCRA requires treatment, storage, and disposal (TSD) facilities to have permits, to comply with operating standards, to meet financial requirements in case of accidents, and to close their facilities in accordance with EPA regulations when they decide to cease operation.

The 1984 amendments imposed several new requirements on TSD facilities to minimize land disposal. Bulk or non-containerized hazardous liquid wastes are prohibited from being disposed of in any landfill. Severe restrictions are placed on the disposal of containerized hazardous liquids, as well as on the disposal of non-hazardous liquids in hazardous waste landfills.

The major non-hazardous solid waste provision in RCRA is the prohibition of open dumps. The prohibition is implemented by the states, using EPA criteria to determine which facilities qualify as sanitary landfills in order to remain open.

Underground storage tanks (USTs) are also monitored — the result of a growing nationwide problem surrounding leakage. To prevent future leaks, the regulation set technical standards for tank design and installation and mandated corrective action and tank closure when the situation necessitates such a response.

In 1984, significant amendments were made to RCRA. In addition to restrictions on land disposal, the regulation of small-quantity generators, and the UST program, these amendments also established very strict requirements for all aspects of treatment, storage, and disposal-facility operation. All TSD facilities must be inspected annually or every other year, depending on the type of facility. Additional regulatory conditions specific to TSDs deal with timetables for issuing and denying TSD permits, time limits for TSD permits, submittal of information regarding the potential for public exposure to hazardous materials, and expanding the EPA's enforcement powers.

TOXIC SUBSTANCES CONTROL ACT (TSCA)

The Toxic Substances Control Act (TSCA) regulates the creation and use of newly discovered extremely hazardous substances or changes to existing manufacturing processes. This act authorizes the EPA to screen existing and new chemicals used in manufacturing and commerce to identify potentially dangerous products or uses that should be subject to federal control. As enacted, TSCA also included a provision requiring EPA to take specific measures to control the risks from polychlorinated biphenyls (PCBs). Subsequently, other sections of the regulation have been added to address concerns about other specific toxic substances: asbestos in 1986 and lead in 1992.

The EPA has the authority to require manufacturers and processors of chemicals to conduct tests and report results from research studies to determine the effects of potentially dangerous chemicals on living things. Should the EPA determine that using these dangerous chemicals does present an unreasonable risk to human health or the environment, the EPA can regulate their use. Regulating the use of such a chemical can range from a total ban on production, to additional training requirements, to proper product labeling at the point of sale.

Federal legislation to control toxic substances was originally proposed in 1971 by the President's Council on Environmental Quality. Its report, "Toxic Substances," defined a need for comprehensive legislation to identify and control chemicals whose manufacture, processing, distribution, use and/or disposal was potentially dangerous

Did You Know?

Asbestos and PCB are the reason the Toxic Substances Control Act was created.

and not adequately controlled under other environmental statues. The House and Senate each passed bills but controversy over these bills precluded enactment.

Then, many waterways became contaminated with PCBs. Chlorofluorocarbons began to threaten the ozone layer. It wasn't until these episodes of environmental contamination occurred and more accurate estimates of the costs of imposing toxic substances controls that the legislation was passed and enacted in 1976.

The TSCA directs the EPA to require manufacturers and processors to conduct tests for existing chemicals if any of the following conditions exist:

- Their manufacture, distribution, processing, use, or disposal may present an unreasonable risk of injury to the health or the environment; or they are to be produced in substantial quantities and the potential for environmental release or human exposure is substantial or significant.
- Existing data are insufficient to predict the effects of human exposure and environmental releases.
- Testing is necessary to develop such data.

Manufacturers must also prevent future risks through pre-market screening and regulatory tracking of new chemical products. They must control unreasonable risks already known or as they are discovered for existing chemicals. Finally, manufacturers must gather and disseminate information about chemical production, use, and possible adverse effects to human health and the environment.

EMERGENCY PLANNING AND COMMUNITY RIGHT-TO-KNOW ACT (EPCRA)

The EPCRA was enacted in 1986, establishing state and local entities and assigning them the responsibility to ready their communities to respond to a hazardous chemical release. Each state created an emergency-response commission and then divided the state into emergency planning districts, which in turn established local emergency planning committees. Local firefighters, health officials, government representatives, and industry management serve on these local committees.

The EPCRA also mandated that facilities that use or manufacture hazardous substances must report their annual releases to the government. Facilities must also notify local authorities should an accident at their plant present a potential threat to their industrial and community neighbors.

The EPCRA's main purpose is to protect the health, safety, and environment of our communities. The major provisions for emergency planning and notification are as follows:

- The EPA established a list of "extremely hazardous substances" and established threshold planning quantities for each substance.
- Facilities that handle these substances are required to have emergency response plans and employee training programs for those called upon to respond to emergency situations.

- The immediate report of any sudden release of these substances is required. In addition, a release of any hazardous substance exceeding the reportable quantity must also be reported to state, local, and federal officials.

Facilities are required to submit Material Safety Data Sheets (MSDS) for each hazardous chemical within the plant's inventory to the agencies and the local fire department. In addition, an inventory must be submitted to the same agencies and provide the following information:

- Estimates of the maximum amount of each inventoried chemical present at any time throughout the year
- Estimates of the average daily amount of each inventoried chemical
- Location of each inventoried chemical

COMPREHENSIVE ENVIRONMENTAL RESPONSE AND LIABILITY ACT (CERCLA) AND SUPERFUND AMENDMENTS AND REAUTHORIZATION ACT (SARA)

The Comprehensive Environmental Response and Liability Act (CERCLA) of 1980 created the Superfund hazardous substances cleanup program. It was enlarged and reauthorized by the Superfund Amendments and Reauthorization Act of 1986 (SARA).

The CERCLA authorizes the federal government to respond to spills and other releases (or threatened releases) of hazardous substances, as well as to uncontrolled hazardous waste dumps. Hazardous substances are identified under the Solid Waste Disposal Act, the Clean Water Act, the Clean Air Act, and the Toxic Substances Control Act, or are designated by the EPA. Response is also authorized for releases of "pollutants or contaminants," which are broadly defined to include virtually anything that can threaten the health of "any organism." Most nuclear materials and petroleum are excluded, except for those petroleum products that are specifically designated as hazardous substances under one of the laws previously mentioned.

The government classifies responses under two categories: short-term removals, where emergency action is necessary, and long-term remedial action taken at sites on the National Priority List (NPL). Removals are limited to a one-year effort with an expenditure of not more than $2 million. Remedial actions are of a longer term, are more expensive, and frequently involve extensive engineering at the sites.

The CERCLA and the SARA also provide for liability and financial responsibilities. Waste generators, transporters who select disposal sites, and disposal facility owners and operators are liable for response costs and for damage to natural resources. The acts do not provide for victims of exposure to hazardous substances. Victims must seek restitution in state court.

Potentially responsible parties (PRPs) are accountable for enforcement and cleanup costs. There are no limits to the liability amount that can be assessed if the release is due to misconduct, negligence, violation of any safety, construction or operating standards or regulations, or when cooperation and assistance requested by a public official in connection with response activities is denied. Should PRPs refuse to comply with a cleanup order, they may be assessed punitive damages that can amount to three times the liability amount. Because some waste sites are abandoned and/or the original owners or dumpers cannot be found, it may be difficult to determine a PRP. In these cases a superfund is created. A **superfund** is a monetary fund that comes from tax dollars paid by the chemical industry to pay for the cleanup of abandoned waste sites in the event no PRP can be found.

The SARA set deadlines in order to expedite cleanup progress. Deadlines were set through 1990 for the following:

- Site inspections
- NPL listing and ranking
- Remedial investigations
- Feasibility studies
- On-site work

This act stipulates that cleanups must assure the protection of health and the environment. Remediation must meet the standards of federal and state environmental laws, specifically the Safe Drinking Water Act's recommended maximum contaminant levels and the Clean Water Act's water quality criteria.

Permanent remedies, such as incineration, rather than non-permanent remedies are mandated. If a non-permanent remedy, such as solidification and burial, is allowed under waiver, the EPA must inspect the site every five years.

Summary

Hazardous chemicals vary in their composition and are classified in a variety of categories. For example, substances can be combustible, flammable, water reactive, pyrophoric, explosive, or oxidizing. These hazardous agents typically do not present a threat to the environment or to the average citizen in the community unless they are released from a controlled environment.

Leaks, spills, and releases all involve the uncontrolled discharge of hazardous process fluids. These types of discharge can be caused by any number of factors, including human error, exceeding operating limits, corrosion and erosion, improperly designed equipment, equipment failure, process changes, and weather conditions.

Process technicians must be familiar with the ways hazardous substances can enter the environment, know how to recognize a discharge or a potential discharge, and know what actions to take to prevent or reduce a toxic discharge. In addition, process technicians must be familiar with the substances they are working with and be aware of the environmental regulations associated with those substances.

Checking Your Knowledge

1. Define the following terms:
   a. Corrosion
   b. Erosion
   c. Explosive
   d. Leak
   e. Oxidizer
   f. Peroxide
   g. Pyrophoric material
   h. Release
   i. Spill
2. *(True of False)* A flammable liquid is any liquid that has a flashpoint at or above 100 degrees F.
3. Describe what causes each of the following materials to react:
   a. Water-reactive materials
   b. Pyrophoric materials
   c. Explosives
4. List five factors that could lead to leaks, spills, and releases.
5. *(True of False)* Operator fatigue has no impact on the likelihood that a leak, spill, or release will occur.
6. Why was the Clean Air Act established?
7. *(True of False)* The Clean Water Act was designed to reduce water pollution and protect water quality.
8. The Superfund site cleanup program was established under:
   a. Comprehensive Environmental Response and Liability Act (CERCLA)
   b. Toxic Substances Control Act (TSCA)
   c. Clean Air Act (CAA)
   d. Water Quality Act (WQA)

Student Activities

1. Research one of following topics and write a one to two-page paper describing what you learned.
   a. Clean Air Act (CAA)
   b. Clean Water Act (CWA)
   c. Resource Conservation and Recovery Act (RCRA)
   d. Toxic Substances Control Act (TSCA)
   e. Emergency Planning and Community Right-To-Know Act (EPCRA)
   f. Comprehensive Environmental Response and Liability Act (CERCLA)
   g. Superfund Amendments and Reauthorization Act (SARA)

2. Pick a pollutant (e.g., ozone). Develop a presentation about its sources and/or how it is formed, its effects on public health and the environment, and methods of control.

3. Attend a Local Emergency Planning Commission (LEPC) meeting to better understand how the Emergency Planning and Community Right-To-Know Act (EPCRA) regulation is implemented in your community. Discuss your findings with your fellow classmates.

4. Request facility hazardous chemical inventory forms for any local facilities by contacting the Local Emergency Planning Commission (LEPC) or the public library. Review the forms to see what chemicals are listed.

5. Contact your local water authority and obtain the latest tests on your drinking water. Report your findings (e.g., water source, which substances are over the limits, and which substances are close to the limits).

6. Research a superfund site (e.g., Love Canal). Write a paper or create a presentation explaining the following:

   a. Where the site is located
   b. What happened
   c. How the site was remediated (cleaned up)
   d. What the site is used for today

CHAPTER 17

Introduction to Hazard Controls

Objectives

This chapter provides an overview of hazard controls in the process industries, which can affect the safety and health of workers.

After completing this chapter, you will be able to do the following:

■ Describe the three major types of hazard controls:

Engineering

Administrative

PPE

■ Discuss why, when, and how these controls are applied.

Key Terms

ABC model—a model used to provide observation and feedback on safety performance to workers: events or activators (A) in an environment often direct performance or behavior (B), which is motivated by the consequences (C) that people expect to avoid or receive.

Audit—a review, typically conducted by people from the company, a hired third party, regulatory agencies, or a combination of these groups, to determine if a particular facility is complying with established safety, health, and/or environmental programs.

Hazard control—the recognition, evaluation, and elimination (or minimization) of hazards in the workplace.

Inspection—a pro-active activity conducted prior to a need for action by plant personnel to ensure that safety, health, and environmental programs are being followed.

Monitor—a process used to gather data to evaluate the work environment using specialized equipment.

Process Hazard Analysis (PHA)—an organized, systematic process used to identify potential hazards that could result in accidents causing injury, death, property damage, or environmental damage.

Introduction

The process industries employ many different types of controls to ensure safety. These controls include engineering controls, administrative controls, personal protective equipment, and government regulations. Process technicians should be aware of the safety controls in place in their facility and how to use them.

This chapter describes the various efforts used to reduce hazards in the process industries.

Hazard Controls Overview

As earlier chapters in this book have shown, the potential for a variety of hazards exists in the process industries, such as exposure to chemicals, fire, biological agents, falls, noise, pressure, and temperature extremes. Process technicians must understand the hazards they might encounter, as well as the controls that companies put in place to eliminate or reduce these hazards.

Hazard control involves the recognition, evaluation, and elimination (or minimization) of hazards in the workplace. The best way to address a hazard is to eliminate it entirely or substitute a less-hazardous alternative. Sometimes, a process can be redesigned to eliminate a hazard completely, such as automating a process that previously exposed workers to mechanical hazards or using non-toxic or less-toxic raw materials in place of a more-toxic material.

Hazard control is addressed using one of three different methods, listed from most preferable to least preferable:

• Engineering controls are physical changes that address the source of the hazard, and then eliminate or reduce exposure to it. This can include equipment changes, process changes, or material changes. For example, sound-dampening materials could be installed to reduce a noise hazard or proper ventilation could be used to significantly reduce a hazardous atmosphere. An engineering control is the preferred method of hazard control; however it can be expensive to implement.

• Administrative controls use policies, procedures, programs, training, and supervision to establish rules and guidelines for workers to follow, thus reducing their exposure to a hazard. Administrative controls reduce but do not always eliminate the hazard.

• Personal Protective Equipment (PPE) is special gear and equipment that workers can use to create a physical barrier between themselves and hazards. For example, a hard hat can reduce the hazard of falling objects. Respirators can keep workers from breathing chemical vapors or particulates. These controls only reduce exposure to the hazard.

The following sections expand on these controls.

Engineering Controls

Engineering controls vary widely, based on factors such as the following:

• Applicable government regulations (e.g., safety, health, environmental)
• Type of industry
• Management decision
• Economics
• Type of process
• Required equipment
• Available materials
• Technology

Although engineering controls vary widely based on these factors, the following are some general forms they can take (Figure 17-1):

• Process containment
• Automation
• Noise abatement
• Sound dampening
• Heat shielding
• Radiation shielding
• Equipment layout
• Ventilation
• Detection systems
• Alarms
• Electrical grounding and bonding
• Automatic shutdown devices
• Machine guarding
• Guardrails
• Spill containment
• Work areas and working surfaces
• Fire suppression equipment
• Redundant (backup) systems

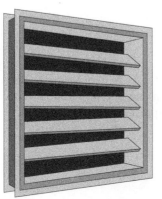

FIGURE 17-1 Proper Ventilation is an Engineering Control Used to Reduce Hazardous Environments

- Flares and pressure-relief valves
- Interlocks (devices designed to prevent an action unless a certain condition is met)
- Improved equipment design (e.g., fails less, leaks less)
- Closed-loop sampling

Engineers, safety and health personnel, management, and purchasing are involved in the process of identifying, evaluating, and implementing engineering controls. Process technicians play a vital part in providing input about the need for and placement of engineering controls, participate in safety committee reviews, and report any potential problems. Chapters 18 and 19, the *Engineering Controls* chapters, describe various engineering controls in more detail.

Administrative Controls

If an engineering control cannot be used to control a safety, health, or environmental hazard, then an administrative control can be sought. Administrative controls can take a wide range of forms (Figure 17-2):

- Programs (e.g., hearing-awareness week and fire-prevention week)
- Procedures
- Policies
- Training
- Supervision
- Monitoring
- Permits
- Process Hazard Analysis
- Documentation
- Housekeeping
- Inspection and audits
- Safe-work observations (watching selected work practices and providing feedback)
- Signs and signals (can be audio, visual, or both)
- Reward programs
- Work-area changes (i.e., limiting workers' exposure to a hazard by moving them to another area temporarily)

In general terms, administrative controls fall into two broad categories: programs and activities. Programs are documented approaches for controlling a hazard, and activities are the programs put into action.

Program design and development are determined by requirements set by one or more of the following:

- Government (regulatory)
- Company
- Site-specific or plant-specific
- Unit or department-specific
- Work environment (union versus non-union)
- Community

FIGURE 17-2 Warning Signs are a Type of Administrative Control Used to Increase Worker Safety

Program elements include:

- Policies—guiding principles
- Procedures—step-by-step instructions for accomplishing a task
- Plans—methods prepared in advance for carrying out actions
- Principles—a set of standards or rules
- Rules—statements about what can and cannot be done, along with possible rewards or consequences
- Agreements—plans for coordinating activities between different organizations
- Systems—organized, interdependent sets of principles developed for industry-wide use

Following are some examples of common programs:

- Regulatory-mandated or those created in response to government regulations such as OSHA and EPA standards. These programs interpret the laws and establish a compliance plan using a written program (e.g., a Hazard Communication, or HAZCOM, program).
- Evacuation and emergency plans, or established responses outlining required emergency actions (e.g., communications, designated safety officials, evacuation routes)
- Mutual aid agreements, or agreements between a facility and the nearby emergency response unit or local government, for rendering community aid in the event of an emergency
- Incident Management System (IMS), formerly the Incident Command System (ICS), or methods for assigned command and control during an incident

Training is a critical activity in the process industries. No matter what job process technicians are assigned to do, they must receive some type of training to perform the job safely, competently, and effectively. The importance of training has grown in recent years. A better-trained workforce is required because of advances in the field of process technology, increased quality standards, and the always-present need for safety.

Companies create a system of permits to ensure safe work practices. Permits can be required for confined-space entry, lockout/tagout, hot work, and other potentially hazardous practices. Safe work permits can involve listing the steps required to complete a task, a hazard assessment, the elimination or minimizing of the hazard, and the provision for maintaining safe working conditions until the task is complete. See Chapter 21, *Permitting Systems,* for details.

A **Process Hazard Analysis (PHA)** is an organized, systematic process used to identify potential hazards that could result in accidents that could cause injury, death, property damage, or environmental damage. Many different methodologies can be used to conduct a PHA. One methodology is called HAZOP, or Hazards and Operability Study. A HAZOP identifies operations problems and hazards to personnel, company property, and the environment. The benefits of a properly done PHA are many; it can identify hazards, optimize productivity, mitigate legal liabilities, reduce operating costs, improve employee morale, and shorten project schedules.

Monitoring is a process by which data are gathered to evaluate the work environment (Figure 17-3). Industrial health monitoring involves sampling various elements in the work environment to determine if hazards exist and to what level. Specialized equipment is used to assist the monitoring process, such as the following:

- A decibel meter to measure noise levels
- Absorbing materials (in the form of badges or sorbent tubes) to capture and measure chemical exposure
- A dosimeter to measure radiation exposure
- A "sniffer" to detect hazardous atmospheres

The data are recorded and tracked to show historical perspective. These data are analyzed and compared to published standards (e.g., OSHA's hearing conservation

FIGURE 17-3 Monitoring is a Process by Which Data are Gathered to Evaluate the Work Environment

program or OSHA's air contaminant standard). Industrial health monitoring is usually designed to determine workers' exposure to hazardous agents and often requires workers to wear monitoring equipment over a period of time.

Safe-work observations provide observation and feedback to workers on safety performance, including safe behaviors or at-risk behaviors. This process involves the basic **ABC model**: the idea that Behavior (B) is directed by Activators (A) and motivated by Consequences (C). In other words, events (activators) in an environment often direct performance (behavior), but people most likely do what they do because of the consequences they expect to avoid or receive. Often, employees observe each other at work and then provide one-on-one coaching to each other. The aim is to make the process part of the work culture.

Housekeeping, or keeping work areas neat and orderly, is also important for maintaining a safe environment. Process technicians must assist with housekeeping tasks; it is a part of the job description. Good housekeeping can prevent accidents, save money, increase productivity, and improve morale.

Inspections and audits are activities that involve conducting regular safety, health, and environmental checks. These activities are pro-active, meaning that they are conducted prior to a need for action. **Inspections** are usually conducted by plant personnel to ensure that safety, health, and environmental programs are being followed. Inspections can be conducted on a regularly timed basis, or can be unannounced (such as a spot-check). Process technicians may be asked to participate on an inspection team. **Audits** are examinations typically conducted to determine if a particular facility is complying with established programs. The group conducting the audit can be comprised of people from the company, regulatory agencies, a hired third party, or a combination. Audits are performed less frequently than inspections.

Investigations occur in reaction to an incident (either an accident or a near miss). A team usually performs an investigation. The team can be made up of individuals with different skills, experience, and viewpoints. Investigations can accomplish the following:

- Decide which problem will be addressed.
- Describe the problem.
- Develop a thorough understanding of the problem.
- Examine the problem.
- Agree on basic causes of the problem.
- Develop an effective solution.
- Implement corrective action.

Process technicians can be required to participate on an investigation team or provide information to a team.

Chapter 20, *Administrative Controls: Programs and Practices* describes various administrative controls in more detail.

Personal Protective Equipment

Personal Protective Equipment (PPE) helps form a protective barrier between workers and a hazard if engineering and/or administrative controls do not sufficiently reduce or eliminate the hazard.

PPE comes in a wide variety of forms:

- Footwear
- Gloves
- Coveralls, lab coats, aprons
- Safety goggles or glasses
- Hearing protection (e.g., ear plugs, ear muffs)
- Respirators
- Headgear and hard hats
- Face shields

FIGURE 17-4 PPE Helps Form a Protective Barrier Between Workers and a Hazard

Selecting the proper PPE for the hazard, then making sure the PPE fits properly, is key to ensuring a worker's safety. For PPE, process technicians must understand the following tasks:

- Selection—choosing the right PPE (or combination of PPE) to address the hazard
- Fit—adjusting the PPE to meet the needs and be as comfortable as possible (called fit testing)
- Inspection—checking the PPE for wear and damage
- Wear—donning the PPE before performing the task associated with the hazard
- Maintenance—caring for (cleaning, repair) the PPE
- Storage—properly putting away the PPE

It is crucial that process technicians wear PPE at all times when they are working around potential hazards. Removing a piece of PPE, even for a moment, can expose a worker to the hazard. Workers must know the limitations of PPE and understand that hazards still exist and not become complacent. Employers are required to train workers on PPE and provide appropriate PPE. Check with your company's safety and/or health professionals for proper selection, fit, use, and maintenance.

Chapter 22, *Personal Protective Equipment and First Aid*, describes this control in more detail.

Summary

The process industries employ many different types of controls to ensure safety. These include engineering controls, administrative controls, personal protective equipment, and government regulations.

Engineering controls are physical changes (e.g., heat shielding, guardrails, alarms, and shutdown devices) that address the source of the hazard and then eliminate or reduce exposure to it.

Administrative controls use policies, procedures, programs, training, and supervision to establish rules and guidelines for workers to follow, thus reducing their exposure to a hazard. Administrative controls reduce but do not always eliminate the hazard.

Personal Protective Equipment (PPE) is special gear and equipment that workers can use as a physical barrier between themselves and hazards (e.g., a hard hat can reduce the hazard of falling objects, and respirators can keep workers from breathing chemical vapors or particulates). PPE does not eliminate a hazard; it simply reduces exposure to the hazard.

Process technicians should be aware of the safety controls in place in their facility and how to use them.

Checking Your Knowledge

1. Define the following terms:
    a. ABC model
    b. Administrative controls
    c. Audit
    d. Engineering controls
    e. Hazard control
    f. Inspection
    g. Monitor
    h. Personal Protective Equipment (PPE)
2. Which of the following is the first you should consider to control a hazard?
    a. Administrative
    b. Elimination
    c. PPE
    d. Engineering
3. What factors make it difficult to eliminate a hazard? *(select all that apply)*
    a. Physical limitations
    b. Economic limitations
    c. Proper PPE is not available.
    d. Government regulations do not address the hazard.
    e. Workers will not approve of the change.
4. _____ controls are physical changes that address the source of the hazard and then eliminate or reduce exposure to it.
5. List at least seven forms that engineering controls can take.
6. What are two broad categories of administrative controls?
    a. PPE and engineering plans
    b. Training and documentation
    c. Behavior and observation
    d. Programs and activities
7. List at least five elements included under administrative controls.
8. What does industrial-hygiene monitoring involve? *(choose the best answer)*
    a. Fugitive emissions levels
    b. Sampling various elements in the work environment to determine if hazards exist
    c. Record-keeping
    d. Noise levels
9. What is a PHA?
10. What do the initials PPE stand for?
11. PPE helps form a(n) _____ _____ between workers and hazards.
12. *(True or False)* The most important requirement of PPE is comfort.

Student Activities

1. Meet with your fellow students to discuss various products or equipment that you have used which presented some type of hazard. Come up with different engineering solutions for controlling the hazard, and have the group vote on the best solution.
2. Think of a potential hazard for a situation, such as replacing a car battery, and write a procedure for how to safely complete the task.
3. Choose a type of PPE. Research the requirements for selecting, fitting, wearing, and maintaining the PPE. Write a paper describing the PPE, the types of hazards the PPE addresses, how to properly fit the PPE, and how to clean and maintain it.

4. Think about engineering controls in everyday life (e.g., a lawn mower's blade brake) that improve safety. Discuss your ideas with your fellow students.

5. What administrative controls do you use at home to prevent accidents (e.g., turning off a breaker before fixing a switch or installing a light fixture, and ensuring ladders have good footing)? Make a list of the tasks.

6. Make a list of PPE that you wear at home when doing repairs, yardwork, housework, or other similar tasks.

18

Engineering Controls: Alarms and Indicator Systems

Objectives

This chapter provides an overview of engineering controls relating to alarm and indication systems.

After completing this chapter, you will be able to do the following:

■ Describe the role of alarms in providing a warning of conditions that can lead to emergencies, leaks, spills, and releases, and discuss the dangers of improper responses or failures of alarms.

■ Identify various engineering controls, specifically alarm and indication systems, used by the process industries to minimize and/or eliminate threats to health, safety, and the environment:

Fire alarms and detection systems

Toxic gas alarms and detection systems

Redundant alarm and shutdown devices

Automatic shutdown devices

Interlocks

■ Know and understand actions required by process technicians when alarms occur.

■ Know the two main types of alarm systems and how they operate.

Key Terms

Alarm—a signal that indicates the existence of an unusual or potentially hazardous situation.

Annunciator—a device that displays alarm conditions through the use of flashing and continuously lit panels.

Audible alarm—an alarm that uses sound to warn workers of a particular condition or hazard.

Detection device—equipment designed to sense a particular condition (e.g., smoke, vapors, flame) and send a signal to an alarm system if the condition exceeds a pre-set limit.

Indicator device—a generic term for a type of equipment that indicates process variables; may be visual (e.g., light), audible (e.g., horn), or both.

Interlock—a type of hardware or software that does not allow an action to occur if certain conditions are not met.

Permissive—a type of interlock that does not allow a process or equipment to start up unless certain conditions are met.

Pressure switch—a mechanical device that uses electrical contacts to complete an electric circuit and generate an alarm signal.

Redundant system—a system that provides a backup in the event the primary system fails.

Visible alarm—an alarm that uses visual means (e.g., lights, motion, color) to warn workers of a particular condition or hazard.

Introduction

Alarms and indicators are designed to alert workers if something should go wrong and/or notify responsible personnel that a potentially dangerous situation exists that requires attention. These systems are crucial to maintaining the safety and health of workers and protecting the community and the environment. Alarms are also used to maintain peak operating efficiency as well as ensure product quality.

This chapter provides an overview of detection devices, alarms and indicators, and automatic responses. The process technician must understand how these systems detect hazardous or unusual conditions and generate alerts or automatic responses.

Hazardous and Emergency Situations

A wide range of potential hazards exist in the process industries, such as fires, explosions, and releases. Dangerous or hazardous situations can occur, threatening workers, equipment, the facility, and the surrounding community and environment.

Hazardous situations can include any of the following:

- Fires
- Explosions
- Spills
- Leaks
- Releases
- Terrorist attacks

Process conditions in a facility must be constantly monitored for changes. If these changes are not detected early, they can rapidly escalate into hazardous situations or emergencies that can result in injuries, loss of life, damage to equipment and facilities, and environmental damage.

Once a hazardous change or situation is detected, it must be communicated to workers using an alarm and indicator system. Finally, workers must know how to properly respond to the alarm.

Detection Devices and Alarms

Detection devices are equipment used to monitor a process, equipment, condition, or area for changes that could lead to potentially hazardous situations. Devices can be designed to detect for various conditions, such as the following:

- Smoke
- Fire
- Explosive atmosphere
- Hazardous atmosphere
- Intrusion (unauthorized entry)
- Vibration
- Radiation
- Noise
- Low or high:
  - Fluid levels
  - Pressures
  - Temperatures
  - Flow
  - Speed or velocity
  - Chemical composition
- Access to restricted areas

A condition can be measured in simple yes or no terms or when a pre-determined setting is reached (e.g., when a concentration of vapors exceeds a set amount). When a device detects a condition, it sends a signal to an alarm that notifies the system of that condition. **Indicator devices** may be visual (e.g., light), audible (e.g., horn), or both.

Alarms are signals that indicate the existence of an unusual or potentially hazardous situation. Alarms warn workers when a hazard or dangerous condition is about to happen, could potentially happen, or is currently happening. For example, a filtration system can use an alarm to warn when a filter must be replaced to avoid equipment damage. Or, a smoke alarm can warn of a potential fire hazard.

An **audible alarm** uses sound to warn workers of a particular condition or hazard, whereas a **visible alarm** may use lights, motion, or color to alert workers. A device that displays alarm conditions through the use of flashing or continuously lit panels is called an **annunciator**. When an alarm goes off, it is called annunciation.

Some alarms can be manually activated, and others have a **pressure switch** that uses electrical contacts to complete an electric circuit and generate an alarm signal. Alarms are used to warn of hazards, dangerous conditions, and/or emergencies such as the following:

- Fires
- Natural disasters (e.g., tornadoes)
- Evacuations

FIGURE 18-1 Detection Devices Monitor for Changes that could Lead to Potentially Hazardous Situations

- Spills
- General states of emergency

Alarms can create an audible (can be heard) signal, visible (can be seen) signal, or a combination of both. In a noisy plant environment, most alarms produce both audible and visual signals.

Alarms are coded with certain meanings that are specific to each facility (e.g., red flashing lights may indicate a hazard or a chemical release). A facility-wide alarm can use a sounder (horn) to generate tones in pre-determined sequences and or pitches (frequencies) that indicate the type of hazard. For example, two long blasts of a horn followed by one short blast could mean a hazardous leak has occurred. Speakers can also be placed throughout a facility so verbal instructions can be provided to workers. It is vital that process technicians learn all the alarm signals, systems, and meanings for their facility. They must also know how to respond to each type of alarm.

FIGURE 18-2 Alarms are Signals that Indicate the Existence of an Unusual or Potentially Hazardous Situation

In some situations, closed-circuit TV is used to allow remote monitoring of areas where detectors and alarms are used but workers are not usually present. A camera is set up in the area so it can transmit video or images back to a central location. TV monitors display the video feed or images so workers can monitor these areas.

Alarm systems can be tied to plant instrumentation and the facility's process control system, typically called a Distributive Control System (DCS). In simple terms, a DCS is a computerized control system that links to various sensors and detection devices that monitor various aspects of a process, called process variables. The DCS allows process technicians to view the real-time status of process variables and operating statistics using computer monitors. The DCS also has a control interface that allows process technicians to automatically control or respond to process variables, making corrective changes as necessary.

Although process control and DCS details are outside the scope of this textbook, the process technician must understand the role of the DCS in an alarm system.

Detectors can link to the DCS, providing updates on the potentially hazardous conditions that they monitor. If a change occurs, the DCS will display an alarm condition (Figure 18-3). The process technician must acknowledge the alarms and then perform an appropriate response. Typical DCS alarms are prioritized using one of the following designations:

- High high
- High
- Low
- Low low

Keep in mind that other designations can be used. Each facility is responsible for prioritizing, categorizing, and naming its alarm conditions.

Alarm systems can also be tied to process upset control systems (e.g., flares, relief valves), automatic shutdown devices, firefighting systems, explosion suppression systems, deluge systems, ventilation, and other similar engineering controls.

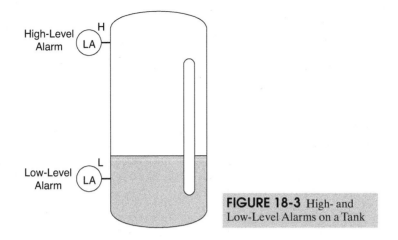

FIGURE 18-3 High- and Low-Level Alarms on a Tank

Alarm-system failure can result in an extremely dangerous situation. Some alarm systems perform self-checks and diagnostics to ensure that they are still working. However, all alarms must be manually tested on a regular basis to make sure they function properly. Alarm-system reliability is imperative for the safe operation of the plant or facility, so redundant (backup) systems are often used. A later section discusses redundant alarm systems.

Detection equipment, by its nature, is often delicate and sensitive to its environment. Potential factors that can cause damage or false readings include the following:

- Dust
- Temperature
- Moisture and humidity
- Corrosive atmospheres
- Mechanical damage
- Chemicals with properties similar to the substance being monitored

Detectors and alarm systems must be tested and maintained regularly to ensure they function properly for both normal and abnormal situations.

Alarm Handling

When alarms are actuated, process technicians should follow company-specified procedures according to their training (Figure 18-4). Each company writes procedures to address emergencies as required by PSM 1910.119.

Some alarms can go off even though no problem exists; this is called a false positive response. Or, sometimes alarms do not go off when there is an actual hazard; this is called a false negative response. False negatives are obviously dangerous, because workers are not alerted to the threat. Hazardous conditions can build, with no advance warning of impending disaster. False positives are also problematic because they waste resources (e.g., firefighters responding to a false alarm) and may cause workers to ignore alarm signals even when an actual emergency exists.

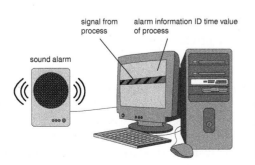

FIGURE 18-4 When Alarms are Actuated, Process Technicians should Follow Company-Specified Procedures According to Their Training

If a process upset occurs, process technicians can be faced with an "alarm flood," or numerous alarms going off at the same time. Disasters in the process industries have demonstrated how alarm flood can overwhelm workers and produce catastrophic results. An alarm management system can help reduce the alarm flood and allow workers to handle alarms effectively and efficiently, potentially averting disaster. A well-designed alarm management system aims to alert, inform, and guide workers in normal and abnormal conditions.

It is essential the workers follow plant emergency procedures. Because alarms can cause a variety of emotional and physical responses in workers, many facilities require operators to know emergency procedures from memory, without referring to manuals.

Alarm management takes into account human factors (i.e., how people interact and respond), such as the following:

- Does the alarm get the attention of workers? (note: alarms that go off too frequently are often ignored – e.g., car alarms).
- Is the alarm message clear?
- Does the worker have the necessary information to respond appropriately?
- Are alarms presented at a proper rate (not too fast or too slow)?
- Can workers take action in a sufficient amount of time?
- Is it easy to make a mistake?

Process technicians can provide feedback about alarms to supervisors, who can make adjustments as necessary. This process can help workers better respond to alarms in abnormal conditions, potentially alleviating dangerous situations.

Some alarms must allow workers to acknowledge them so that more-important alarms can take precedence. For example, if a worker shuts down a pump, a low-level alarm for a tank might go off. The worker must be able to acknowledge the low-level alarm, indicating an awareness of what caused the condition.

Facilities must set policies, procedures, and standards for their alarms. Each alarm should have a well-defined and planned response, including roles and responsibilities of workers to address normal and abnormal conditions. Each response procedure must be well documented (i.e., provide enough information for the process technician to act appropriately).

Using such an approach, alarms should meet the following standards:

- Clearly communicated (effectively annunciated with clear meaning)
- Prioritized based on order of importance and urgency
- Organized (categorized into logical groupings)
- Presented in a timely way

Fire Alarms and Detection Systems

Fire and/or explosive atmosphere detectors can be installed throughout a facility. Fire detection devices include the following:

- Smoke detectors
- Flame detectors
- Temperature switches
- Carbon monoxide detectors
- Explosive gas detectors

Each is activated when a predetermined level, such as high temperature, an abundance of smoke particles, or an excessive amount of carbon monoxide, is reached. This triggers the fire alarm automatically, providing audible, and sometimes visual, warnings to workers. Because a worker might see flames before the detector can sense them, so fire alarms can also be set off manually.

Fire alarms and explosive atmosphere alarms can be tied to a deluge system, which will rapidly dump water or a fire- or explosion-suppression substance (Figure 18-5). Firefighting systems can use water sprinklers, halon systems, carbon dioxide systems,

FIGURE 18-5 Fire Alarms and Explosive Atmosphere Alarms can be Tied to a Deluge System, Which will Rapidly Dump Water or a Fire- or Explosion-Suppression Substance

foam, or other systems based on the type of fire that must be suppressed. Water is used to extinguish fires and cool down hot equipment. Halon is used around computer systems, telephone switch rooms, and other electrically sensitive equipment. Carbon dioxide is used for fires involving electrical and electronic equipment. See Chapter 24, *Fire, Rescue, and Emergency-Response Equipment* for details.

It is possible to prevent the buildup of an explosive atmosphere by deluging the area with foam or dry-powder suppression agents. In areas where explosions are potential hazards, other engineering controls can be used to minimize or contain the explosion (e.g., special walls or building design to contain an explosion and prevent it from doing damage to other areas).

Toxic-Gas Alarms and Detection Systems

Toxic-gas alarms and detection systems are used for processes that handle toxic materials, which can produce fumes, vapors, or other hazardous atmospheres (e.g., hydrogen sulfide, cyanide).

Toxic-gas detection devices are installed throughout the process area to detect a release regardless of wind conditions. Sniffers, or sensors configured to detect certain toxic gases, continually sample the ambient air for traces of the specific toxic substance. When the sensor detects a predetermined concentration of the substance, it will send a signal to an alarm.

The sniffer devices are located in enough locations in the area that a leak will be detected no matter which direction an air current is moving. In certain circumstances, a water or foam deluge system can automatically be activated upon alarm conditions to suppress and/or disperse the toxic gas.

Devices can be permanently mounted, or be portable for use in areas such as a confined space that is not normally occupied. Operators may also be required to wear personal monitors, as a standard operating procedure, while in a unit that has a toxic gas such as cyanide.

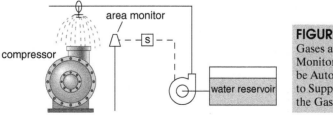

FIGURE 18-6 When Toxic Gases are Detected by an Area Monitor, Water or Foam may be Automatically Dispensed to Suppress and/or Disperse the Gas

Redundant Alarm Systems

Many critical processes and equipment have two or more alarm systems linked to them, which provide a backup in case the primary system fails. The need for

redundant system (backup systems) is based on not just how critical a process is, but how devastating a major accident as a result of the hazardous condition could be. For most critical alarms, at least two separate and totally independent sets of detection devices and alarms are used. Some systems will use two or more different power sources, such as the main electrical system and a battery backup.

Automatic Shutdown and Response Devices

Some systems link detection devices and alarms to an automatic shutdown or response device. These devices ensure that processes and equipment are maintained in a safe state during failures or hazardous conditions.

When a detection device senses a potentially hazardous condition, an alarm will generate a signal to another device, which initiates a process shutdown and/or automatic response. The following are some examples:

- A smoke detector shuts down a compressor and initiates a fire-suppression system.
- A level detector shuts down a pump and liquid product is routed to a relief system to prevent overflow.
- A blocked conveyor automatically stops.
- An automatic valve shuts down feed valves to block the reactor if conditions of runaway reaction are met.

The following are some types of conditions that can be monitored and linked to an automatic shutdown and response system:

- Temperature (e.g., high temperature in a reactor)
- Pressure (e.g., high pressure in a vessel)
- Flow (e.g., high or low flow in a line)
- Level (e.g., high or low level in a storage tank)
- Hazardous atmospheres (e.g., oxygen-deficient, flammable materials present)
- High radiation (e.g., X-rays)
- Chemical leak (e.g., high hydrogen sulfide levels in the atmosphere)
- Flames, smoke, and explosion (e.g., fire)
- Jams, stoppages, or foreign objects (e.g., removal of a guard)
- Speed and/or velocity (e.g., over-speed trips on a turbine)
- Product quality (e.g., product off specifications)

Interlocks

An **interlock** is a control system that does not allow an action or change unless a certain condition or conditions are met (e.g., you can't take a key out of the ignition unless the transmission is in park). Interlocks are used to ensure that a proper sequence is followed or, if it is important enough, to shut down a process.

Additionally, when operating a furnace, certain conditions (e.g., proper fuel pressure, air flow, fan is running, the damper is open) must be met before the burners can be lit (Figure 18-7). This is an example of a **permissive** interlock. Another example of an interlock is a sensor that detects a foreign object in a process and sends a signal that causes an interlock to halt that process before the object can cause damage. Equipment capable of generating radiation often contains interlocks to make sure that the equipment cannot start until the appropriate conditions are met.

In the case of an interlock with two or more conditions, all conditions must be met before the interlock will allow an action to proceed (e.g., starting up or shutting down a process). Interlocks can also prevent two incompatible events from occurring at the same time. For example, a reversible electric motor control can have an interlock that prevents the forward and reverse contactors from operating at the same time.

Interlocks can be part of the software used in a process control system, or a piece of hardware or equipment (which typically use electrical relays). Hardwired and

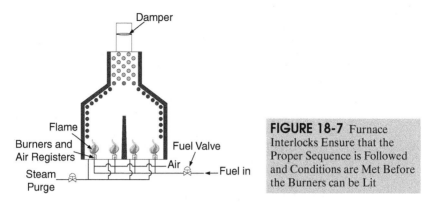

FIGURE 18-7 Furnace Interlocks Ensure that the Proper Sequence is Followed and Conditions are Met Before the Burners can be Lit

softwired interlocks can be bypassed, but doing so removes a layer of safety protection. Hardwired and softwired interlocks can operate independently of each other, or can be used together to provide redundancy.

Administrative Controls

The following are administrative controls that relate to alarm systems. Companies must perform the following tasks:

- Develop emergency action plans (many are required by regulating agencies like OSHA).
- Train workers how to respond to alarms.
- Conduct regular drills and critiques.
- Establish monitoring procedures.
- Provide Personal Protective Equipment (PPE) and train workers on its use.
- Mark evacuation routes.

Process Technician Requirements

Following are some recommendations that process technicians should follow when dealing with detection and alarm systems (Figure 18-8):

- Always pay attention to alarms, even if one goes off regularly.
- Remain calm and logical.
- Understand what the facility's various alarms mean.
- Be familiar with emergency plans and operations associated with various alarms.
- Know the evacuation route for your area and meeting or check-in locations.
- Be alert to conditions around you (e.g., wind direction or air current, which determines the direction a hazardous atmosphere might travel).
- Report alarms; do not be deterred by panic or the consequences of a false alarm.
- Know which alarms are not functioning and the backup plan for that alarm.
- Know what backup plans exist if alarm systems fail.
- Participate in emergency drills.

FIGURE 18-8 Process Technicians should Know Emergency Evacuation Routes and Participate in Emergency Drills

- Never tamper with or disable an alarm.
- If an alarm seems to go off inadvertently or too often, inform your supervisor.
- Know how to troubleshoot alarm conditions and determine the root cause.
- Participate in regular testing processes, including backup systems.
- Assist with any alarm system repairs.
- Be able to locate emergency PPE quickly and use it properly.
- Perform housekeeping tasks regularly to keep a work area clean and make sure evacuation routes are uncluttered.

Summary

A wide range of potential hazards exist in the process industries, such as fires, leaks, spills, explosions, and releases. Dangerous or hazardous situations can occur, threatening workers, equipment, the facility, and the surrounding community and environment. Because of this, process conditions in a facility must be constantly monitored for changes. If these changes are not detected early, they can rapidly escalate into hazardous situations or emergencies that can result in injuries, loss of life, damage to equipment and facilities, and environmental damage.

Alarm, detection, and indicator systems are used to warn workers about emergencies or abnormal situations. Alarms can be audible or visible. Process technicians must understand how these systems detect hazardous or unusual conditions and generate alerts or automatic responses. They must also know what each alarm means and how to respond to it.

In addition to alarms, some systems contain automatic shutdown devices or interlocks. These devices are engineered to protect workers and equipment and should not be tampered with or overridden.

Checking Your Knowledge

1. Define the following terms:
   a. Administrative controls
   b. Alarm
   c. Annunciator
   d. Audible alarm
   e. Detection device
   f. Engineering controls
   g. Indicator device
   h. Interlock
   i. Permissive
   j. Pressure switch
   k. Redundant system
   l. Visible alarm
2. An alarm can be audible, _____, or both.
   a. Virtual
   b. Variable
   c. Vital
   d. Visible
3. Some alarm systems can perform self-checks and _____.
4. Name at least three environmental factors that can damage detection equipment.
5. *(True or False)* Fire alarm systems are always linked to fire-suppression equipment.
6. Toxic gas detectors _____ the air to detect toxic substances.
   a. Sniff
   b. Moisten
   c. Recirculate
   d. Heat
7. *(True or False)* A redundant alarm system provides a backup in case the primary system fails.
8. Name five different conditions that can be monitored by an automatic shutdown system.
9. Describe what an interlock does.
10. A(n) _____ interlock does not allow a process to start until certain conditions are met.
    a. Prohibitive
    b. Permissive
    c. Negator
    d. Hardware

11. What happens when a process technician acknowledges an alarm?
    a. The alarm condition is corrected.
    b. The horn is silenced.
    c. The alarm starts to flash.
12. What is the basic function, or reason, for alarms?
13. What role do alarms play in preventing or mitigating unplanned releases of hazardous materials? What can result from false-positive or false-negative alarms?
14. How should workers respond to false-positive alarms?

Student Activities

1. Check your workplace, school, home, and car for detectors, alarms, and indicators. Make a list of the alarms you find, what they detect, how they detect (if you know), and what type of alarm they generate (audible, visible, or both).
2. Research detectors and alarms on the Internet, using manufacturer sites or safety organizations. Write a two-page paper about why detectors and alarms are important, their various uses and limitations, how to test and maintain them, or other related topics.
3. Think of a process or equipment that you are familiar with that could benefit from an interlock device. Discuss with your fellow students the process and equipment, the hazard it presents, and how an interlock could lessen the hazard.
4. Choose a monitor and demonstrate how the monitor works and is calibrated.
5. Research process safety or environmental incidents using the Internet or newspaper archives and find an example of an incident in which alarms were involved. Describe how the alarm system helped mitigate the hazard or why it did not.

CHAPTER 19

Engineering Controls: Process Containment and Process Upset Controls

Objectives

This chapter provides an overview of engineering controls used to contain process materials and control process upsets.

After completing this chapter, you will be able to do the following:

- Recognize various engineering controls, specifically process containment and control systems, used by the process industries to minimize and/or eliminate threats to health, safety, and the environment.

- Describe various engineering controls, specifically process upset control systems, used by the process industries to minimize and/or eliminate threats to health, safety, and the environment.

- List common process fluids used in the process industries and describe the potential safety and health hazards posed by these materials.

Key Terms

Closed-environment drain system—a system of devices such as pumps, piping, and scrubbers to prevent the release of liquids, gases, and vapors to the atmosphere.

Effluent—liquid wastewater discharge from a process facility.

Flare—an environmentally-approved device that burns waste gases collected from various process sources to reduce pressure.

Fugitive emission—an intentional or unintentional release of a gas.

Ground-Fault Circuit Interrupter (GFCI)—a device that protects personnel from the possibility of electrical shock by shutting off the power to electric tools when a small amount of current-to-ground is detected.

Positive pressure control—a system used to keep external air, which may contain airborne toxic substances, from entering the building.

Pressure-relief valve—a safety valve that automatically opens at a set pressure to protect process vessels or piping from excessive pressure.

Process fluid—any material that flows; it can be either liquid or gas. When under pressure, both gases and liquids transmit force equally. Process gases are compressible and liquids are not.

Introduction

When working with hazardous and flammable substances, leaks and spills are always a concern. The process industries use a number of process containment controls and process upset control systems to minimize and/or eliminate threats to health, safety, and the environment.

In addition to hazardous chemicals, process fluids such as air, nitrogen, steam, and water have the potential to cause serious injury.

Process Containment and Control Systems

Fugitive emissions are leaks or releases that occur when containment controls fail. Sources of fugitive emissions include pump and compressor seals, storage and processing vessels, loading facilities, flow-control and pressure-relief valves (including valve packing), flange gaskets, and pipelines carrying materials from one process to another.

Fugitive emissions are monitored at potential leak points in the plant. Although a specific leak point level might be low, all the small amounts around the facility can result in a significant total emission. These emissions must be monitored to determine if hazardous agents are present at unacceptable levels set by the EPA. Inline sensors (part of the process) or portable detection equipment can be used. See Chapter 23, *Monitoring Equipment,* for more details.

This section describes examples of engineering measures and devices used to control common sources of fugitive emissions in the following:

- Closed systems
- Closed-loop sampling
- Floating roof tanks
- Ventilation
- Effluent control
- Waste treatment
- Noise-abatement devices
- Safeguards against electrical shock

CLOSED SYSTEMS

A major concern of the process industries is how to contain or dispose of waste byproducts. In the past, chemical sewers were used to drain and collect hazardous waste byproducts. However, these sewers sent toxic fumes to the atmosphere. To help minimize exposure to toxic fumes, plants now implement **closed-environment drain systems**. These systems use devices such as pumps, piping, and scrubbers to prevent the release of liquids, gases, and vapors into the atmosphere.

CLOSED-LOOP SAMPLING

Closed-loop sampling is a method for controlling fugitive emissions from the material being sampled. Using inline-sampling containers such as sample bombs or small draw-off points, closed-loop sampling systems help reduce liquid waste and air emissions.

The loop is first purged or flushed with the material being conveyed. The valve drain to the container is then opened and the sample is collected. To prevent sample contamination, the drain-valve connection to the sampling loop is short and the amount of "dead" space is small.

After the sample analysis is complete, the sample stream is returned to the process via piping to eliminate the release of waste or vapors into the atmosphere. The purged material required to get an accurate sample is then recovered.

The safest way to collect a sample from a toxic stream is to use an in-line automatic analyzer sampling system. Although not entirely free of fugitive emissions, this method greatly reduces potential exposure problems posed by manual sampling.

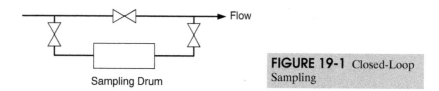

Flow

Sampling Drum

FIGURE 19-1 Closed-Loop Sampling

FLOATING ROOF TANKS

A floating roof tank is an atmospheric tank (one at ambient temperature and pressure), equipped with a roof that floats on the surface of the liquid stored in the tank. The floating roof and the seal between the roof and the sides of the tank prevent vapors from being emitted into the atmosphere.

An external fixed roof tank has an external fixed roof and an internal floating roof, which helps prevent hazardous vapors from leaking into the vapor space. This design helps eliminate toxic vapors from being released into the atmosphere. The vapor space can be equipped with a nitrogen blanket system so that, when the tank does vent, only nitrogen is released to the atmosphere.

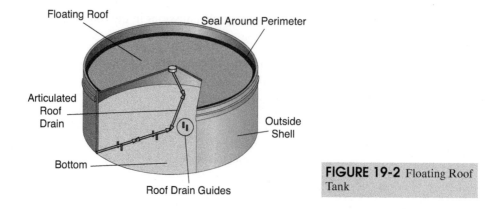

Floating Roof

Seal Around Perimeter

Articulated Roof Drain

Outside Shell

Bottom

Roof Drain Guides

FIGURE 19-2 Floating Roof Tank

VENTILATION

Ventilation systems are installed in plants to protect employees from airborne contaminants, such as vapors and dusts. Such systems can include air movers, such as electric fans, forced-air fans, intake fans, and exhaust fans, as well as filters, such as bag filters.

Vent hoods are used in process and laboratory settings. They allow technicians to work in a controlled environment around vapors, fumes, and/or dusts. Vent hoods work by drawing hazardous materials into the ventilation system and away from personnel. These materials are then removed from the exhaust air before it is discharged to the environment.

Positive pressure control is used to keep external air, which may contain airborne toxic substances, from entering the building. Positive pressure is achieved by maintaining the air pressure within a building at a slightly higher pressure than ambient pressure.

FIGURE 19-3 Ventilation Systems are Installed in Plants to Protect Employees from Airborne Contaminants, such as Vapors and Dusts

EFFLUENT CONTROL

Process plants use **effluent** control to ensure that wastewater leaving the facility complies with the requirements of any type of discharge permit under which the facility operates. Effluent control can involve monitoring, analysis, and adjustment of the wastewater stream. Wastewater streams can be monitored continuously by the use of inline analytical instruments and/or by manual monitoring and analysis by process technicians.

WASTEWATER TREATMENT

Wastewater treatment is the process of removing contaminants from a plant's process sewer prior to discharging the effluent. Depending on the type of discharge permit the facility has, all or some of the following types of treatment can be required:

- Oil and solid separation
- Biological treatment
- Chemical treatment
- pH control
- Waste sludge removal (e.g., incineration, land fill disposal).
- Additional treatment as required

NOISE-ABATEMENT DEVICES

There are many sources of noise in a process unit (e.g., fluid flow, atmospheric venting, equipment, and machinery) that can contribute to noise pollution and cause hearing damage. To help reduce the noise level associated with process units, plants employ noise-abatement devices.

Noise-abatement methods are used to reduce the level of noise, and include acoustical insulation or reducing the flow velocity of fluids. Examples of noise-abatement devices include equipment mufflers and noise insulation between walls and within doors.

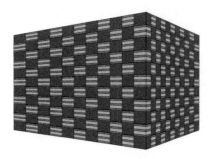

FIGURE 19-4 Noise-Abatement Walls are Used to Reduce the Level of Noise

SAFEGUARDS AGAINST ELECTRICAL SHOCK

Electricity takes the path of least resistance in order to ground itself, including going through a human body or a piece of equipment that happens to be in its path. The resulting shock may cripple or kill. Safeguards to prevent such an occurrence employ the principles of bonding and grounding.

Grounding and Bonding Mechanisms

Grounding involves connecting an object to the earth using metal in order to provide a path for the electricity to travel. The metal is usually a piece of copper wire connected to a grounding rod or underground water piping (grounding mechanisms). This path allows electricity to dissipate harmlessly into the ground.

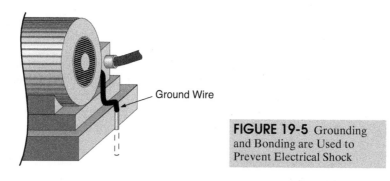

Ground Wire

FIGURE 19-5 Grounding and Bonding are Used to Prevent Electrical Shock

Bonding involves connecting two objects together, usually with a copper wire. An ungrounded object is grounded by bonding it to a grounded object. Care must be taken so equipment is not unintentionally bonded, causing an unexpected electrical hazard.

Static-dissipation devices are permanent wiring systems used for grounding process equipment.

Ground-Fault Circuit Interrupters

A **Ground-Fault Circuit Interrupter (GFCI)** is a device that protects personnel from the possibility of electrical shock when using electrical tools. The device will shut off the power when a small amount of current-to-ground is detected.

Process-Upset Control Systems

Engineering controls can be implemented to help prevent any process upsets from turning into a hazardous situation.

Process-upset controls come in a variety of forms, some of which include the following:

- Flare systems
- Pressure-relief valves
- Deluge systems

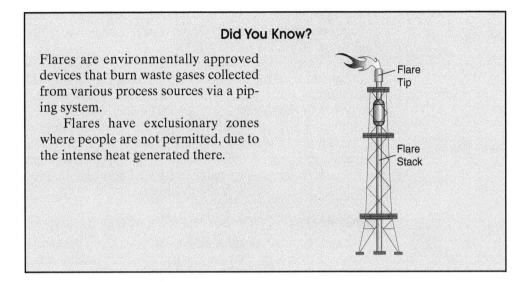

Did You Know?

Flares are environmentally approved devices that burn waste gases collected from various process sources via a piping system.

Flares have exclusionary zones where people are not permitted, due to the intense heat generated there.

Flare Tip

Flare Stack

- Explosion-suppression systems
- Explosive-proof designs (control rooms)
- Spill containment

FLARE SYSTEMS

Flares are environmentally-approved devices that burn waste gases collected from various process sources via a piping system. Sometimes, a process can generate excess pressure. Gases are released to reduce pressure; but, instead of releasing to the atmosphere, the gases are piped into a flare system. The gases are burned and released into the atmosphere as carbon dioxide, thereby controlling the process upset. Some flare systems include methods for trapping entrained liquids before gases are sent to the flare.

A common type of flare on a process unit is elevated on a tower several hundred feet in the air, to disperse fumes and due to the heat that can be generated. Flares must be designed to provide smokeless burning of the gases. This is usually accomplished with the addition of steam during the combustion process.

PRESSURE-RELIEF VALVES

Pressure-relief valves are safety valves that automatically open at a set pressure to protect process vessels or piping from over-pressuring. These valves do not rely on external sensing devices; instead, they rely on spring tension settings adjusted to a set pressure and are inherently very reliable.

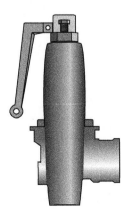

FIGURE 19-6 Pressure-Relief Valves are Safety Valves that Automatically Open at a Set Pressure to Protect Process Vessels or Piping from Over-Pressuring

For example, the spring tension of a relief valve may be set at 50 psi. If the pressure in the process exceeds 50 psi, the valve will automatically open. When the pressure drops below 50 psi, the valve will close.

The material being released through the relief valve may be piped to discharge to a flare system or to the atmosphere, depending on the substance involved. For highly toxic materials, relief valves typically discharge to a closed-vent system, either to the flare system or through emission control equipment.

A rupture disc can be installed upstream of the pressure-relief valve to minimize fugitive emission. The disc is a thin metal dish designed to burst at a specified pressure. It separates the process fluid from the safety relief valve to prevent leakage through the valve. A pressure gauge can be provided between the rupture disc and the relief valve.

DELUGE AND EXPLOSION SUPPRESSION AND DETECTION SYSTEMS

Deluge systems dump or spray an extinguishing agent to suppress a fire, a toxic release, or a hydrocarbon spill. They may be automated or require manual activation.

Explosive-gas alarms and detectors operate in a similar manner to toxic-gas alarms and detectors, but are designed specifically to detect explosive gas. In addition to an alarm they may also provide automatic activation of a deluge system.

Explosion-suppression systems are walls erected to contain an explosion. The walls can be erected around a particular process area, or an entire building (e.g., a process control room) can be constructed with explosion suppression capability.

SPILL CONTAINMENT

Spill containment is designed to contain hazardous materials that have escaped the controlled environment of their vessels. When spills are contained, large areas are protected from pollution and the spills can be cleaned up and disposed of easily. Dikes and containment basins are used around storage tanks and are sized to contain the full contents of the tanks within their enclosure. They also prevent contaminated rainwater from entering public waterways. Curbs are used for process units and usually drain into a chemical sewer.

Process technicians are responsible for securing the area around a spill (marking off it using barricading, tape, or rope). They can also help contain the spill using absorbent pads, sandbags, booms, and other devices.

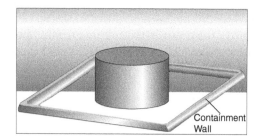

Containment
Wall

FIGURE 19-7 Spill Containment is Designed to Contain Hazardous Materials that have Escaped their Vessels

Instrumentation and Process Upsets

Instrumentation and other devices such as alarms, permissives, interlocks, and Emergency Shutdown Devices (ESDs) are important engineering features to assure safe and environmentally sound operation of potentially hazardous equipment and processes. Such devices must never be bypassed without operations and safety management approval and appropriate permits.

Instrumentation senses critical process or equipment variables such as temperature, pressure, flow, levels, analyticals, motion, and vibration. If these variables approach or exceed predetermined limits, an alarm is generated. The predetermined limits can be associated with quality control, safe operation, or both. Alarms typically continue until acknowledged by the process technician and safe, proper control conditions are regained. In some instrumentation systems, alarm signals can indicate that an action was automatically taken.

Permissive devices require that one or more specific conditions be met before a piece of equipment or process can be safely started (these conditions can be sequential). For example, igniting a furnace could require that the flow of process fluid be heated first, followed by an analytical check that explosive conditions do not exist in the furnace, followed by adequate fuel feed pressure being available. Instrumentation can perform the analytical check and alert the process technician.

Interlock devices monitor critical process variables and provide logic that permits automatic adjustments or overrides of the primary controls. The out-of-control critical parameter is either corrected or the equipment or process is safely shut down. For example, a distillation column pressure that exceeds safe limits can override the steam flow control to the column reboiler, until the pressure returns to a normal range. Low lubricant pressure to a compressor bearing can shut down the compressor.

Emergency Shutdown Devices (ESDs) are used to secure equipment and processes in an orderly and safe manner when a critical condition occurs that cannot otherwise be corrected. An ESD system includes the following:

- Sensors for critical variables
- Safe shutdown logic
- Interlock controls
- Process energy source interrupters (e.g., electric breakers, hydraulic fluid diverters, steam turbine trips)
- Uninterruptible power supply (UPS) to the critical instrumentation and controls
- Environmental control systems activators

Alarms, permissives, interlocks, and ESDs are discussed in Chapter 18, *Engineering Controls: Alarms and Indicator Systems*. Additional information about instrumentation is outside the scope of this textbook.

Process Fluids

Process fluids, even fairly inert fluids such as water, can be hazardous in certain industrial situations. Process fluids can be deemed hazardous because they are toxic or flammable, or because they are under high pressure or used at extreme temperatures. Process fluids become even more hazardous when they escape from their containers in an uncontrolled manner. Leaks, spills, and releases are examples of such uncontrolled discharges.

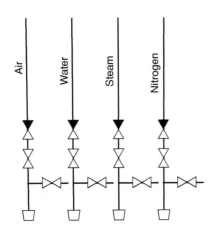

FIGURE 19-8 Utility Drops Supply Process Technicians with Substances such as Air, Water, Steam, and Nitrogen

The following are common process fluids:

- Air
- Nitrogen
- Steam
- Condensate
- Water

AIR

Air is used for powering pneumatic equipment and instrumentation, along with pressure testing. Air can also be used as part of either waterblasting or sandblasting cleaning processes for vessels and equipment, although this type of task is often handled by outside contractors. Process technicians should be aware of the hazards of air. Within a process plant, air is typically used in a compressed state (under pressure). Often, the air pressure can reach between 100-175 psi. High-pressure air can generate flying debris, which could puncture or be pushed into the eyes and skin. If air is inadvertently introduced into a process line and mixed with hydrocarbons, an explosive mixture could be created.

NITROGEN

Nitrogen is often used to purge or sweep vessels and often serves as a backup for compressor air. Nitrogen can be found at medium to high pressure (30 psi to 1,500 psi) in the process industries.

Nitrogen is hazardous to the process technician because it displaces oxygen. If nitrogen is introduced into a confined space, an oxygen-deficient atmosphere can be created. Process technicians working in oxygen-deficient atmospheres must use an air-supplied respirator.

When exposed to pure nitrogen, a person passes out, can stop breathing, and may even die. As the human body takes in oxygen, it uses the oxygen and creates carbon dioxide. The brain senses the build-up of carbon dioxide, which triggers the need to breathe. It is the excess of carbon dioxide, not the lack of oxygen, which causes a person to breathe. An excess of nitrogen displaces the oxygen in the body so no carbon dioxide is formed.

STEAM

When heat and water combine, steam is created. Steam is water in vapor form. In the process industries, steam is used to drive steam-power engines (turbines); to heat other process fluids; to clean equipment; to power pumps and turbines; to heat reboilers, kettles, tanks, and buildings; to purge air or hydrocarbons from lines and equipment; to blanket containers to prevent flammable conditions; and to snuff fires.

One pound of water will expand 1,600 times when converted to steam. In the process industries, steam is found at high temperatures (212 degrees F and higher) and high pressures (600 psi to 1,500 psi). As pressure is increased, the temperature elevates. For example, at 100 psi, the steam temperature is 338 degrees F, while at 600 psi, steam temperature is 488 degrees F.

Hazards associated with steam are its temperature and expansion ratio, which, if uncontrolled, can result in the following:

- Severe burns
- Severe cuts from a high-pressure jet of steam
- Thermal expansion
- Thermal shock
- Water hammer, which occurs when steam is introduced into a unit too rapidly
- Ruptured equipment

CONDENSATE

Condensate is condensed steam that becomes extremely hot water. Condensate is usually found in vessels that have been steam-cleaned, but have not yet been purged with air. It can also be found in steam traps and condensate drains; some facilities collect condensate and return it as a source of boiler feed water. Condensate can cause severe burns.

Condensate can also be used as a seal flush. Use caution around pumps using hot condensate as a seal flush.

If steam condenses in a closed container (e.g., vessel) not rated for vacuum service, it will create a vacuum because steam condenses at a significantly reduced volume when it returns to the water phase. This can result in the container collapsing. The same effect can also potentially damage turbines, compressors, pumps, and other equipment.

WATER

In the process industries water is used for cooling, purging, firefighting, cleaning, decontaminating, and as a raw material for many end-products. If water is under pressure and then rapidly vaporized, it can cause vessels to rupture. If water freezes, it can rupture pipes or create other hazardous working conditions.

There are many different types of water used in the process industries. These include process water, fire water, drinking water, potable water, deionized water, and waste water. Table 19-1 compares the different types of water:

TABLE 19-1 Types of Water

Type of Water	*Description*
Process water	Water that serves in any level of the manufacturing process of certain products
Fire water	Water that is used to extinguish a fire
Drinking water	Water which is intended for human use and consumption and considered to be free of harmful chemicals and disease-causing bacteria, cysts, viruses, or other microorganisms
Potable water	Water that is safe for human consumption
Deionized water	Water which has been specifically treated to remove minerals
Waste water	Any water which has been used at least once and cannot be used again without being treated

In the pharmaceutical industry there are other types of specialized water that are not found in other industries. These include Purified Water, Water for Injection, Sterile Purified Water, Sterile Water for Injection, Sterile Bacteriostatic Water for Injection, Sterile Water for Inhalation, and Sterile Water for Irrigation. The United States Pharmacopeia (USP) sets the qualifications for sterility and packaging methods that delineate between these various types of water.

Summary

When working with hazardous and flammable substances, leaks and spills are always a concern. The process industries use a number of process containment controls and process upset control systems to minimize and/or eliminate threats to health, safety, and the environment.

Process containment and control systems include a variety of engineering measures and devices used to control common sources of fugitive emissions. These engineering controls include closed systems, closed-loop sampling, floating roof tanks, ventilation systems, effluent control, waste treatment, noise abatement, static-dissipation devices, ground-fault circuit interrupters, and grounding mechanisms.

Occasionally something will go wrong during a process. Engineering controls are implemented to help prevent any process upsets from turning into a hazardous situation. Process-upset controls come in a variety of forms, such as flare systems, pressure-relief valves, deluge systems, explosion-suppression systems, explosive-proof building design, and spill-containment devices.

In addition to hazardous chemicals, process fluids such as air, nitrogen, steam, condensate, and water have the potential to cause serious injury. Process fluids become even more hazardous when they escape from their containers in an uncontrolled manner. Process fluids can be deemed hazardous because they are toxic or flammable, or because they are under high pressure or used at extreme temperatures. For example, steam at high pressure and high temperature can cause severe burns or cuts. And, because nitrogen displaces oxygen, it can create an oxygen-deficient atmosphere in which a technician without an air-supplied respirator could suffocate.

Checking Your Knowledge

1. Define the following terms:
   a. Bonding
   b. Closed-environment drain system
   c. Effluent
   d. Flare
   e. Fugitive emissions
   f. Grounding
   g. Pressure-relief valve
   h. Process fluid
2. _____ emissions are leaks or releases that occur when containment controls fail.
3. Sample bombs or small draw-off points are used for:
   a. Closed systems
   b. Open-loop sampling
   c. Closed-loop sampling
   d. Inline loop controls
4. *(True or False)* The safest way to collect a sample from a toxic stream is to use an inline automatic analyzer sampling system.
5. What is the purpose of a nitrogen blanket system on a tank?
6. What type of control is used to keep external air from entering a building?
   a. Positive pressure
   b. Negative pressure
   c. Mixed pressure
   d. Low pressure
7. Name three types of wastewater treatments.
8. Noise-abatement methods or devices include all of the following EXCEPT:
   a. Reduction in flow velocity of fluids
   b. Acoustical insulation
   c. Equipment mufflers
   d. Venting
9. *(True or False)* Grounding involves connecting two objects together, usually with a copper wire.
10. Describe what a GFCI is and how it works.
11. Name at least three process-upset control systems.
12. What is the purpose of a flare system?
13. A(n) _____ disc is a thin metal dish designed to burst at a specified pressure.
14. What gas is often used to purge or sweep vessels and often serves as a backup for compressor air?
    a. Helium
    b. Hydrogen
    c. Nitrogen
    d. Methane
15. Name at least four hazards associated with steam.

Student Activities

1. Think of process containment and control systems that are part of everyday life. Make a list of those systems and describe how they keep people safe.
2. Using the Internet or other resources, research different types of flare systems. Write a three-page report on the different types of flares, including a description of each and how it is used.
3. Obtain a Material Safety Data Sheet (MSDS) for a chemical, and then create a sample spill using water as a substitute for the chemical. DO NOT ACTUALLY USE THE CHEMICAL. Use the MSDS to respond to the spill and properly clean it up.

CHAPTER

20

Administrative Controls: Programs and Practices

Objectives

This chapter provides an overview of administrative controls found in the process industries.

After completing this chapter, you will be able to do the following:

■ Describe various administrative controls, in the way of specific company SHE programs, used by the process industries to eliminate and/or minimize threats to safety, health, and the environment.

■ Discuss various administrative controls, in the way of company SHE practices, used by the process industries to minimize and/or eliminate threats to safety, health, and the environment.

■ Participate in safe work observations and provide feedback to coworkers on safe and potentially unsafe work practices.

■ Conduct a site safety inspection and/or audit to identify potential workplace hazards.

■ Describe general procedures for how to safely handle materials.

■ Discuss the impact of government regulations and industry organization guidelines.

Key Terms

Agreement—a plan for coordinating activities between different organizations.

Industrial hygiene monitoring—monitoring the health and well-being of workers exposed to chemical and physical agents in their work environment.

Near miss—an unsafe act that does not result in an incident or accident.

Plan—a method, prepared in advance, for carrying out an action.

Policy—a guiding principle.

Principle—a set of rules or standards.

Proactive—a preventive activity conducted prior to a need for action.

Procedure—a step-by-step set of instructions for accomplishing a task.

Reactive—a corrective activity conducted in response to a need.

Rule—a statement describing how to do something or stating what may or may not be done.

System—an organized, interdependent set of related principles or rules.

Introduction

According to the hazard-control hierarchy, when elimination, substitution, and engineering controls cannot completely protect the process technician from possible exposure to hazardous agents, administrative controls are then employed. Administrative controls can be broken down into two broad categories: programs and activities. This chapter introduces process technicians to the various types of administrative controls (programs and activities) used in the process industries.

Common Administrative Programs

Administrative hazard-control programs can consist of any combination of the following:

- **Policies**—guiding principles
- **Procedures**—step-by-step instructions for accomplishing a task (Figure 20-1)
- **Plans**—methods, prepared in advance, for carrying out actions
- **Principles**—sets of rules or standards
- **Rules**—statements describing how to do something or stating what may or may not be done
- **Agreements**—plans for coordinating activities between different organizations
- **Systems**—organized, interdependent sets of related principles or rules

Exactly what is included in a written program depends upon regulatory requirements, company-specific requirements, site- or plant-specific requirements, and unit- or department-specific requirements.

FIGURE 20-1 Procedures are Step-by-Step Instructions for Accomplishing a Task

Regulatory-mandated programs are those created in response to government regulations. Most of the EPA regulations and OSHA standards that affect the process industries require industrial facilities to capture their interpretation of the law and their plans for complying with the law in a written program. An example of such a program would be a facility's Hazard Communication Program.

Evacuation and accountability plans are established procedures that outline evacuation routes and assembly areas designed to promptly evacuate and account for on-site personnel in the event of an emergency.

Mutual-aid agreements are agreements between industrial facilities, local government-directed emergency responders, and other outside emergency-response teams to render aid to industrial neighbors in the event of a release or other hazardous situation.

Incident Command Systems (ICS) are pre-established methods, documented in writing, for assigned command and control of any incident.

FIGURE 20-2 Mutual-Aid Agreements are Agreements between Industrial Facilities, Local Government-Directed Emergency Responders and Other Outside Emergency-Response Teams

COMMON ADMINISTRATIVE ACTIVITIES

Recall that programs are written documents that explain how hazards are to be controlled. Activities are the programs put into action. Administrative hazard-control activities can consist of any combination of the following:

- Documentation and shipping papers—listing all required product and shipment information
- Training—providing instruction on job tasks or concepts
- Permits—preparing a job site to ensure that conditions are safe for work to begin
- Inspections and audits—conducting regular, scheduled checks to ensure a safe and healthy workplace
- Investigations—taking action to determine the cause or cases of an incident or accident
- HAZOPs—conducting a study to identify possible hazards in a system
- Monitoring—taking samples to determine if toxic or hazardous substances are present in the workplace
- Safe-work observations—looking at selected work practices and providing feedback on the safe and potentially unsafe behaviors observed
- Housekeeping—maintaining a clean, neat, and orderly workplace
- Community awareness—implementing policies and procedures to inform the surrounding community about the nature of the process and operations ("Right to Know" law)

Documentation and Shipping Papers

Documentation is required to identify hazardous materials (Figure 20-3). This documentation includes Material Safety Data Sheets (MSDSs) as well as paperwork for shipping hazardous materials. By identifying the materials as hazardous, employees and community members can better understand the hazards associated with the materials as well as appropriate emergency response for the material in the event of an incident or accident.

FIGURE 20-3 Appropriate Documentation is Required When Shipping Hazardous Materials

Shipping papers for hazardous materials must include the following:

- Description of the materials
- Emergency-response telephone numbers and other information
- Number of pages if there is more than one
- Shipper's signed certification indicating that the shipment was prepared according to rules

Training

Training is one of the most critical activities that occur in the process industries. No matter what job a process technician is assigned to do, some type of training will be required in order to acquire the knowledge and skill to do the job competently and safely.

Training has always been important in the process industries, but it has grown even more important over the last 10 years. Advances in the field of process technology, the heightened demand for quality, the increasing necessity for safe behavior in the workplace, and increased government regulations have brought about the need for a better-trained workforce. Regulations now stipulate that process technicians must be trained and qualified to perform their jobs.

Permits

Safe-work permits are used in the process industries to ensure that appropriate precautions are taken to prevent possible harm to workers when they will be engaging in potentially dangerous activities (Figure 20-4). Issuing safe-work permits involves several phases:

- A listing of all steps required to complete the task
- An assessment of each step to determine any possible threat to worker safety
- The elimination or minimization of each threat
- The provision for maintaining safe working conditions until the task is complete

The following activities commonly require safe-work permits:

- Lockout/tagout
- Hot work

FIGURE 20-4 Safe-Work Permits are Used in the Process Industries to Ensure that Appropriate Precautions are Taken to Prevent Possible Harm to Workers

- Confined space
- Line opening and blinding
- Radiation
- Critical lifts
- Scaffold tags

For more information on permit systems and how they work, refer to Chapter 21, *Permitting Systems*.

Inspections and Audits

Inspections and audits are similar in that both activities involve conducting regular safety, health, and/or environmental checks. These activities are **proactive** in nature, which means that they are conducted prior to a need for action. Investigations, which will be discussed next, are considered **reactive** because they are conducted in response to a need.

Inspections and audits are different in the following ways:

- Inspections are usually conducted locally by plant personnel to ensure that safety, health, and environmental procedures are being followed. Inspections tend to be conducted on a regular and frequent schedule, such as daily, weekly, monthly, quarterly, and yearly.

- Audits are typically conducted by outside personnel (company or corporate employees) to determine if the plant is in compliance with company and regulatory requirements. Corporate audits are usually performed according to a schedule but tend to be done less frequently than inspections.

- Process technicians are often required to conduct inspections and/or audits within their unit or plant.

Investigations

In order for an investigation to take place, an incident, an accident, or **near miss** (an unsafe act which did not result in an incident or accident) must have occurred. A team of individuals often performs an investigation so that people with varying viewpoints and areas of expertise can look at the same problem and bring different insights to the situation.

Investigations are conducted the same way one would approach solving a problem:

1. Decide which problem will be addressed.
2. Arrive at a statement that describes the problem (e.g., what happened, when it happened, and the extent of the problem).
3. Develop a complete picture of all the possible causes of the problem.
4. Agree on the basic causes of the problem.
5. Develop an effective solution that can be implemented.

FIGURE 20-5 Investigations Occur When an Incident, Accident, or Near Miss Occurs

6. Create an action plan.
7. Implement the solution.

Process technicians often participate in investigations. They may be called upon to participate because they were directly involved in the incident or because someone with their expertise is needed on the team.

HAZOPs

A Process Hazard Analysis (PHA) is an organized and systematic process used to identify potential hazards that could result in accidents causing injury, property or environmental damage, production loss, and/or third-party liability.

Many different methodologies can be used to conduct a PHA, depending on the goals of the study, the stage in a facility's life cycle, and the time available. One of the most comprehensive PHA methodologies that can be used at any stage in a facility's life cycle is a HAZOP (Hazards and Operability Study).

A HAZOP identifies operability problems as well as hazards to personnel, company property, and the environment. The benefits of using a HAZOP include the identification of hazards, the optimization of productivity and profitability, the mitigation of employee morale, and the shortening of project schedules.

Monitoring

Two types of monitoring activities conducted in the process industries are industrial hygiene monitoring and fugitive emissions monitoring.

Industrial hygiene monitoring is conducted by an industrial hygienist, and involves sampling the working environment to determine if any hazardous agents are present at unacceptable OSHA levels. Industrial hygienists use a variety of tools to capture data on workplace hazards. They then analyze these data and compare them to published standards. Common industrial hygiene sampling activities include noise monitoring, toxic substances sampling, and ergonomic studies.

Fugitive emissions monitoring is very similar to industrial hygiene monitoring in that samples are taken, analyzed, and compared to government or company standards. However, fugitive emissions monitoring involves a sampling of the working environment to determine if any hazardous agents are present at unacceptable EPA levels. Also, emissions monitoring can be done either by inline sensors or manually with detection devices. For more information on monitoring equipment, refer to the appropriate chapter.

FIGURE 20-6 Industrial Hygiene Monitoring Involves the Sampling of a Work Environment to Determine if Any Hazardous Agents are Present

Safe-Work Observations

A behavior-based observation and feedback process provides visibility and control over upstream indicators of safety performance—namely safe and at-risk behaviors.

Most behavior-based observation and feedback systems follow the ABC model of behavior change, which reflects the following basic principle:

- Behavior (B) is directed by
- Activators (A) and motivated by
- Consequences (C).

In other words, stimuli or events in the environment often direct performance, but people do what they do because of the consequences they expect to receive or avoid. Using simple but effective observation techniques, employees observe each other and then give appropriate one-on-one coaching regarding the observed behavior. As employees become more comfortable with the informal observation process, they begin to observe and coach informally and the process becomes a natural part of the work culture.

Housekeeping

Housekeeping in a process plant is basically the same as housekeeping in the home. In order to maintain a safe, orderly environment, workers must keep their work areas tidy and uncluttered.

Good housekeeping accomplishes the following goals:

- Prevents accidents
- Saves money
- Increases productivity
- Improves worker morale

Community Awareness

Community awareness programs are established in order to build relationships between industrial sites and the surrounding communities. A partnership is established to show that industry is committed to operating in a safe and responsible manner. Many companies will encourage and support employee participation in community volunteer efforts and contribute to local organizations focusing on education and environment programs.

Companies can work with local authorities to address emergency-response and planning issues, along with Right-To-Know information (part of an EPA regulation), through an American Chemical Council (ACC) program called Community Awareness Emergency Response (CAER). See Chapter 16, *Recognizing Environmental Hazards* for more details on Community Right-To-Know. See also the Responsible Care section later in this chapter for more information.

SAFE MATERIAL HANDLING

Regulations and administrative programs are not enough to protect workers from exposure to hazardous agents. Process technicians must share in this responsibility. To do this, they must be able to recognize hazardous agents within the facility in which they work.

Employees must always avoid unsafe acts while on the job. An unsafe act is any behavior within the unit, such as smoking, that increases the likelihood of a worker experiencing an accident or injury.

Finally, process technicians are expected to follow all safe operating policies and procedures and wear the appropriate personal protective equipment for the tasks being performed.

GOVERNMENT REGULATIONS AND INDUSTRY ORGANIZATION GUIDELINES

There are a number of regulations and guidelines that process industries companies follow as part of their administrative controls. The major regulations and guidelines include the following:

- Process Safety Management
- HAZWOPER
- Responsible Care

- ISO 14000
- DOT Hazardous Materials Handling: Loading and Unloading
- DOT Hazardous Materials Packaging and Marking
- NFPA Hazardous Materials Storage Requirements

Process Safety Management (PSM)

OSHA's PSM requirements, implemented in 1993, apply to companies that process certain hazardous chemicals. PSM Regulations are in place to prevent or minimize the consequences of catastrophic releases of toxic, reactive, flammable, or explosive chemicals. In order to remain in compliance, these companies are required to have the following:

- Written Operating Procedures: Employers must provide clear instructions for safely conducting activities within the covered process area. The procedures must include steps for each operating phase, operating limits, safety and health considerations, and safety systems and their functions. Written procedures, which may also be maintained electronically, must be accessible by all employees who work on or maintain a covered process. The procedures must be reviewed as often as necessary to ensure they reflect current operating practice. They must also include safe work practices where needed to provide for special circumstances, such as lockout/tagout and confined-space entry.

- Employee and Contractor Training: Employers must provide training on all covered processes to ensure employees are trained on an overview of the process, all required operating procedures, safety and health hazards, emergency operations, and safe work practices. Employees receive initial training and certification and then must receive periodic refresher training. Contract employers are also required to train their employees to safely perform their jobs around highly hazardous chemicals and to document that employees received and understood training. In addition, they are responsible for ensuring that contract employees know about potential process hazards and the work-site employer's emergency action plan.

- Pre-Startup Safety Review: Employers are required to perform a safety review of new or modified equipment or facilities prior to starting up operations. This helps ensure that equipment is constructed to meet design specifications, procedures are developed and in place, training is completed, and all required PHAs are performed and changes implemented.

- Hot-Work Permits: Hot-work permits must be issued for hot-work operations conducted on or near a process covered under the PSM standard. Hot-work operations include electric or gas cutting or welding, brazing or soldering, grinding, hot-tar projects, any portable gas procedures, and steam-generating work.

- Mechanical Integrity: Employers are required to establish and implement written procedures to ensure the ongoing integrity of process equipment that contains and/or controls a process covered under the PSM standard. This section of the standard does not apply to contract employers; however, contract employees are required to follow the written procedures.

- Management of Change Process: This ensures that when changes are made to a process, those changes will provide employees with the same protection from highly hazardous chemicals as the original equipment or process. Employers are required to provide training to all on-site employees prior to starting up the renovated process or equipment. Contract employers must train their own contract personnel on the new procedures. Management of change also ensures that process safety information and operating procedures are updated correctly as needed. Typically, all changes to procedures are reviewed by supervisors and engineers to ensure they are accurate. These changes are then documented and implemented into the existing or new procedures and employees are trained on the changes.

- Incident Investigations: The PSM standard requires employers to investigate as soon as possible following an incident but no later than 48 hours after an incident. Covered incidents include those that either resulted in or could have resulted in a catastrophic release of covered chemicals. The standard requires that an investigation team, which includes at least one person knowledgeable in the process and others with knowledge and experience in investigations and analysis of incidents, work together to develop a written incident report. These reports must be kept by the employer for five years.

• Emergency Planning and Response: Employers must develop and implement an emergency action plan. This plan must include procedures for handling small releases of highly hazardous chemicals. Employees must be trained to follow these procedures and the procedures must be accessible to all employees who may be affected.

• Compliance Audits: Internal audits are required every three years in facilities with covered processes. These audits must certify that employers have evaluated their compliance with process safety requirements. Employers are required to respond promptly to audit findings and must document how deficiencies were corrected. Employers must retain the most recent internal audit.

HAZWOPER

OSHA's HAZWOPER standard applies to any facility that has employees involved in the following:

• Cleanup operations involving hazardous substances
• Cleanup operations at sites covered by the Resource Conservation and Recovery Act (RCRA)
• Voluntary cleanup operations at sites recognized by governmental agencies as uncontrolled hazardous waste sites
• Operations involving hazardous wastes that are conducted at treatment, storage, and disposal facilities licensed under RCRA
• Emergency-response operations for release of, or substantial threat of release, of hazardous substances

The following requirements apply to these sites:

Safety and Health Program: The program must be designed to identify, evaluate, and control safety and health hazards. It must also provide a documented plan for emergency response in the event of a release of the hazardous materials.

Site-Control Program: This program must include a site map, site work zones, site communications, safe work practices, and identification of the nearest medical assistance. Employers are also required to implement a buddy system as a protective measure in particularly hazardous situations. This would allow one employee to keep watch on another to ensure quick aid could be provided if needed.

Employee Training: There are two levels of HAZWOPER training. For example, employees involved in hazardous waste cleanup are required to have more intensive training than an equipment operator with little potential for hazardous waste exposure. Most sites require initial training and periodic refresher training on hazardous waste operations or emergency responses.

Medical Surveillance: This is required for all employees exposed to any hazardous substance at or above established exposure levels. It is also required for those who wear approved respirators for more than 30 days on site in a year and for workers exposed by unexpected or emergency releases. In addition to annual medical checks, these employees must have a medical check at the termination of their employment.

Reduction of Exposure Levels: Engineering controls, work practices, personal protective equipment, or a combination of all three must be implemented to reduce exposure below established levels for any hazardous substances. In other words, employers must make all possible efforts to reduce exposure levels to acceptable levels to protect their employees.

Air Monitoring: On-site air monitoring is required to identify and quantify levels of hazardous substances. This monitoring must be performed periodically to ensure the proper protective equipment is used on site.

Information Program: Employers must provide the names of key personnel responsible for site safety and health, names of alternate personnel responsible for site safety and health, and a listing of the HAZWOPER standard requirements.

Decontaminating Procedures: Employees and equipment must be decontaminated before leaving an area where they may have been exposed to hazardous materials. These operating procedures must minimize exposure through contact with exposed equipment, other employees, or used clothing. Showers and changing rooms must be provided where needed.

Emergency Response Plans: Plans are required for handling possible on-site emergencies as well as off-site emergencies. These plans are often performed as drills and may involve non-employees, such as emergency response technicians, firefighters, and medical personnel.

Responsible Care® Guiding Principles

The American Chemical Council provides Responsible Care guiding principles to respond to public concerns about the manufacture and use of chemicals. Through Responsible Care, member chemical companies are committed to support the continuing effort to improve the industry's responsible management of chemicals.

Some guiding principles are listed below:

- Recognize and respond to community concerns about chemicals and operations.
- Develop and produce chemicals that can be manufactured, transported, used, and disposed of safely.
- Make health, safety, and environmental considerations a priority in planning for all existing and new products and processes.
- Promptly report information on chemical-related health or environmental hazards and recommended protective measures to officials, employees, customers, and the public.
- Counsel customers on the safe use, transportation, and disposal of chemical products.
- Operate plants and facilities in a manner that protects the environment and the health and safety of employees and the public.

ISO 14000

The International Organization for Standardization (ISO) created the ISO 14000 standards for environmental management. ISO-registered organizations are required to do the following:

- Minimize harmful effects on the environment caused by activities
- Achieve continual improvement of environmental performance

DOT Hazardous Materials Handling: Loading and Unloading

The Hazardous Materials Transportation Act was created to improve the Secretary of Transportation's regulatory and enforcement authority and thereby protect all citizens against risks associated with the commercial transport of hazardous materials. This regulation places specific requirements on the activities of loading and unloading hazardous materials, and in some cases, tracking of those materials.

DOT Hazardous Materials Packaging and Marking

DOT requires that hazardous materials that are being transported be packaged and labeled in such a way as to clearly identify them as hazardous. Labels and placards are required to clearly identify the container contents as hazardous. Labels must be attached directly to the container. If the materials are then placed inside freight containers or tanks, larger placards must be attached to both ends of the larger containers. The labels and placards identify the material and its hazard class, a four-digit identification number.

NFPA Hazardous Materials Storage Requirements

The National Fire Protection Association (NFPA) has requirements about how hazardous materials can be stored. Following are the major requirements:

- Ensure the compatibility of materials being stored in the same area.
- Ensure proper ventilation within the storage container as well as in the immediate area.
- Allow adequate traffic routes and escape routes.
- Ensure that heat or ignition sources are not in the area.
- Ensure proper labeling of hazardous materials.
- Maintain MSDS for all chemicals at the facility.

Summary

According to the hazard-control hierarchy, when elimination, substitution, and engineering controls cannot completely protect the process technician from possible exposure to hazardous agents, administrative controls are then employed.

Administrative controls are policies, procedures, programs, training, and supervision to establish rules and guidelines for workers to follow in order to reduce the risk of exposure to a hazard. Administrative controls can be broken down into two broad categories: programs and activities.

Common administrative activities include documentation and shipping papers, training, permits, inspections and audits, investigations, monitoring, safe-work observations, housekeeping, and community awareness.

Administrative controls are used in the process industries to help keep employees safe while on the job. They also help protect the environment and the community by ensuring that hazardous materials are handled in as safe a manner as possible. Process technicians should be familiar with the government regulations that pertain to the various administrative controls.

Checking Your Knowledge

1. Define the following terms:
    a. Administrative controls
    b. Agreements
    c. Fugitive emissions
    d. Industrial hygiene monitoring
    e. Near miss
    f. Plans
    g. Policies
    h. Principles
    i. Proactive studies
    j. Procedures
    k. Reactive activities
    l. Rules
    m. Systems

2. Procedures are defined as:
    a. Sets of rules or standards.
    b. Organized, interdependent sets of related principles or rules.
    c. Methods, prepared in advance, for carrying out actions.
    d. Step-by-step instructions for accomplishing a task.

3. _____-mandated programs are those created in response to government regulations.

4. Name at least five common administrative hazard control activities.

5. List ways in which inspections and audits are similar and different.

6. What do the initials PHA stand for?
    a. Protection Hazard Assessment
    b. Process Hazard Analysis
    c. Preventive Hazard Action
    d. Process Hazardous Acquisition

7. *(True or False)* Two types of monitoring activities are industrial hygiene and fugitive emissions.
8. Name three benefits of good housekeeping in the work environment.
9. Describe OSHA's PSM Management of Change process.
10. *(True or False)* Decontamination procedures are part of the OSHA Process Safety Management regulation.
11. Describe the purpose of the Responsible Care program.
12. Which ISO standard addresses environmental management?
    a. 9000
    b. 9001
    c. 1910.119
    d. 14000
13. List at least four NFPA Hazardous Material Storage Requirements.

Student Activities

1. Select one of the following topics: ISO 14000, DOT Hazardous Material Handling, or NFPA Hazardous Material Storage Requirements. Using the Internet or other resources, research the topic and write a three-page paper describing the topic and how it relates to safety, health, and the environment.
2. Brainstorm a list of tasks that might fall under housekeeping duties in the process industries. Share the list with your fellow students.
3. Perform a Process Hazard Analysis (PHA) on an activity or task that you do at school, work, or home (e.g., mowing the lawn, using a hand tool, or cleaning at heights).

Permitting Systems

Objectives

This chapter provides an overview of permitting systems found in the process industries. After completing this chapter, you will be able to do the following:

- Report the function and purpose of permitting systems found in local plants.
- Conduct a job-safety analysis and complete a safe-work permit to ensure the work environment is safe prior to beginning a job.
- Use locks, tags, and blinds to isolate a piece of equipment.
- Describe government regulations and industry guidelines that address permitting.

Key Terms

Affected and other employees—employees who are responsible for recognizing when the energy control procedure is used and understand the purpose of the procedure and importance of not starting up locked/tagged equipment.

Authorized employees—employees who are responsible for implementing the energy-control procedures and performing service or maintenance work. They receive the most detailed training on procedures.

Cold work—any work performed in an area that contains bulk quantities of combustible or flammable liquids or gases or bulk quantities of liquids, gases, or solids that are toxic, corrosive, or irritating.

Critical lift—any lift that could result in death, injury, health impacts, property damage, or project delay if there is an accident.

Fire watch—a trained employee who monitors the conditions of an area for a specified time during and after hot work to ensure that no fire danger is present.

Hot work—any fire or spark-producing operation (e.g., welding, burning, riveting).

Job-safety analyses (JSAs)—a method of analyzing how a job is performed in order to identify and correct undesirable conditions.

Lockout device—a device placed on an energy source, in accordance with an established procedure, that ensures the energy is isolated and the equipment cannot be operated until the lockout device is removed.

Opening/blinding permit—a permit used to help ensure that accidental leaking from pipes does not occur.

Safe-work permit—a permit used to ensure the area is safe for work to be performed and for communicating that information.

Scaffold tags—tags used to clearly label the status of a scaffold and communicate its approved use.

Introduction

Permit-to-work systems are used in the process industries to control hazardous conditions in the workplace. Employers are required to maintain written procedures for work duties that may place an employee in a hazardous situation. Permits protect employees from hazards and potential hazards. They provide written authority for an employee to complete work for a non-routine task, such as maintenance on a specific piece of equipment.

Written permits normally indicate the task to be performed, the potential hazards associated with performing that task, and the protective controls to be used while the task is performed. The permits are approved by appropriate personnel for a specified duration in accordance with company procedure.

Permits communicate information between the issuer and the recipient about a potentially hazardous work assignment. The permit signifies that both the issuer and the recipient have recognized their responsibilities to ensure a safe work environment. The responsibilities of each person are identified in the list that follows.

The permit issuer performs the following tasks:

- Tests the area for hazardous conditions
- Frees the area of hazards
- Informs the recipient of potential hazards that exist
- Communicates the job responsibilities to the recipient
- Identifies safety equipment that is near the work site

The permit recipient performs the following tasks:

- Checks the work area, following all permit procedures
- Shares all details with the work crew
- Notifies the issuer if the scope of work changes
- Monitors the worksite for hazards
- Cleans up the area after completing the job duties
- Returns the signed permit to the issuer after the work is complete

Permits become invalid when conditions become unsafe, an emergency alarm sounds, an accident occurs at the worksite, the scope of work changes, or the work exceeds the expiration date and time of the permit.

Permit Procedures

Typically, a permit process involves multiple employees and possibly multiple departments. Regardless of the type of permit being obtained, the basic steps of the permit process are generally the same.

1. Determine the need for the permit and discuss permit needs with a supervisor or other authorized person.
2. Determine the type of permit to be issued.
3. The permit is completed and the precautions are noted.
4. The permit is reviewed with all workers who will be affected by or involved in the work.
5. The authorized person and the employee completing the work sign the permit. The supervisor maintains the original and the employee completing the work maintains the copy.
6. The job is completed or the permit becomes invalid or expires.

Types of Permits

There are several types of permits that are commonly used in the process industries. They include the following:

- Confined Space
- Lockout/Tagout
- Hot Work
- Cold Work
- Opening/Blinding
- Radiation
- Critical Lifting
- Scaffold Tags
- Safe Work

CONFINED-SPACE PERMITS

Confined spaces are work areas not designed for continuous employee occupancy, which restrict the activities of employees who enter, work inside, and exit the area, and provide a limited means of egress. There is typically a limited amount of oxygen and/or air flow in these spaces. These spaces may have open tops but restrict the movement of air, or they may be enclosed with limited entry openings. In the process industries, it

Did You Know?

According to the National Institute for Occupational Safety and Health (NIOSH), more than 60% of confined-space fatalities occur among would-be rescuers.

Process technicians should NEVER attempt a confined-space rescue without proper training, equipment, and authorization.

☐ Other
☐ Rescuer

may be necessary to enter these confined spaces for non-routine activities, such as inspection, repair, cleaning, and painting. When these types of non-routine activities are necessary, confined space permits are required. Figure 21-1 shows an example of a confined-space permit.

An area qualifies for a permit if at least one of the following characteristics is present:

- The area contains or could potentially contain a hazardous atmosphere.
- The area contains material that could potentially engulf an entrant.
- The area has an internal design that could trap or cave in on an entrant.
- The area contains any other serious safety or health hazard.

FIGURE 21-1 Confined Space Permit

CONFINED SPACE ENTRY PERMIT			867534

Date and Time Issued: _____ Job Site / Space ID: _____ Equipment to be worked on: _____
Date and Time Expires: _____ Job Supervisor: _____ Work to be performed: _____

Entry Signature IN: _____   _____   _____
Entry Signature OUT: _____

Stand-by personnel: _____   _____   _____

Back-up personnel: _____   _____   _____

Atmospheric Check Requirements: Time: _____Oxygen %: _____Explosive %: _____Toxic PPM _____
Tester's Signature: _____

Ventilation Required: Yes _____ No_____
      Type: Blower ____ Fan ____ Ambient ____ Other _____

Atmospheric Check After Iso & Vent: Time:_____Oxygen %: _____Explosive %: _____Toxic PPM _____
Tester's Signature: _____

Entry, standby, and back up persons currently trained: Yes____No____

Equipment:	N/A	Yes	No
Breathing Apparatus			
Communications			
Gas Monitor			
Hoisting Equipment			
Protective Clothing			
Safety Harness			
Other:			

Periodic Atmospheric Check :   Time: _____Oxygen %:_____Explosive %: ____Toxic PPM _____
                                 Time: _____Oxygen %:_____Explosive %: ____Toxic PPM _____
                                 Time: _____Oxygen %:_____Explosive %: ____Toxic PPM _____
                                 Time: _____Oxygen %:_____Explosive %: ____Toxic PPM _____
                                 Time: _____Oxygen %:_____Explosive %: ____Toxic PPM _____

Permit prepared by: _____   Approved by: _____

Mechanical equipment can also directly or indirectly cause hazards within a confined space. If a possible equipment startup could cause injury, the equipment must be isolated to prevent an accidental startup while the workers are in the confined space. If the mechanical equipment could cause flammable vapors or gases, or a buildup of static charge, it is also necessary to isolate the equipment. This type of isolation is completed through a lockout/tagout permit, discussed later in this chapter.

When using a confined-space permit, it is essential that the vessel attendant (sometimes referred to as "standby watch," "fire standby," or "hole watch") remain in communication with the worker inside the confined space. An injury could quickly become a fatality if there is a lack of communication between the worker and the standby watch whose primary duty is to help guard the safety of the worker. In addition, entry areas must be easily accessible if a rescue is required.

Additional hazards may be present as a result of working in the confined space area. These hazards include thermal effects (heat and cold), noise, vibration, and radiation. Some of these hazards are more difficult to control, but for the sake of the worker proper protective equipment must be used to limit the exposure to the hazard.

Confined-space permits may vary from business to business, but the following information is generally required on the permit:

- Test results
- Tester's signature
- Supervisor's name and signature
- Identification of the space to be entered, employees authorized to enter the space, and authorized supervisors
- Purpose of entry and known hazards
- Precautions taken to isolate the space and control hazards
- Emergency-response name and telephone numbers
- Date and length of entry
- Acceptable entry conditions
- Communication equipment required and communication procedures to be followed
- Additional required permits for work to be performed
- Special procedures for required equipment
- Other information to ensure safe working conditions

Safe work permits are used on a wide variety of jobs. These may cover jobs from a pump seal replacement to painting a hand rail. This permit is used to ensure that the area is safe for work to be performed and for communicating the work being done to other personnel.

LOCKOUT/TAGOUT PROCEDURES

Lockout/tagout procedures are used in the process industries to isolate energy sources from a piece of equipment. Lockout/tagout procedures help protect employees from hazardous energy when they are performing maintenance on equipment. In addition to isolating the energy source, it prevents accidental startup on machinery with rotating equipment that could also harm the employee. Whenever there is a potential for exposure to energy, stored energy, or the point of operation of a machine that could place an employee in a hazardous situation, lockout and tagout devices are required.

The **lockout device** can be a key or combination-type lock, a cable tie, or some other positive means for isolating energy from the equipment. A blank flange or bolted slip blind can also be considered a lockout device. Figure 21-2 illustrates an example of a lockout device.

The tag-out device provides warning to all that the equipment is not to be operated until the tag-out device is removed through the proper energy control procedure steps. Figure 21-3 illustrates an example of a tag-out device.

OSHA has requirements for the lockout/tagout devices that may be used. These devices must be durable and able to withstand the environments in which they are

FIGURE 21-2 Lockout Device With Tag

used. If they will be exposed to corrosive materials, they must be constructed and printed in a way that will not deteriorate or become illegible. The devices must also be standardized by color, shape, and size. A third requirement is that they be substantial enough to minimize early or accidental removal. Locks must be strong enough that they cannot be removed without excessive force or special tools. Finally, the tags must identify the name of the employee who applied them and must indicate warnings such as "DO NOT OPERATE."

DO NOT OPERATE

TO BE REMOVED ONLY BY

OPERATOR

FIGURE 21-3 Lockout Tag

Authorized employees are responsible for implementing the energy control procedures and performing the service or maintenance work. They receive the most detailed training on the procedures. **Affected and other employees** are responsible for recognizing when the energy control procedure is used and understanding the purpose of the procedure and the importance of not starting up locked or tagged equipment.

The following procedure provides a guideline of the major activities involved in isolating equipment during lockout/tagout.

Placement of Lockout/Tagout Devices:

1. Prior to shutdown, the authorized employee must have knowledge of:
   * The type and magnitude of the energy (note: energy can be pressure, electrical, or anything else that powers the equipment)
   * Hazards of the energy
   * Methods or means to control the energy
2. The machine or equipment must be shut down using the procedures established for the equipment.
3. Equipment must be totally isolated from its energy sources with the necessary isolating devices.

4. Lockout or tagout devices must be affixed to each energy-isolating device.
5. All potentially hazardous stored or residual energy must be relieved, disconnected, restrained, and made safe.
6. Prior to work on equipment, the authorized employee must <u>verify</u> the isolation and de-energization of the equipment.

Removal of Lockout or Tagout Devices:

1. The authorized employee must check the work area to ensure that nonessential items have been removed and that equipment components are operationally intact.
2. The authorized employee must check to ensure that all employees have been safely positioned or moved, and notify affected employees that lockout or tagout devices have been removed.
3. Each lockout or tagout device must be removed in accordance with procedures.

HOT-WORK PERMITS

Hot work is any fire or spark-producing operation, including welding, riveting, and flame cutting. Because of the potential for fire hazards, hot work requires the issuance of a permit to ensure the hazards are controlled. Many employers use a **fire watch** for a specified time during and after hot work to ensure that a fire does not begin as a result of the hot work that was performed. Figure 21-4 shows an example of a hot-work permit.

No. 42658

HOT WORK PERMIT

Date Issued	Time
Permit Expires:	Time

Location:

Equipment:

Description of Work:

Required Fire Equipment:
☐ Fire Extingusher ☐ Fire Hose (Water)
☐ CO₂ ☐ Dry Chemical
Other:

Special Precautions:

Required PPE:
☐ Breathing Air ☐ Face Shield
☐ Fire Suit ☐ Goggles
Other:

Maintenance Signature

Operations Signature

Job Completed	Maintenance Signature
☐ Yes ☐ No	

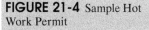

FIGURE 21-4 Sample Hot Work Permit

SAFE-WORK PERMITS

Safe-work permits are permits acquired to do "**cold work**," which is work that does not utilize spaces containing bulk quantities of combustible or flammable liquids or gases, or bulk quantities of liquids, gases, or solids that are toxic, corrosive, or irritating. Cold work requirements state the following:

- Liquid residue of hazardous materials be removed from workspaces as thoroughly as practicable before cold-work operations are started.
- Testing must be conducted to determine the concentration of flammable, combustible, toxic, corrosive, or irritant vapors prior to conducting cold work.

- Continuous ventilation must be provided to keep flammable and toxic vapors at acceptable levels.
- Testing of the environment must be conducted to ensure that air concentrations are safe.
- Spills must be cleaned up as work progresses.
- No ignition sources must be present.

OPENING/BLINDING PERMITS

Piping can also present a hazard due to pressurized and/or toxic fluids. To help ensure that accidental leaking from pipes does not occur, permits for **opening and blinding** are used. OSHA defines blanking or blinding as, "the absolute closure of a pipe, line, or duct by the fastening of a solid plate (e.g., a spectacle blind or a skillet blind) that completely covers the bore and that is capable of withstanding the maximum pressure of the pipe, line, or duct with no leakage beyond the plate" (29 CFR 1910.147).

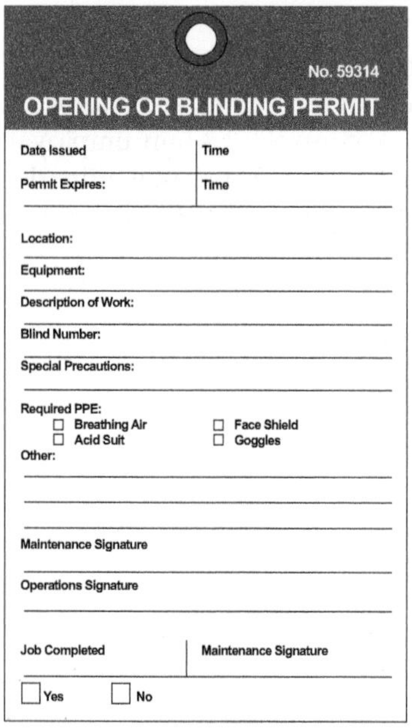

FIGURE 21-5 Sample Blinding Permit

RADIATION PERMITS

Ionizing radiation is an extreme hazard and is usually confined to restricted areas. To work with or near radioactive materials, employees must follow strict guidelines. The standard practice is to keep exposure to ionizing radiation "as low as reasonably achievable (ALARA)." This is achieved through the use of permits, which require continuous personal monitoring and time limits for exposure to certain grades of ionizing radiation (Figure 21-6).

CRITICAL LIFTING PERMITS

Critical lifts are those that could result in death, injury, health impacts, property damage, or project delay if there is an accident. Critical lift plans are necessary to ensure the lift is planned in such a way as to prevent an accident. These plans, implemented by trained personnel, effectively help identify potentially unsafe conditions that could lead to accidents.

The person in charge of the lift typically prepares a pre-lift plan. Within the plan, the items being lifted are identified, the operating equipment is described, rigging sketches (as appropriate) are created to serve as a guide, and operating procedures are identified.

It is essential that experienced, trained operators and signalers be assigned to operate the equipment and give required signals. The required equipment must be used, and the lift must be completed as identified in the plan. Those involved in the lift

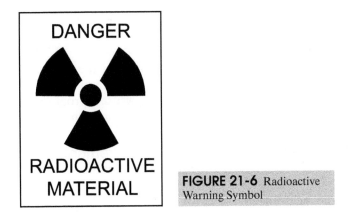

FIGURE 21-6 Radioactive Warning Symbol

typically participate in a pre-lift meeting to review the plan and get answers to any questions.

SCAFFOLD TAGS

Scaffold tags are used in the process industries to clearly indicate the status of a scaffold. Some tags are used to indicate that the scaffold has not been inspected and should not be used. Other tags indicate that it has been inspected and is considered safe to use. Some tags are used to communicate that while the scaffold has passed inspection, specific fall protection should be used by the workers.

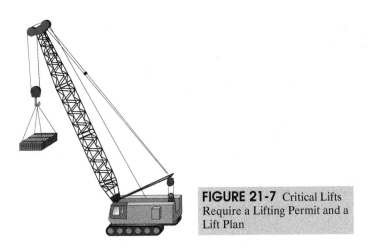

FIGURE 21-7 Critical Lifts Require a Lifting Permit and a Lift Plan

Job Safety Analyses

Job Safety Analyses (JSAs) are conducted to analyze the company's performance in identifying and correcting hazardous conditions that could result in a serious accident, injury, or near miss. Within most companies, the first-line supervisors are primarily responsible for the development of JSAs because of their knowledge of the process and experience in recognizing and eliminating hazards and potential hazards. They involve the employees who perform the job in order to make the JSA as effective as possible.

A variety of methods are used in the process industries to perform a JSA. All of the commonly used methods, however, contain the same basic components:

1. Identify the specific steps performed to complete a job.
2. Identify hazards and potential hazards associated with performing the job.
3. Determine solutions for removing and/or minimizing the hazards and potential hazards.

Statistics show that employers with a strong JSA program are able to reduce the number of accidents and incidents that occur at the location.

Government Regulations and Industry Guidelines

OSHA STANDARDS

OSHA's standards for General Industry and for the Construction Industry (29 CFR 1910) specifically identify the requirements for employers in protecting their employees. In most instances, OSHA requires written programs, thorough training for employees, and strict adherence to the written programs.

Confined-Space Permit Program Requirements

Employers are required to develop and implement a written program for employee entrance into permit-required confined spaces. OSHA requires the following components to be included:

- Identification and evaluation of hazards before entrance
- Testing conditions before entry; monitoring during entry
- Performing atmosphere, oxygen-combustible, and toxicity tests
- Establishing safe work practices and procedures to eliminate or control hazards
- Identifying employee job duties
- Providing proper personal protective equipment to perform the job
- Using a trained standby watch outside the space while the job is performed
- Coordinating activities of multiple employees in the space
- Implementing rescue and emergency procedures
- Implementing procedures for permit programs
- Performing an annual review of the permit program; revising the program as needed to ensure employee safety

Lockout/Tagout Permit Program Requirements

The written program requirements for lockout/tagout include the following:

- Documented energy-control procedures for shutting down and re-starting equipment
- An employee training program on the control procedures
- Regular inspections of the use of the procedures
- Correction and documentation of procedural limitations
- Employees are retrained as needed to ensure procedure changes are followed

An energy control program ensures that whenever the possibility of unexpected equipment startup or energization occurs or when the possibility of unexpected release of stored energy could occur and cause injury, the equipment is isolated from its energy source and rendered inoperative.

TRAINING

Employees are required by OSHA to undergo training on permit programs. Re-training may be required if job performance, job duties, or permit processes change. Training must be documented and records retained for at least three years.

Summary

Permit-to-work systems are used in industry to control hazardous conditions in the workplace. Employers are required to maintain written procedures for work duties that may place an employee in a hazardous situation. Permits are used to protect employees from hazards and potential hazards. They provide written authority for an employee to complete work for a non-routine task, such as maintenance on a specific piece of equipment.

Written permits indicate the task to be performed, the potential hazards associated with performing that task, and the protective controls to be used while the task is performed. The permits must be approved by appropriate personnel for a specified duration in accordance with company procedure.

Permits communicate information between the issuer and the recipient about a potentially hazardous work assignment. The permit signifies that both the issuer and the recipient recognize their responsibilities to maintain a safe work environment.

Permits become invalid when safe conditions become unsafe, an emergency alarm sounds, an accident occurs at the worksite, the scope of work changes, or the work exceeds the expiration date and time of the permit.

Permit procedures are required by OSHA to safeguard employees from hazardous conditions that can be controlled. These permit programs must be written and employees must be trained on how to follow the procedures. In addition, permit programs must be reviewed on an annual basis and retraining conducted as needed to ensure that proper procedures are followed.

Checking Your Knowledge

1. Define the following terms:
   a. Cold work
   b. Confined space
   c. Critical lift
   d. Fire watch
   e. Hot work
   f. Lockout device
   g. Lockout/tagout
   h. Opening/blinding permits
   i. Scaffold tag
   j. Tagout device
2. What is the purpose of permits?
3. Permits typically contain which of the following?
   a. A description of the task to be performed
   b. Protective controls to be used when performing the task
   c. Potential hazards associated with the task
   d. All of the above
4. When working with permits, both the issuer and the recipient are responsible for certain tasks (e.g., testing the area for hazardous conditions). List these tasks in the table below.

Issuer Responsibilities	*Recipient Responsibilities*
1.	1.
2.	2.
3.	3.
4.	4.
5.	5.
	6.

5. *(True or False)* Permits become invalid if the scope of work changes, if conditions become unsafe, or if the work exceeds the expiration date and time of the permit.
6. List the seven basic steps of the permitting process.

Match the permit type with the appropriate description.

Permit Type
   7. Confined Space
   8. Lockout/Tagout
   9. Hot Work
   10. Cold Work
   11. Opening/Blinding
   12. Radiation
   13. Critical Lifts
   14. Scaffold Tags

Description
   a. Required for work that could produce a spark
   b. Used to isolate energy from a piece of equipment
   c. Required for lifts that could cause health impacts, injury, or property damage
   d. Used to prevent exposure to excessive amounts of radiation
   e. Required when performing work in a small area that contains large quantities of flammable, toxic, corrosive, or irritating substances
   f. Used to prevent accidental leakage from a pipe
   g. Required for work areas that restrict activities
   h. Used to communicate the status of a scaffold and its intended use

Student Activities

1. Select one of the permit types discussed in this chapter. Use the Internet to identify at least one incident, accident, or fatality that could have been prevented if proper permitting procedures had been followed. Write a one-page report that explains your findings.

CHAPTER

22

Personal Protective Equipment and First Aid

Objectives

Upon completion of this chapter you will be able to do the following:

■ Understand basic first-aid responses.

■ Discuss the function and purpose of personal protective equipment (PPE) in the process industries:

Respiratory protection

Eye protection

Hearing protection

Head protection

Hand protection

Foot protection

Skin protection

■ Describe the levels of protection and how to select the proper PPE.

■ Explain the use and care of PPE.

■ Describe government regulations and industry guidelines that address medical and first-aid responses and PPE:

OSHA 1910 Subpart K: Medical and First Aid

OSHA 1910.132 – Personal Protective Equipment (PPE)

OSHA 1910.133 – PPE: Eye and Face Protection

OSHA 1910.134 – PPE: Respiratory Protection

OSHA 1910.138 – PPE: Hand Protection

Key Terms

CPR—cardiopulmonary resuscitation, which is an emergency method to assist a victim whose heart has stopped beating properly.

First-degree burns—burns that affect only the outer layer of skin and cause pain, redness, and swelling.

Second-degree burns—burns that affect both the outer and the underlying layer of skin.

Third-degree burns—burns that affect deeper tissues and cause white or blackened, charred skin that may be numb.

Introduction

OSHA requires that employees be given a safe and healthy workplace that is reasonably free of occupational hazards. However, process technicians can be exposed to chemical, biological, physical, and ergonomic hazards inherent to working in a process facility. Although every effort is made to prevent accidents and emergencies, they can still occur on the job, because a process industry worksite can pose a variety of potential hazards.

Company policies and procedures determine whether process technicians can provide first aid to victims. Often, an emergency-response team (or first responders) and/or a resident medical staff will handle emergencies and accidents, including first-aid treatment. At the very least, the process technician must promptly report emergencies and accidents to the proper authorities. Your company will train you on how to report emergency situations and accidents, along with other basic emergency procedures (potentially including first-aid training).

Note: This book is not intended to be a first-aid or medical guide. Its goal is to inform process technicians of potential emergencies and accidents they might encounter, so they can report useful information to trained first responders or medical staff.

To prevent or minimize hazards, government and the process industries have implemented engineering controls, administrative controls, and the use of personal protective equipment (PPE). Process technicians must understand the proper selection, use, care, and maintenance of PPE. This chapter describes different types of PPE, levels of PPE protection, proper fit and use, and upkeep.

Potential Injuries

In a typical process facility, workers routinely perform tasks that, if not performed safely or properly, might cause injuries. Physical hazards are common; they come from environmental factors such as excessive levels of noise, temperature, pressure,

FIGURE 22-1 Some of the Most Common Injuries in the Process Industries Include Cuts, Pinches, Scrapes, Bruises, Burns, Strains, or Splinters

vibration, radiation, electricity, or machinery. This chapter discusses some potential types of injuries that can occur, although the nature and severity of an injury varies due to the hazard agent (the substance, method, or action by which damage can happen to personnel) and site-specific conditions (e.g., work environment, processes, materials, equipment).

Often, injuries result in minor cuts, pinches, scrapes, bruises, burns, strains, or splinters. (Figure 22-1) However, in rare cases more-serious injuries can occur. The following are some types of injuries and emergency situations that can occur in the process industries:

- Eye injuries
- Bleeding
- Impact injuries (bruises, sprains, or fractures)
- Back injuries
- Burns
- Head injuries
- Electric shocks
- Breathing problems
- Injuries from contact with chemical or biological substances

Regardless of the type of injury that you or a fellow worker sustains, you must follow your facility's procedures for obtaining trained help and reporting the injury (OSHA requires certain types of injuries to be recorded).

Following are some general recommendations for dealing with injuries in the workplace:

- Report the situation immediately to the proper authorities, as outlined in your company's policies and procedures. Provide as much accurate information as possible about the location, situation, and victim's status and symptoms.
- If properly trained and company policies and procedures permit, administer first aid.
- Do not move a victim unless a critical situation threatens the victim (e.g., fire, threat of explosion, hazardous atmosphere).
- Remember to use proper PPE and follow safety procedures while dealing with the situation, so you do not become a victim also.
- Remain calm and observant.
- Provide a full and accurate report on the situation and/or assist with any investigation.

The following are general descriptions of each type of injury. To learn more about these injuries and their treatment, check with your employer about available first-aid courses. The American Red Cross and National Safety Council also offer courses to the public. For additional training, you can enroll in an Emergency Medical Technician (EMT) or paramedic program.

EYE INJURIES

Eye injuries can occur when a foreign object or substance comes into contact with the eye. Flying particles or falling objects, such as chips, metal shavings, dust, and other similar hazards are common causes of injury, according to the Bureau of Labor Statistics. Often, these objects are tiny and are moving fast (e.g., being thrown by a moving equipment part). Sparks (e.g., from tasks such as welding and grinding) can also strike the eye and cause damage.

FIGURE 22-2 Safety Glasses, Safety Goggles, and Face Shields Protect the Eyes from Injury Caused by Flying Debris

Contact with chemicals and other hazardous substances (e.g., molten metal, biological agents) is another common cause of eye injuries. Other accidents are caused by swinging objects (e.g., ropes, chains) that strike the eye.

The eyes can be protected from injury using PPE such as safety glasses, safety goggles, or face shields. Eyewash stations (like drinking fountains, but with two streams) can be used to help with eye injuries. See the PPE section in this chapter for details.

BLEEDING

Bleeding can be mild or severe. Most bleeding occurs from minor cuts, scrapes, punctures, or gashes. The hands are a common place for these types of injuries. If you step on a sharp object, your foot can be punctured. Thrown or falling objects can cause impact injuries (bruises, fractures), sometimes accompanied by bleeding. Severe cuts (such as to an artery) can cause a victim to bleed profusely, pass out, and even die.

A variety of hazards in the process industries can cause injuries that bleed: using tools, coming into contact with moving equipment parts, getting hit by a thrown or fallen object, or getting a body part caught between two hard surfaces.

The hands can be protected from injury by using gloves. Feet can be protected by wearing safety footwear. Other body parts can be protected by other types of PPE. See the PPE section in this chapter for details.

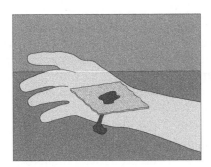

FIGURE 22-3 Most Bleeding Occurs from Minor Cuts, Scrapes, Punctures, or Gashes

IMPACT INJURIES

Impact can occur in a variety of ways:

- Thrown or falling objects
- The worker dropping materials he or she is carrying (e.g., pipes, drums, bags)
- The worker slipping or tripping
- The worker falling from a height
- A vehicle accident

Impact can result in injuries such as the following:

- Bruises
- Bleeding
- Strains
- Fractures
- In extreme cases, death

Even a short fall of three to four feet can cause a major injury or even death. Also, if a person is working at heights and using fall protection, a fall can still cause impact injuries (e.g., from the harness violently jerking against a body part). (Figure 22-4)

FIGURE 22-4 Process Technicians should Always Wear Proper Protective Equipment When Working at Heights

Different PPE can be used to protect workers against impact injuries, such as the following:

- Hard hats
- Gloves
- Safety shoes
- Fall protection
- Safety glasses, goggles, or face shield

See the PPE section in this chapter for details.

BACK INJURIES

Back injuries can range from minor to serious. Some require rest and other simple treatments, while severe ones can result in surgery, permanent disability, or death. Back injuries can occur in any situation, whether the work environment is a process unit or an office environment. They are not always caused by lifting heavy objects or performing laborious tasks; even something as simple as bending over to pick up a dropped object can cause an injury. (Figure 22-5)

The following are some situations that can potentially result in a back injury:

- Lifting or handling materials incorrectly
- Using tools improperly
- Slips, trips, and falls
- Falling from heights
- Impact (e.g., falling objects, getting caught between two hard surfaces)

Improper Heavy Lift

FIGURE 22-5 Many Back Injuries Occur as a Result of Improper Lifting Technique

- Twisting the body in an unnatural position
- Vehicle accidents
- Bending over

Although a back belt can be used, following proper ergonomic practices is the better approach to preventing or minimizing a back injury. See Chapter 15, *Recognizing Ergonomic Hazards* for more details. Check your company's policy on the use of back belts. Proper footwear can help prevent slips and falls, while fall protection can minimize injuries due to falls from heights.

BURNS

There are three levels of burns. A **first-degree burn** affects only the outer layer of skin and causes pain, redness, and swelling. A **second-degree burn** affects both the outer and the underlying layer of skin. These burns cause pain, redness, swelling, and blistering. A **third-degree burn** affects deeper tissues and causes white or blackened, charred skin that may be numb. (Figure 22-6)

Burns can be caused by fire, hot steam, or liquids, radiation, friction, heated objects, the Sun, electricity, or chemicals. Thermal burns are the most common type of burn in the process industries. They occur when hot metals, liquids, steams, or flames come in contact with the skin. Because there is a large amount of hot piping and steam releases within a plant, thermal burns occur frequently.

Various types of PPE can be used to prevent or minimize the hazards of burns, such as the following:

- Flame-resistant clothing (FRC)
- Specially cooled clothing
- Gloves
- Safety footwear
- Helmets or masks
- Tinted face shields, safety goggles, or safety glasses
- Protective (barrier) creams

Respiratory protection can provide a barrier against flame-related hazards such as smoke or vapors.

See the PPE section in this chapter for details.

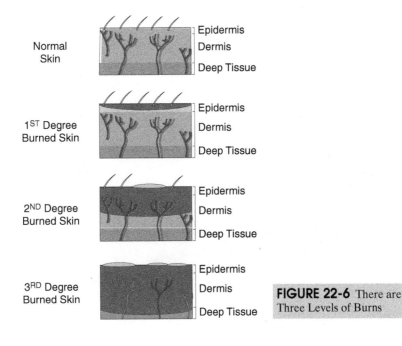

FIGURE 22-6 There are Three Levels of Burns

HEAD INJURIES

Head injuries can occur in the plant due to situations such as the following:

- Impact (e.g., thrown or falling objects, getting caught between two hard surfaces)
- Slipping or tripping
- Falling from a height
- A vehicle accident

Head injuries can be minor or major, but even a minor head injury can result in a more severe condition, such as concussion, blackout, or disorientation. Major head injuries can result in permanent disability or death. Head injuries can also include neck or spine injuries. All head injuries should be considered as major until a professional medical person can provide diagnosis and treatment.

The most common type of PPE to prevent or minimize head injuries is the hard hat. See the PPE section in this chapter for details.

ELECTRICAL SHOCKS

Electrical hazards can result from situations such as the following:

- Improper wiring or grounding
- Short circuits (Figure 22-7)
- Cracked, degraded, or wet insulation
- Surges and overloads
- Equipment failure
- Static electricity buildup
- Downed power lines
- Lightning strikes

An electrical shock occurs when a person is exposed to electric current. Electrical shock and other electrical hazards occur when a person contacts a conductor carrying electricity while also touching the ground or an object that has a conductive path to the ground. The person completes the circuit as the current passes through his or her body.

Electrical shock can cause any of the following conditions:

- Burns
- Cardiac arrest
- Ventricular fibrillation (rapid, irregular contractions of the heart)
- Muscle damage
- Cessation of breathing
- Death

Following are some symptoms of electrical shock:

- Unconsciousness
- Cessation of breathing

FIGURE 22-7 Electrical Shorts can Produce Fire and Electrical Shock Hazards

- Weak
- Weak or absent pulse
- Stopped heart
- Burned skin
- Stiffness of muscles in the body

If you see a victim of an electrical shock, DO NOT touch the victim. Make sure the power source is turned off immediately. If you cannot turn off the power, use a non-conductive material (e.g., wood or plastic) to push, pull, or roll the victim away from the source of electricity.

PPE can prevent or minimize the hazard of electrical shocks:

- Hard hats
- Insulated gloves
- Safety footwear (non conductive)

See the PPE section in this chapter for details.

BREATHING PROBLEMS

Breathing problems can result from a variety of causes, including exposure to hazardous atmospheres, air pollution, allergens, extreme temperatures, pressure changes, and overexertion.

Breathing troubles can result in the following conditions:

- Coughing and wheezing
- Sneezing
- Shortness of breath or gasping
- Gagging
- Vomiting
- Loss of consciousness
- Turning blue (a condition called cyanosis)
- Cessation of breathing
- Death

The following are some hazardous atmospheres the process technician might encounter:

- Smoke
- Airborne substances (toxic or non-toxic)
- Oxygen-deficient environment
- Oxygen-enriched environment

Respiratory protection can provide a barrier between a worker and hazardous atmospheres. (Figure 22-8)

FIGURE 22-8 Respiratory Protection can Provide a Barrier between a Worker and Hazardous Atmospheres

CONTACT WITH CHEMICAL OR BIOLOGICAL SUBSTANCES

Workers handling chemicals and/or biological substances can come into contact with these substances. The effects will vary, based on factors such as the type of substance, the amount of the substance, the type of exposure (e.g., inhalation, absorption), and the length of exposure. See the *Recognizing Chemical Hazards* and *Recognizing Biological Hazards* chapters for more details.

Various types of PPE can be used to prevent or minimize exposure to these substances:

- Chemical suits, biohazard suits, or similar clothing and gear
- Respiratory protection
- Hard hats
- Gloves
- Safety footwear
- Face shields, safety goggles, or safety glasses

Eye-wash stations can be used to remove any chemical or biological substances in or near the eyes, while safety (deluge) showers can be used to remove such substances from the body.

See the PPE section in this chapter for details.

OTHER INJURIES

The following are some other potential injuries and emergency situations:

- Cessation of breathing
- Stopped heartbeat
- Shock
- Choking

Trained personnel can administer Cardio-Pulmonary Resuscitation **(CPR)** to victims who have stopped breathing and/or have no heartbeat. Administering CPR properly and promptly can make the difference between life and death. You can become CPR-certified through your company's program or other outside agency (e.g., Red Cross).

Shock is different from an electrical shock. This life-threatening medical emergency results in the bodily collapse or near collapse that comes from an inadequate delivery of oxygen to the body. Shock victims require immediate medical attention, or else permanent disability or death can happen. Shock symptoms can appear in a range of ways:

- Weakness
- Pale, cold and/or clammy skin
- Cyanosis (turning blue)
- Chills
- Vomiting
- Shallow or rapid breathing
- Change in body temperature (high or low)
- Change in heart rate (fast or slow)
- Chest pain
- Restlessness
- Personality change
- Mental confusion
- Coma

Choking is caused by food or an object blocking a person's upper airway, which prevents proper breathing. A first-aid technique called the Heimlich maneuver can be performed on a choking victim to dislodge the trapped object from the airway.

Choking signs include the following:

- Inability to talk
- Clutching the throat with one or both hands (this is the universal choking sign)
- Wheezing or coughing
- Panic
- Wild gestures
- Turning blue
- Loss of consciousness

As stated previously, only trained personnel should attempt any first-aid treatment, including CPR or the Heimlich maneuver. CPR and first-aid skills should be updated regularly because treatment methods change and technology provides better ways to help people (e.g., Automatic Electronic Defibrillators, or AED).

FIGURE 22-9 The Heimlich Maneuver can be Performed on a Choking Victim to Dislodge a Trapped Object from the Airway

Personal Protective Equipment

Personal protective equipment is specialized clothing and equipment worn or used by workers to minimize the risk of injury from exposure to various hazards in the workplace, including chemical, physical, biological, and ergonomic hazards. OSHA regulations require "the use of personal protective equipment (PPE) to reduce employee exposure to hazards when engineering (technological and engineering improvements may be used to isolate, diminish, or remove a hazard) and administrative controls (e.g., policies, procedures, and activities) are not feasible or effective in reducing these exposures to acceptable levels." PPE is considered the last line of defense against workplace hazards.

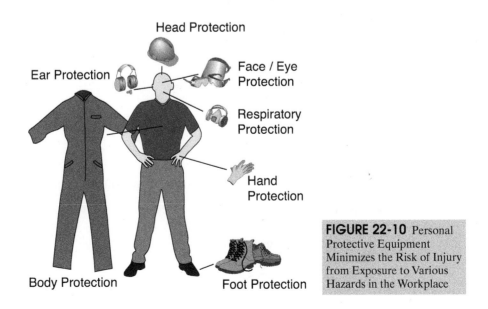

Head Protection

Ear Protection

Face / Eye Protection

Respiratory Protection

Hand Protection

Body Protection

Foot Protection

FIGURE 22-10 Personal Protective Equipment Minimizes the Risk of Injury from Exposure to Various Hazards in the Workplace

The following are different types of PPE that may be provided to the process technician:

- Respiratory protection: air-supplying respirators and air-purifying respirators
- Hearing protection: earmuffs, earplugs, helmets
- Eye protection: safety glasses, safety goggles, face shields
- Face protection: face shields, helmets, masks
- Head protection: hard hats, bump caps
- Body and skin protection: chemical suits, flame-resistant clothing, aprons, slickers
- Hand protection: chemical and liquid-resistant gloves, leather or canvas gloves, insulated gloves
- Foot protection: safety shoes or boots, rubber boots, non-conductive footwear

Two main groups are responsible for promoting standards that PPE must meet to provide proper protection:

- The National Institute for Occupational Safety and Health (NIOSH) is a government agency that carries out research and training, and recommends new standards and criteria to OSHA. NIOSH standards are used for respirators.
- The American National Standards Institute (ANSI) is an organization that develops and promotes standards in a wide variety of areas, including PPE.

OSHA requires that many categories of PPE meet or be equivalent to ANSI standards:

- Eye and Face Protection—ANSI Z87.1-1989
- Head Protection—ANSI Z89.1-1986
- Foot Protection—ANSI Z89.1-1991

No ANSI standards are available for hand protection, so gloves should be selected based on the task to be performed and the glove's performance and construction characteristics.

When PPE is required, a PPE program should be implemented according to OSHA, which covers the following:

- Hazards present
- Selection, maintenance, and use of PPE
- Training of employees
- Monitoring of the program to ensure its ongoing effectiveness

Employers perform a hazard assessment, evaluating potential hazards in the facility based on basic categories such as the following:

- Impact
- Penetration
- Compression (roll-over)
- Chemical
- Heat or cold
- Harmful dust
- Light (optical) radiation
- Biologic

The following are hazards that are surveyed during the assessment:

- Electricity sources
- Motion sources
- High-temperature sources
- Radiation sources
- Harmful dust sources
- Types of chemicals used or present

- Types of biological materials used or present
- Sharp objects
- The potential for falling or dropped objects

Based on this assessment, employers determine the proper types of PPE required. OSHA recommends that employers provide a level of protection greater than the minimum required.

Companies provide PPE to workers, along with training on how to select, fit, use, and maintain it. If a company does not provide certain types of PPE, they can reimburse the employee for purchasing the appropriate items on their own (e.g., safety shoes, prescription eye protection).

EXPECTATIONS OF PROCESS TECHNICIANS

OSHA mandates that employers check that workers demonstrate an understanding of PPE training, along with the ability to properly wear and use the PPE. This must be done before workers perform the task that requires the PPE.

The following are some general expectations of process technicians relating to PPE:

- Know when PPE is necessary.
- Select the proper PPE.
- Understand the limitations of PPE.
- Inspect PPE before use and make sure it fits.
- Use PPE properly, especially when using multiple types of PPE together.
- Take off PPE when done, and then inspect it again.
- Clean, maintain, and store the PPE.
- Report damaged PPE and replace it.

A wide range of PPE types, many of which are described in this chapter, are used to protect process technicians from head to toe in a variety of situations and hazards. Because the types of PPE vary and there are numerous PPE manufacturers, it is outside the scope of this textbook to cover all possible PPE information.

OSHA regulations require employers to supply different types of PPE to workers. They must then train workers on hazards, along with proper PPE selection, limitations, fit, use, care, and maintenance. Process technicians must receive this training when they first join a company and periodically throughout their employment.

Thus, this section will provide only general tips related to PPE:

• Understand the hazards in your workplace, including chemical, physical, biological, and ergonomic. Consider the relationship hazards have with each other (e.g., fire hazards with chemical hazards, hot weather with fired equipment). Hazards can change based on materials used, equipment maintenance, new processes, environmental conditions, or other factors. Keep updated on these hazards.

• Know what effects hazards can cause, and how they relate to you, your co-workers, the facility, the local community, and the environment. For example, understand what chemicals cause cancer, damage skin, or produce respiratory troubles.

• Follow all government regulations, company policies and procedures, and unit-specific requirements relating to safety, health, and the environment. Read and follow documentation, Material Safety Data Sheets (MSDSs), labels, signs, and other important information. Also, remember that common sense is a good tool for all workers when it comes to safety.

• Choose the right PPE. Make sure it fits properly and comfortably. PPE must be used correctly and continuously to be effective. For example, if you remove hearing protection for even a moment, you can be exposed to damaging high noise levels. When finished using PPE, make sure to inspect it for damage and repair or replace it as necessary. Thoroughly clean PPE following manufacturer's recommendations and company procedures. Then, store PPE so it is ready for its next use.

• Practice good personal hygiene (e.g., wash hands, shower, keep ears free of wax) and grooming (no beard, long hair secured). Proper diet, exercise, and rest are vital also, along with periodic medical examinations (e.g., health, vision, and hearing). Make sure you are up to date on all vaccines (e.g., tetanus).

FIT AND COMFORT

Workers are more likely to properly wear PPE that fits well and is comfortable. Properly fitting PPE can mean the difference between being protected or exposed. Make sure you select the proper-size PPE.

Once PPE has been selected, remember the 3 Cs:

• Correct—adjust the PPE so it fits properly.
• Comfortable—make sure the PPE is comfortable and does not significantly hamper normal motions.
• Compulsory—make sure you wear the PPE at all times during hazard exposure.

Often, multiple pieces of PPE are worn together. For example, a process technician might need to wear earmuffs, a hard hat, safety goggles, gloves, safety footwear, and other PPE to perform a task. Make sure that the PPE is compatible and does not cause problems when worn together. All PPE must be adjusted to fit together, so that protection is not reduced or compromised in any way. Once all PPE is adjusted, practice some common motions of work tasks (e.g., squatting, lifting an arm, walking) to get the feel of the PPE and see if any additional adjustments are required.

LIMITATIONS

It is crucial that process technicians understand not just the protection, but also the limitations. Companies will provide training on specific limitations, but the following is a general list:

• PPE can take time to put on correctly. This can be critical in emergency situations. The process technician must understand how to put on and take off PPE quickly, especially respirators. In fact, donning a respirator should be practiced.
• Wearing PPE can limit mobility and hinder the wearer, especially if multiple pieces are worn at the same time. Gloves can reduce dexterity, full-face respirators can limit vision, and full chemical suits can reduce arm and leg motion.
• Communications (speech and hearing) can be impaired when wearing a respirator or hearing protection.
• The likelihood of heat stress increases when wearing PPE. Some body protective suits do not "breathe" properly, so heat and sweat is trapped inside the suit.
• PPE adds to the total weight of the wearer, making normal tasks harder and resulting in increased exertion.
• Some PPE is limited in its use. The wearer can only be exposed to a hazard for a certain amount of time before the risk of exposure increases (e.g., respirators with filters).
• PPE can be constrictive and cause the wearer pyschological stress, especially in situations such as confined-space entry or working at heights.
• Improper fit and use can lead to exposure.
• The wearer can experience a feeling of overconfidence that results in lax safety habits or not following safety procedures.

PPE LIFESPAN

PPE can lose its effectiveness due to the following factors:

• Penetration—hazardous materials pass through the PPE barrier, due to an opening caused by a tear, PPE slippage, incorrect fit, or some other factor
• Permeation—hazardous material crosses through the PPE barrier. This depends on the properties of the PPE, the nature of the hazard, and duration of exposure

- Degradation—hazardous material or other forces (sunlight, heat, moisture) break down the PPE properties through contact over time
- Contamination—the wearer is exposed to hazardous materials and the PPE is not thoroughly cleaned

TYPES OF PPE

The following is an illustration of different types of PPE, followed by specific descriptions of each type.

Respiratory Protection

Respirators protect workers from exposure to hazardous atmospheres that can result in acute or chronic health hazards. Respirators fall into one of two types:

- Air supplying—provides breathable air to the wearer through a mask and hose connected to a clean-air source. Examples include an air line, hose mask, self-contained breathing apparatus (SCBA)
- Air purifying—filters contaminants out of the air, protecting the wearer from particles as small as 0.3 microns. They cannot be used in oxygen-deficient atmospheres, those with less than 19.5% oxygen (O_2) content. Examples include a dust mask, half or full face mask, gas mask

If you wear prescription glasses or contacts, check with your company about the impact these have on respirator use. Special adjustments might need to be made for full-face respirators (e.g., SCBA masks, air line masks).

See Chapter 8, *Hazardous Atmospheres and Respiration Hazards* for details on respirator descriptions, selection, fit, use and care.

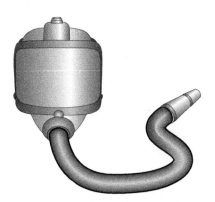

FIGURE 22-11 Air-Supplying Respirators Provide Breathable Air to the Wearer Through a Mask and Hose Connected to a Clean-Air Source

HEARING PROTECTION

Hearing protection is used when excessive noise is present in the workplace. Following are factors that determine excessive noise:

- Loudness of the sound (in dBs)
- Duration of worker's exposure
- Whether the noise is generated by one or multiple sources
- If the worker moves between work areas with different noise levels

Hearing protection only reduces the amount of noise that reaches the ears; it does not eliminate it entirely. Hearing protection has an associated noise reduction rating (NRR).

The main types of hearing protection are earplugs and earmuffs:

Earplugs– made of materials such as foam, rubber, waxed cotton, fiberglass wool, or plastic, and inserted into the ear canal. Some types are pre-formed or molded (fitted for a specific individual by a professional) while other types are self-forming (i.e., they expand in the ear). Earplugs can be reusable or single-use.

Earmuffs– cover the entire ear with a cushioned cup held in place with a headband.

Earplugs and earmuffs can reduce noise levels by up to 20-30 dBA. Earmuffs generally have a higher noise reduction rating than ear plugs. Worn together, they can reduce the noise level up to an additional 5 dBA (this is required if the worker is exposed to noise levels of 110 dBA or higher).

In some cases, special helmets must be used that dampen (reduce) noise and protect against vibrations. Some helmets also provide head protection, like a hard hat. These vary by the type of hearing hazard and work situation.

Hearing protection is selected based on the following criteria, in order of importance:

1. Intensity of noise
2. Personal comfort
3. Availability

Fingers or cotton do not provide adequate protection against noise. Use proper hearing protection.

Earplugs must be inserted properly in the ear canal to be effective. Consult your company's procedures and/or manufacturer's instructions to ensure a good fit. Clean reusable earplugs regularly, based on the manufacturer's recommendations. Replace earplugs if they cannot be thoroughly cleaned. Dispose of single-use earplugs after use.

Make sure earmuffs fit snugly over your ears and form a perfect seal. Clean them thoroughly after use. Glasses, long hair, facial movements (e.g., yawning or chewing) can reduce the protective value of earmuffs.

See Chapter 10, *Hearing and Noise Hazards* for more details.

Eye Protection

Eye protection is PPE that reduces the risk of hazards to the eyes from the following:

- flying objects (e.g., metal chips, dust)
- falling objects
- extreme temperatures
- chemical splashes (e.g., corrosive liquids)
- irritating mists
- hot fluid splashes (e.g., steam, molten metals)
- glare
- sparks
- radiation burns

Did You Know?

Noise-induced hearing loss (NIHL) can be caused by a single loud impulse noise (e.g., an explosion) or by loud, continuous noise over time (e.g., noise generated in a woodworking shop).

Hearing protection should always be worn when working in environments with sounds louder than 80 decibels (normal conversation is around 60 decibels).

Other sounds that can cause NIHL include motorcycles, firecrackers, and firearms, all of which range from 120 to 140 decibels.

FIGURE 22-12 Proper Eye Protection can Prevent Injury or Blindness

Eye protection PPE comes in a variety of types, based on the hazard:

• Safety glasses—protective eyewear with metal or plastic frames and impact-resistant lenses. They are more shatter resistant than normal eyewear. They can come with or without side shields.

• Safety goggles—tight-fitting protective eyewear that completely cover the eyes, eye sockets, and face around the eyes. They protect your eyes from the front and sides against impact, dust, and splashes.

• Face shields—a sheet of transparent plastic attached to a headband, which extends from the eyebrows to below the chin and across the width of the face. Face shields protect the face and front of the neck from flying particles, dust, sprays, or splashes. They do not protect against impact. They can be worn with safety goggles or safety glasses to provide extra protection. They can be clear or polarized (tinted) for glare protection.

• Helmets and masks (e.g., welding shields)—these types of eye protection vary based on the specific hazard.

Other types of specialized eye protection are available, depending on the hazard (radiation, laser light, UV light). Your company will provide these.

All eye protection must meet or be equivalent to the ANSI standard (ANSI Z87.1-1989). Eye protection should allow for air to circulate between the eye and the lens, and fit properly. Your company can help you select the proper eye protection for the work situation. You must understand how to use it, along with when and where to use it.

For people who wear corrective eyewear (prescription glasses or contacts), eye protection should either include the prescription in the design or properly fit and not interfere with the prescription glasses or contacts (i.e., the worker's vision must not be inhibited or limited). Eyeglasses are never a substitute for proper eye protection. Contacts present a hazard in that they can trap a hazardous substance (e.g., chemical, metal shaving, dust) between the wearer's eye and the lens. Some companies have policies specifying where tinted-lens glasses and/or sunglasses cannot be worn (e.g., indoors or in dimly lit areas).

The following are some tips for using and maintaining eye protection:

• Follow company guidelines and manufacturer's recommendations for cleaning eye protection.
• When cleaning eye protection, never use harsh abrasives that could scratch the lenses or remove any protective coatings.
• In cases where workers must share eye protection, it must be disinfected after each use.
• Prescription eye protection should not be shared.

In case of accidental exposure to a hazard, eyewash stations are located around the facility. Process technicians should know where the closest eyewash station is to their work area.

Face Protection

Face protection is used to reduce hazards to the face (including the head and neck), such as impact, chemical or hot metal splashes, heat, radiation, and other hazards.

Some typical types of face protection include the following:

• Face shields—protect the face and eyes from flying particles, dust, sprays, or splashes. They do not protect against impact. (Figure 22-13)

FIGURE 22-13 Face Shields Protect the Face and Eyes from Flying Particles, Dust, Sprays, or Splashes, but do not Protect Against Impact

- Acid-proof hoods—protect the head, face, and neck against splashes from corrosive chemicals.
- Welding helmets—protect against splashes of molten metal and radiation burns.

Head Protection

Head protection is used to protect against the following hazards:

- Impacts from falling objects or fixed objects (e.g., low-hanging beams or equipment such as pipes)
- Penetration
- Electrical shock
- Burn hazards

Hard hats must be impact-resistant and meet the ANSI standard (ANSI Z89.1-1986) for protective headwear. A hard hat consists of a hard outer shell and a web lining suspension system that absorbs and spreads the shock of an impact. Most hard hats have a bill across the front, but some have a brim all around it, similar to a traditional safari helmet.

Hard hats fall into one of three categories:

- Class A—provides impact and penetration resistance along with limited voltage protection (up to 2,200 volts).
- Class B—provides the highest level of protection against electrical hazards, with high-voltage shock and burn protection (up to 20,000 volts); this type also provides protection from impact and penetration hazards by flying or falling objects.
- Class C—provides lightweight comfort and impact protection, but does not protect against electrical hazards.

FIGURE 22-14 Hard Hats Must be Impact-Resistant and Meet the ANSI Standard for Protective Headwear

The following are some tips for properly using and maintaining a hard hat.

- Never modify the hard hat (e.g., add stickers, paint it, drill holes, or remove webbing).
- Make sure it is adjusted to fit on your head properly (refer to the manufacturer instructions with the hard hat):
  - Adjust the headband to fit your head, while still allowing sufficient space between the outer shell and the web lining.
  - The hat should not bind, fall off, or irritate the skin.

- Wear your hard hat with the bill turned to the front (do not place it on your head backward).
- If you have long hair, secure it tightly under the hard hat.
- Follow the manufacturer's recommendations for cleaning. Some cleaning materials might damage the shell and/or reduce electrical resistance.
- Store the hard hat out of direct sunlight and heat, which can weaken the shell.
- Replace the liner regularly (about once a year).
- Inspect the hard hat daily. Replace it when any of the following occur:
  - You notice any cracks, holes, breaks, flaking, scratches, brittle spots, discoloration, or loss of gloss.
  - It receives a significant impact.
  - The date stamp has expired (hard hat material can age and lose its integrity, especially if exposed to harsh conditions frequently).

Some hard hats are designed for use with other PPE or optional accessories, such as earmuffs, safety glasses, face shields, and mounted lights. Accessories should not compromise the safety elements of the hard hat.

Another type of protective headwear is called a bump hat or cap. These do not meet ANSI standards, and are intended only to protect against bumping into an obstruction and not against impact (e.g., a falling object).

Body and Skin Protection

In some situations, workers must shield most or all of their bodies against hazards in the workplace such as exposure to the following:

- Chemicals
- Hot metals
- Hot liquids
- Biological hazards
- Radiation
- Hazardous material or waste
- Impacts
- Cuts, abrasions

Body and skin protection is used chiefly to protect most of the body (e.g., torso) or all of it, including the arms, legs, and head. Skin provides a large surface area for chemicals (chemical hazards can come from a solid, liquid, or gas element, compound, or mixture) and biological hazards (from a living or once-living organism — e.g., viruses, mosquitoes, or snakes) to be absorbed or attack so protecting it is vital. The torso

FIGURE 22-15 Flame-Resistant Clothing (FRC) Helps Protect the Wearer Against Flames and Heat

includes major organs, including the heart, lungs, kidneys, liver, and intestines, all of which must be protected from chemical and biological hazards.

Various materials are used to create body protection, including fire-retardant wool and cotton, plastic, rubber, leather, neoprene synthetics, paper-like fiber (typically for disposable items) and other materials. Body and skin protection comes in various forms, depending on the hazard:

• Chemical protective suit—protects the wearer from hazardous chemical spills and splashes. Multiple layers of different materials can be used to increase the level of protection. These suits do not protect against all types of chemicals; the type of chemical hazard will determine the type of chemical suit to use. They do not protect against heat and flames. Chemical hats, hoods, gloves, and boot covers can be added to provide full protection. Sometimes, tape is used to seal potential entry points (e.g., sleeves, cuffs).

• Totally encapsulating chemical protective (TECP) suit—provides the highest level of protection from hazardous chemicals (e.g., spills, splashes, vapors), covering the wearer from head to toe (full body suit); this suit is air-tight and used with an air-supplying respirator. There are various types available for specific chemicals and situations. A similar type of suit, commonly called a biohazard suit, protects against biological hazards. Another type shields against radioactivity.

• Aprons and smocks—protect a major portion of the wearer against chemical splashes and spills; gloves, face shield and goggles, boots, respirator, and other PPE can be added to improve protection.

• Flame-resistant clothing (FRC)—protects the wearer for a limited time against flames or heat. (Figure 22-15) They can also protect against sparks and bursts of electric arcs. These types of garments use specially treated materials to provide protection that will resist bursting into flames for a brief period of exposure. FRC does not protect against lengthy exposure to flames or heat. It also does not provide protection from chemical exposure.

• Slickers—protect workers from wet conditions; they consist of separate pants and a jacket that can be worn over another garment. Slickers are typically not fire resistant. They can provide limited protection against certain types of chemicals.

• Reflective clothing—protects against radiant heat. See the Special PPE section of this chapter for more details on temperature extremes.

• Barrier creams—provide skin protection from irritants; they are also called protective ointments. They can be used in cases where gloves or other hand and arm protection cannot be safely or effectively used. Some barrier creams (similar to sunscreen) can be used to protect against certain heat or light exposure. Typically, they must be reapplied periodically during a shift. Check with your company about barrier cream use. When conditions allow proper PPE use, barrier creams cannot be used as a substitute.

Other types of body and skin protection can be used in various forms (e.g., lab coats, vests, jackets, coveralls, and full body suits). Your company will provide this PPE as the hazard or situation requires.

Hand and Arm Protection

For protection against chemicals, glove selection must take into account the chemicals encountered, the chemical resistance, and the physical properties of the glove material.

Hand and arm protection is used to protect against hazards such (Figure 22-16) as the following:

• Chemical exposure
• Cuts, abrasions, and scratches
• Impact (fractures, bruises, strains)
• Penetration and punctures
• Burns (chemical or temperature)

FIGURE 22-16 Gloves Help Protect the Hands from Injury and Exposure to Hazardous Substances

- Hot or cold temperatures
- Electrical shock

Hand and arm protection comes in a variety of types:

- Gloves (extending to the wrists, mid-arm, or elbow)
- Hand pads
- Finger guards
- Wristlets
- Arm coverings or sleeves

To select the proper hand and arm protection, consider the following factors:

- Hazards present (chemical, physical, biological, ergonomic)
- Specific work activities planned
- Body part requiring protection (hand only, forearm, arm)
- Grip requirements (dry, wet, oily conditions)
- Duration of contact with the hazard
- Glove performance characteristics
- Size and comfort

OSHA lists four broad categories of protective gloves:

- Leather, canvas, or metal mesh
- Fabric or coated fabric
- Chemical and liquid-resistant
- Insulating rubber gloves

This section focuses on gloves, the primary type of hand and arm protection.

Leather, canvas, or metal mesh materials are used to create sturdy gloves that provide protection against physical hazards, (e.g., punctures, cuts, abrasions, and burns). Leather or canvas gloves also provide protection against sustained heat, sparks, and blows. Metal mesh protects against knives and sharp objects. You should not use any of these types of gloves when working with chemicals.

Fabric and coated fabric gloves are made of cotton or other fabrics. Fabric gloves protect against dirt, chafing, abrasions, and slivers. They do not work well with rough or sharp materials, or heavy materials. Coated fabric gloves typically combine cotton and plastic, which provide some hand protection and offer slip-resistant qualities.

For chemical and liquid protection, select hand and arm protection based on the chemicals or liquids to be handled, the nature of the contact (e.g., splash, immersion), and duration of exposure. Consult company information (e.g., a Material Safety Data Sheet and manufacturer's recommendations) for proper selection. The Department of Energy rates various protective gloves for their effectiveness against specific chemicals in its Occupational Safety and Health Technical Reference manual.

Chemical and liquid-resistant gloves can be made of materials such as neoprene, butyl, nitrile, latex, plastic, or combinations of materials. The following are descriptions of various glove materials and their properties:

• Neoprene—synthetic rubber gloves provide good flexibility and permit dexterity, are high density and tear resistant. OSHA states that neoprene can protect against alcohols, organic acids, alkalis, gasoline, and hydraulic fluids. The chemical and wear-resistance properties are better than natural rubber.

• Butyl—synthetic rubber gloves remain flexible at low temperatures and resist oxidation, ozone corrosion, and abrasion. OSHA states that butyl protects against peroxide, rocket fuels, highly corrosive acids (nitric acid, sulfuric acid, hydrofluoric acid, and red-fuming nitric acid), strong bases, alcohols, aldehydes, ketones, esters, and nitrocompounds. Butyl is not recommended for use with aliphatic and aromatic hydrocarbons and halogenated solvents.

• Latex—natural rubber gloves are comfortable to wear, stretch well without breaking (tensile strength), elastic, somewhat temperature resistant, and can protect against abrasive tasks (e.g., grinding). OSHA states that latex protects against most water solutions of acids, alkalis, salts, and ketones. Latex gloves can cause allergic reactions (see Chapter 4, *Recognizing Biological Hazards*). Hypoallergenic gloves, glove liners, and powderless gloves are good alternatives for workers with allergies to latex.

• Nitrile—copolymer gloves permit dexterity and sensitivity, while standing up to heavy use (even after exposure to substances that cause other glove materials to disintegrate). OSHA states that nitrile protects against chlorinated solvents (e.g., trichloroethylene and perchloroethylene), oils, greases, acids, caustics, and alcohols. Nitrile is not recommended for use with strong oxidizing agents, aromatic solvents, ketones, or acetates.

Chemical and liquid-resistant gloves must be long enough to prevent liquids from entering the top. Generally, the thicker the gloves, the greater the chemical resistance. The tradeoff is impaired grip and dexterity. Do not use these types of gloves when handling rough or sharp objects.

Before using chemical and liquid-resistant gloves, inspect them for tears, punctures, discolorations, stiffness, or other damage or defects. One way to test a glove's protection is to fill it with water and then roll the top (cuff) toward the fingers to check for leaks. Reuse of chemical-resistant gloves should be evaluated carefully, considering the absorptive quality, toxicity of chemicals handled, duration of exposure, and glove storage temperature.

Insulating rubber gloves minimize exposure to electrical hazards. OSHA CFR 1910.137 details requirements for the selection, use, and care of these gloves.

Gloves for extreme temperatures (hot and cold) are discussed in the Special PPE section.

Check your company's policies regarding whether gloves can be worn around rotating equipment and/or power tools.

Foot and Leg Protection

Foot and leg protection is used to minimize hazards such as the following:

• Impacts and/or crushing
• Falling or rolling objects
• Penetration by sharp objects
• Hot surfaces
• Exposure to hazardous substances (e.g., chemical, biological)
• Slippery surfaces

A variety of footwear and leg protection can be worn, based on the type of job or task. For example, you might be required to wear safety shoes or boots, rubber boots, flat-soled shoes, high-tops, or other types of protective footwear. (Figure 22-17) Some types of footwear are not permitted (e.g., open-toed shoes, sandals, high heels).

FIGURE 22-17 Safety Shoes and Boots Help Prevent Slipping, Impact Injuries, and Exposure to Hazardous Substances

Safety shoes or boots must meet ANSI minimum compression and impact performance standards (Z41-1991). ANSI-approved safety footwear provides toe protection and impact or compression protection. The type and amount of protection can vary. Shoe soles are typically non-slip and heat resistant. Some safety footwear includes steel toe boxes and/or metal insoles to prevent penetration. Leather is a common material used for the upper parts of safety shoes or boots.

Metatarsal guards, made of aluminum, steel, fiber, or plastic, can be strapped to the outside of shoes to protect the instep area from impact and compression hazards.

Toe guards also fit over the toes of shoes to provide protection from impact or compression. They can be made of aluminum, steel, or plastic.

Combination foot and shin guards protect the feet and lower legs. They can be used along with toe guards to provide added protection.

Legging, made of leather, aluminized rayon, or other materials, can provide leg protection against a variety of hazards (e.g., falling or rolling objects, molten metal, sharp objects, hot surfaces).

The following are some other types of hazards and the proper footwear to use:

- Rubber or vinyl boots can provide protection against chemical splashes and spills.
- Proper footwear and shoe inserts can also protect against ergonomic hazards.
- Non-conductive footwear should be worn around electrical hazards. This type of safety shoe can prevent the wearer's feet from completing an electrical circuit to the ground. In dry conditions, they can protect against open circuits of up to 600 volts (depending on the type). These should be used along with other insulating PPE and additional precautions. However, the insulating protection can be reduced if the shoes get wet, the soles are worn down, or metal objects become embedded. And, following electrical safety precautions, the worker must not touch conductive, grounded objects. Non-conductive footwear must not be used in explosive or hazardous locations.
- Conductive footwear might be required in some situations to prevent the buildup of static electricity. Foot powder should not be used with conductive footwear, because it provides insulation and reduces the conductivity of the shoes. Socks made of nylon, wool, or silk can produce static electricity and should not be worn with conductive footwear. If working around electrical hazards, you should never wear conductive footwear.

The following are some tips for using and maintaining protective footwear and leg protection:

- Inspect them prior to each use, checking for wear and tear: cracks, holes, separation of materials, discoloration, thin spots, broken buckles, and broken laces.
- Look over the soles of shoes for embedded metal objects or other items.
- Follow the manufacturers' recommendations for cleaning and maintenance.

In some situations, chemical or biological substances can be spilled onto safety footwear. These substances can be absorbed by the shoe material or leak inside the top of the shoe. Use a safety shower to deluge yourself and the footwear.

Special Types of PPE

Special types of PPE can be required, based on the type of hazard and other conditions. The following are some common types of special PPE:

Fall protection (or fall arrest) devices can be used to prevent injuries from falls of six feet or greater. Fall protection consists of a full body harness, an anchor point, and a lanyard connecting the two. The harness is designed to evenly distribute forces of a fall to strong muscle groups that can better absorb these forces than other body parts. See Chapter 9, *Working Area and Height Hazards* for more about falls.

Temperature protection includes both hot and cold-temperature gear. For high-temperature environments, PPE can include the following:

- Reflective clothing
- Insulated suits
- Aramid-fiber gloves (protect against hot or cold, and are cut or abrasive resistant)
- Synthetic gloves (protect against hot or cold, are cut or abrasive resistant, and can handle some diluted acids)
- Aluminized gloves (provide reflective and insulating protection)
- Tinted face shields
- Ice vests and cooling bandanas
- Water-cooled garments
- Cooling inserts for hard hats

For low-temperature environments, PPE can include the following items:

- Polyester or polypropylene underwear
- Polypropylene liner socks
- Outer jackets that can be opened at the waist, neck, and wrists to control the release or retention of heat; these can also include side vents and underarm vents
- Heated protective clothing
- Gloves or mittens
- Hard hat liners
- Leather-upper boots with rubber bottoms and felt lining and insoles
- Eye protection fitted to prevent exhaled moisture from causing frost or fog on the eye piece(s); eye protection can also be tinted to prevent glare

Safety Showers and Eyewashes

Eyewash stations and safety (or emergency) showers can provide on-the-spot decontamination using a large quantity of water or other flushing fluid, serving as a backup to PPE if hazardous material exposure occurs. An exposed worker can flush away the hazardous substance using an approved fluid such as potable (drinking) water or treated water. Remember that water does not neutralize any contaminants, but it does dilute and wash them off the affected area.

Eyewash stations are either fixed units or portable and provide a stream of flushing fluid directly to the eyes and face. Some eyewash stations look like drinking fountains with two streams, while others are containers filled with liquid. Safety showers

FIGURE 22-18 Eyewash Stations are Used to Flush Hazardous Substances from the Eyes

provide a high rate and high-pressure flow of water to rinse away contaminants from the face or body; they can also be used to put out clothing fires.

Both eyewash stations and showers are designed to turn on quickly and operate continuously with little interaction from the worker. Some eyewash stations and safety showers are a combined unit. Eyewash stations and safety showers can include an alarm to alert others that the station or shower is in use.

When a worker is exposed, time is critical. Eyewash and safety showers should be located no more than ten seconds from potential hazards. Locations of eyewash stations and safety showers should be marked with signs and well lit. If used quickly and properly, eyewashes and safety showers can greatly reduce the severity of a hazardous material exposure.

Workers must be familiar with the materials they are working with and read the appropriate Material Safety Data Sheets before they actually are exposed. Some materials react with water and should not be removed using an eyewash station or safety shower (e.g., sodium reacts with water to produce hydrogen, which is an extremely flammable gas). Process technicians must be familiar with the location and operation of eyewashes and safety showers in their work area.

A potential drawback to eyewash stations and safety showers that use potable water is that the water can contain chemicals and other substances that interact with the contaminant and aggravate the situation. Portable eyewash stations may not contain enough fluid to properly flush the contaminant.

Another type of device is called a drench or deluge hose, which is typically attached to a sink or faucet. These devices are not recommended for use as eye-wash stations or safety showers. However, if necessary they can be used to drench a worker's head and body; they should not be used on the eyes due to the high water pressure. They can also be used to spot-rinse an area or to assist a victim who cannot stand or is unconscious.

The following are some tips for eyewash stations and safety showers:

- Flush (or irrigate) the affected area for the recommended amount of time. Refer to the accompanying chart.
- If you wear contacts, wash your hands thoroughly and then remove the lenses to prevent contaminants from being trapped between the eye and the lens.

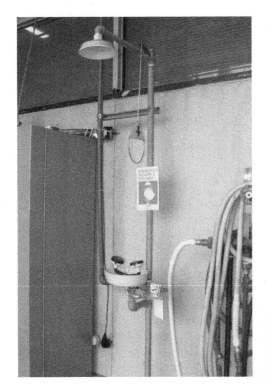

FIGURE 22-19 Safety Showers are Used to Wash Hazardous Substances Off the Body

TABLE 22-1   Minimum Recommended Flushing Times for Eye-Wash Stations and Safety Showers

Contaminant	Minimum Recommended Flushing Times
Mild chemical irritant	5 minutes
Unknown	20 minutes
Moderate to severe chemical irritant	20 minutes
Non-penetrating corrosive chemical	20 minutes
Penetrating corrosive chemical	60 minutes

- Avoid using a safety shower as an eye wash (the water pressure can damage the eyes).
- Seek medical evaluation after using an eyewash station or safety shower; they are not a substitute for medical care.

ANSI has created a standard for Emergency Eyewash and Shower Equipment (ANSI Z358.1-2004), and recommends that the affected body part be flushed immediately and thoroughly for at least 15 minutes using a lot of clean fluid (e.g., water). Flushing times can vary, based on the chemical and its properties.

Table 22-1 illustrates some recommended minimum times, based on the type of contaminant.

TABLE 22-2   Four Classes of PPE

Level	Designation	Probability of Contact with Hazardous Materials	Examples of PPE Used*
A	Highest	Highly probable	Highest level of protection, usually level D plus an air-supplying respirator (e.g., SCBA), totally encapsulating chemical protective (TECP) suit, rubber boots, and chemical gloves
B		Probable	Added protection, usually level D plus a full chemical protective suit, air-supplying respirator, face shield, goggles (in place of safety glasses), and rubber boots
C		Possible	Added protection, usually level D plus PPE (e.g., a face shield, chemical resistant gloves, a chemical protective jacket or slicker jacket, and potentially a respirator)
D	Lowest	Low possibility	Minimal protection. This is typically the daily work uniform and PPE that process technicians must wear (e.g., FRC, safety boots, hard hat, safety glasses, hearing protection)

*Company policy dictates the exact PPE requirements.

Levels of PPE Protection

OSHA and EPA regulations have determined four classes or levels of PPE protection, (Table 22-2) based on the type of work being performed and the probability of the worker coming into contact with hazardous materials. Companies must identify the hazard(s) and determine the concentration (typically parts per million, or PPM). Then appropriate PPE is selected and a list is developed, including skin and respiratory protection, for specific tasks and work areas. These lists are generally posted in appropriate areas around the facility and included in Standard Operating Procedures (SOPs).

Level A provides full protection of skin, the respiratory system, eyes, membranes, and the entire body. It is used in situations where contact with hazardous materials (e.g., sulfuric acid) is very likely, and inhalation and absorption hazards exist.

Level B provides protection in situations where contact with hazardous materials is likely, and absorption of chemicals through the skin is the chief hazard.

Level C PPE provides the same level of skin protection as level B, with a lower level of respiratory protection. The potential absorption hazard is lower than levels A and B.

Level D typically provides minimum skin protection and no respiratory protection.

Government Regulations

The following are OSHA regulations relating to medical or first-aid requirements and personal protective equipment.

OSHA 1910 SUBPART K: MEDICAL AND FIRST AID AND RELATED REGULATIONS

Employers are responsible for determining their own medical and first-aid requirements. They must develop a plan for handling the safety hazards to which their employees are exposed on the job. These plans do not require approval by OSHA; however, during an inspection, the plan is evaluated for accuracy.

Recognition and Evaluation

To determine their needs, employers must evaluate the workplace for requirements. They must consider the following areas when determining what those needs are:

- Location and availability of medical facilities and emergency services
- Availability of medical personnel to consult on occupational health issues
- Types of accidents that could reasonably occur at the workplace
- Response time of external emergency services
- Number and locations of employees at the plant
- Corrosivity of materials used at the facility
- First-aid supplies that should be available
- Level of training required for employees who will render first aid

Medical and First-Aid Plan

After evaluating the workplace, employers must design and implement a program that is tailored to the specific worksite. Elements of the program must include the availability of medical personnel for consultation with employees. Employers must provide the names and telephone numbers for the medical professionals with whom the employer has agreements. Emergency telephone numbers must be conspicuously located by each

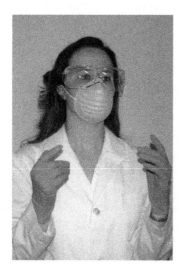

FIGURE 22-20 Proper PPE should Always be Worn When Working Around Blood or Other Body Fluids

plant telephone. In addition, sufficient ambulance service must be available to handle an emergency. This requires ensuring that ambulance services are familiar with the plant location, access routes, and hospital locations.

First-Aid Responders

Employers are required to have at least one person trained in first aid at the worksite if serious or life-threatening injuries can be reasonably expected. The trained personnel must be available within 15 minutes for serious injuries and within four minutes for life-threatening injuries.

Multiple individuals should be trained in order to provide coverage when other responders are unavailable. The responder(s) should be designated and the other workers should know who they are and how to contact them in an emergency. Trained responders must have a current first-aid certificate.

Bloodborne Pathogens

"Bloodborne pathogens" refers to pathogenic microorganisms that are present in human blood or bodily fluids that are capable of causing disease in humans. These pathogens include, but are not limited to, the hepatitis B virus (HBV) and the human immunodeficiency virus (HIV). In most process facilities, it is highly unlikely that employees will be exposed to bloodborne pathogens while performing their normal job tasks. However, they are at risk of exposure if they are in close proximity to an accident victim. Special training is required when exposure is more likely, as is the case for a first-aid provider. Various training courses are required for different types of responders.

Employees who are exposed to waste-treatment systems may require vaccinations against infectious diseases, such as HBV. First-aid kits typically include PPE to protect responders from bloodborne pathogens. (Figure 22-20)

See Chapter 4, *Recognizing Biological Hazards* for more information.

Emergency Equipment

First-aid kits must be available at the worksite. The contents of the kits must be determined through direct consultation with a physician. The contents will vary from facility to facility (or even in different parts of a facility, based on the types of hazards at the facility or area). Eyewash stations and emergency showers are required in locations where corrosive materials are used. The equipment must provide large amounts of clean water and be pressure-controlled and clearly identified.

Training and Recordkeeping

First-aid training is recommended by OSHA during initial job training and on an ongoing basis. The Red Cross and National Safety Council provide first aid and CPR courses.

Employers are required to maintain first aid and medical treatment records for all employees who receive treatment on the job. These records are subject to review in the event of a workplace audit.

PERSONAL PROTECTIVE EQUIPMENT

OSHA 1910.132 – Personal Protective Equipment (PPE)

The OSHA Personal Protective Equipment (PPE) - 29 CFR 1910.132 standard aims to prevent worker exposure to potentially hazardous substances through the use of equipment that establishes a barrier between the hazardous substance and the individual's eyes, face, head, respiratory system, and extremities.

This standard requires employers to assess workplace hazards to do the following:

- Determine what PPE is necessary.
- Provide required PPE to their employees.
- Train employees in the proper use and care of the PPE.
- Ensure that employees use the PPE appropriately.
- Determine the limitations of the PPE.

OSHA 1910.133 – PPE: Eye and Face Protection

The OSHA Personal Protective Equipment (PPE) – 29 CFR 1910.133 standard is intended to prevent worker exposure to potentially hazardous substances through the use of appropriate eye or face protection. The PPE will establish a barrier between the hazardous substance and the individual's eyes and face.

OSHA 1910.134 – PPE: Respiratory Protection

The OSHA Personal Protective Equipment (PPE) – 29 CFR 1910.134 standard is intended to prevent worker exposure to potentially hazardous substances through the use of appropriate respirators when they are deemed necessary to protect the health of the worker.

OSHA 1910.138 – PPE: Hand Protection

The OSHA Personal Protective Equipment (PPE) – 29 CFR 1910.138 standard is intended to prevent worker exposure to potentially hazardous substances through the use of appropriate hand protection when workers' hands are exposed to hazards such as absorption of chemicals through the skin, punctures, chemical burns, temperature extremes, and thermal burns.

Summary

OSHA requires that employees be provided with a safe and healthy workplace that is reasonably free of occupational hazards. However, process technicians can be exposed to chemical, biological, physical, and ergonomic hazards inherent to working in a process facility. Although every effort is made to prevent accidents and emergencies, they can still occur on the job because a process industry worksite can pose a variety of potential hazards.

Company policies and procedures determine whether process technicians can provide first aid to victims. Often, an emergency-response team (or first responders) and/or a resident medical staff will handle emergencies and accidents, including first-aid treatment. At the very least, the process technician is required to report emergencies and accidents in a timely way to the proper authorities. Your company will train you on how to report emergency situations and accidents, along with other basic emergency procedures (potentially including first-aid training).

To prevent or minimize hazards, government and the process industries have implemented engineering controls, administrative controls, and the use of personal protective equipment (PPE). Process technicians must understand the proper selection, use, care, and maintenance of PPE.

Checking Your Knowledge

1. Define the following terms:
   a. Administrative controls
   b. Biological hazard
   c. Chemical hazard
   d. Engineering controls
   e. Hazardous agent
   f. Physical hazard
   g. PPE
2. Name two types of eye protection that can prevent eye injuries.
3. Describe several types of situations that can result in impact injuries.
4. _____ resistant clothing can be worn to minimize the hazards of burns.
5. Name at least three signs of electrical shock.
6. List four types of PPE that can be used to minimize exposure to chemical hazards.
7. Describe eight different types of PPE.
8. Describe at least five limitations of PPE.
9. _____ is when hazardous materials or other forces (sunlight, heat, moisture) break down the PPE properties through contact over time.
   a. Penetration
   b. Contamination
   c. Permeation
   d. Degradation

10. Describe the two main types of hearing protection.
11. *(True or False)* Safety glasses or goggles can be worn with a face shield.
12. What class of hard hat provides impact and penetration resistance along with limited voltage protection (up to 2,200 volts)?
    a. A
    b. B
    c. C
    d. D
13. *(True or False)* Hard hats have a date stamp, after which they should be replaced.
14. Define the acronym TECP.
15. List four types of chemical and liquid-resistant glove materials.
16. *(True or False)* Safety shoes or boots must meet ANSI minimum compression and impact performance standards.
17. Eyewash stations and _____ showers can provide on-the-spot decontamination using a large quantity of water or other flushing fluid.
18. Which of the following PPE levels provides the highest amount of protection?
    a. A
    b. B
    c. C
    d. D
19. Which of the following is addressed by OSHA regulation 1910.132?
    a. Protective personal equipment
    b. Eye and face protection
    c. Respiratory protection
    d. Hand protection

Student Activities

1. Research at least three different types of PPE that protect the body. Make a list, describing the PPE, limitations, proper fit, use, and care and maintenance requirements.
2. Obtain at least three different types of liquid-resistant gloves, then use them to perform various tasks. After repeated use, perform a leak test (place water in the glove, roll the top toward the fingers, and then check for leaks). Did of the any gloves leak? If so, what were they made of?
3. Using at least three different types of PPE (e.g., safety glasses, hearing protection, hard hat), correctly adjust the PPE so it can be worn together comfortably. Demonstrate the process to your fellow students.
4. Attend a course and learn how to properly perform CPR or first aid. Write a three-page paper describing the value of your experience.

23

Monitoring Equipment

Objectives

This chapter provides an overview of monitoring equipment found in the process industries.

After completing this chapter, you will be able to do the following:

■ Explain the function and purpose of testing equipment found in local plants:

LEL/O_2 meters

Gas-detection equipment

Personal monitoring devices

Detector tubes

■ Use an LEL/O_2 meter to test a confined space prior to entry.

■ Describe government regulations and industry guidelines that address usage and permitting of monitoring equipment.

Key Terms

Curb—a method of spill containment for process units. It is sloped to a sewer collection system.

Deluge system—a system of dumping or spraying water to extinguish a fire or suppress a toxic release or hydrocarbon spill.

Dike—a trenchlike structure built around storage tanks to hold the contents of the tanks in the event of a leak.

Explosion suppression barrier—a wall or other device built around a specific process area or the control room to suppress explosions.

Fugitive-emissions monitoring—measuring emissions from sources (e.g., flange connection around packing) to determine if any hazardous agents are present at unacceptable EPA levels.

Gas-detection equipment—equipment that detects the components within a gaseous atmosphere.

Gas-detector tube—an instrument that provides instant measurement of a specific gas, usually indicated by a color change within the tube.

LEL meter—Lower Explosive Limit (LEL) meters; measures the combustible content of an atmosphere.

Personal monitoring device—an instrument that measures an individual's exposure to airborne contaminants.

Process upset controls—devices and systems used in the process industries to respond to and control upsets.

Relief valve—a safety device designed to open if the pressure of a liquid in a closed vessel exceeds a preset level.

Shutdown devices—equipment that ensures that processes and/or equipment are secured from failures that may result in hazardous conditions.

Spill containment—a combination of procedures and devices designed to contain spilled materials in the immediate area of the spill and minimize the effects of a spill on the environment.

Introduction

Testing equipment has a wide variety of applications in an operating facility. Detecting leaks, determining the safety of a work environment, or personally measuring exposure to hazardous chemicals are some of the practical ways in which testing equipment is used.

There are several types of testing equipment commonly used in most operating facilities. They include the following:

- LEL/O_2 meters
- Gas-detection equipment
- Personal monitoring devices
- Detector tubes

LEL Shutdown

A Lower Explosive Limit **(LEL) meter** measures the LEL of a gas in air. LELs are usually expressed as a percentage. For example, a 10% LEL reading means the sample contains 10% of the gas required for combustion. A 100% LEL reading means the sample contains enough flammable gas for combustion. Table 23-1 lists some examples of upper and lower explosive limits.

TABLE 23-1 Examples of Lower and Upper Explosive Limits (LELs and UELs)		
Substance	*Lower Explosive Limit (LEL)*	*Upper Explosive Limit (UEL)*
Acetone	3%	13%
Acetylene	2.5%	100%
Benzene	1.2%	7.8%
Butane	1.8%	8.4%
Ethanol	3%	19%
Ethylene	2.7%	36%
Diesel fuel	0.6%	7.5%
Gasoline	1.4%	7.6%
Hydrogen	4.1%	74.8%
Kerosene	0.6%	4.9%
Methane	5%	15%
Propane	2.1%	9.5%

Lower Explosive Limit (LEL) meters are used to detect leaks in work areas where flammable substances are present. LEL monitors work by drawing a sample of the surrounding atmosphere, heating it, and reading the heated sample to determine the LEL of the atmosphere.

These devices have two distinct components. The LEL portion of a monitor measures the concentration of flammable gas in the sample by one of three methods:

- Oxidation—the detector measures readings based upon the heat that is released when the combustible gas or vapor is burned.
- Metal oxide semiconductor (MOS)—the detector absorbs the combustible gas, causing a change in the electrical conductivity of the sensor, which produces a reading.
- Thermal conductivity—the detector measures readings due to a change in thermal conductivity of the atmosphere in the presence of a combustible gas.

O_2 METERS

O_2 meters measure the oxygen content of the atmosphere. Air normally contains 21% oxygen. Oxygen levels below 19.5% can become dangerous, and an oxygen content below 14.5% can be fatal.

O_2 meters are used to monitor oxygen levels in work areas where flammable substances are present and monitor oxygen levels in work areas where breathing air may be limited (especially in confined-space entry). The O_2 section of a monitor measures the level of oxygen in the sample. Following are the two most commonly used types of O_2 detector cells:

• Coulometric detectors use a coulometric cell to measure oxygen. The coulometric cell functions by selectively allowing oxygen molecules to pass through it. When the sample passes through the cell, any contaminants that are present will produce a reaction that creates an electric current. This current passes through an electrolyte between two electrodes, producing an oxygen measurement.

• Polargraphic detectors are used to measure oxygen and carbon monoxide in circulating air.

GAS-DETECTION EQUIPMENT

Gas-detection equipment detects the components within the gaseous atmosphere. The equipment can either be portable (e.g., a hand-held device) or it can be stationary (e.g., at remote monitoring stations). Even though stationary detectors are used at plants, portable hand-held detectors are most commonly used in the process industries. Types of gas-detection equipment include the following:

• Mercury vapor monitors use either ultraviolet analyzers or direct-reading instruments to determine the amount of mercury vapor contamination in ambient air.

• Direct-reading colorimetric tubes and badges change color due to a reaction with airborne contaminants.

FIGURE 23-1 Mercury Vapor Monitor

• Flame ionization detectors read ionized particles of chemical compounds when exposed to an air or hydrogen flame. The detector collects the ionized particles and charges them with a current to read the concentration of the contaminant.

• Photonionization detectors read particles of chemical compounds that have been ionized due to exposure to ultraviolet light. The detector collects the ionized particles and produces an electric current that the detector reads.

• Electron capture detectors read the current flowing from a chemical compound exposed to a small dose of a radioactive element.

• Thermal conductivity detectors detect changes in thermal conductivity of the atmosphere in the presence of a combustible gas to identify contaminants in the atmosphere.

• Infrared analyzers detect various gases and vapors based on the amount of infrared radiation that they absorb.

FIGURE 23-2 Pump and Tube Monitoring Device

FIGURE 23-3
Photonionization Detector

• Photoacoustic spectrometers detect pressure changes in the instrumentation cell caused by the heated expansion of gas contained within the spectrometer.

• Ultraviolet analyzers work in a similar manner to that of infrared analyzers but detect chemicals that become charged by ultraviolet radiation absorption.

• Gas chromatographs separate volatile organic compounds through a multi-phased process. The separation is based on the behavior of the compounds within the chromatograph.

FIGURE 23-4 Gas Chromatograph

• Ion mobility spectrometers read particles of airborne contaminants ionized by exposure to small doses of a radioactive element.

• Particulate monitors directly read aerosol concentrations or the behavior of charged crystals, depending upon the type of particulate monitor.

PERSONAL MONITORING DEVICES

A **personal monitoring device**, or dosimeter, is used to measure an individual's exposure to airborne contaminants over a short period of time (usually between eight and 12 hours). The device is small so that it can be attached to the worker's clothing within the worker's breathing zone. Usually, the dosimeter is attached to the lapel, where it is closest to the nose and mouth.

Following exposure, the monitoring device is sent to a lab for analysis to determine the amount of contaminant collected.

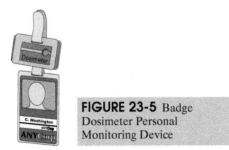

FIGURE 23-5 Badge Dosimeter Personal Monitoring Device

GAS DETECTOR TUBES

Gas detector tubes are easy to read and highly portable. Each tube is designed to detect a specific substance (e.g., chlorine, benzene, and carbon monoxide). When exposed to the target substance, the tube produces a distinct color change, indicating the present of the substance and its concentration.

Alarm Systems and Indicators

Alarms and indicators are a vital part of plant safety. These systems are designed to alert everyone in the facility or work area should something go wrong and/or notify those in control that a potentially dangerous situation exists that requires attention. Several types of alarm systems and indicators are used in the process industries, including the following:

- Fire alarms and detection systems
- Toxic gas alarms and detection systems
- Redundant alarm and shutdown systems
- Automatic shutdown devices
- Interlocks

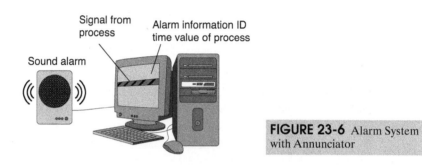

FIGURE 23-6 Alarm System with Annunciator

FIRE ALARMS AND DETECTION SYSTEMS

Fire alarms and detection systems are installed throughout manned process areas and in buildings that are not normally staffed by individuals, such as those that house electrical or computer equipment. These systems can sound audible alarms and provide alarm indication in the control room to alert operations personnel.

In conjunction with an alarm system, other components that may also be activated during an emergency include water sprinklers, deluge systems, halon systems, or carbon dioxide systems. Water is used to extinguish standard fires and cool down hot equipment. Halon is used around computer systems, telephone switch rooms, and other electrically sensitive equipment. Carbon dioxide is used for fires involving electrical and electronic equipment.

Detection devices that are associated with fire alarms include smoke detectors, temperature switches, and LEL detectors. Each device is activated when a predetermined level, such as a high temperature or excessive amount of carbon monoxide, is reached.

Did You Know?

Cameras are highly restricted in plants because of proprietary information and the fact that the flash can interfere with fire-detection sensors or ignite hydrocarbons.

Because of this, process technicians should always ask permission before bringing a camera into a process facility or taking flash photography.

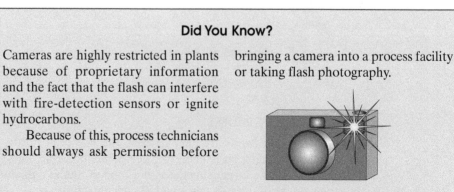

Toxic-Gas Alarms and Detection Systems

Toxic-gas alarms and detection systems are used for processes that involve toxic materials, such as hydrogen sulfide, carbon monoxide, and cyanide. Toxic-gas detection devices are installed in strategic locations throughout the process area and sound when a predetermined concentration of the toxic substance is detected. The devices are located in such a manner that a leak will be detected no matter which direction the wind is blowing.

In some circumstances, a water or foam deluge system will be activated in order to suppress the toxic gas.

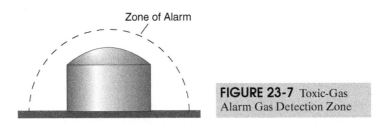

FIGURE 23-7 Toxic-Gas Alarm Gas Detection Zone

REDUNDANT ALARM AND SHUTDOWN SYSTEMS

Redundant devices are used to provide multiple means for warning. **Shutdown devices** ensure that processes or equipment are secured from failures that may result in hazardous conditions.

The need for redundancy is based on the criticality of the process and how devastating a major event would be to the process. Redundancy is provided through two or more separate switches or detection devices and two or more separate alarms. The redundancy may consist of wired electrical devices along with a computer control system that uses programming or software to activate an alarm or warning. Often, relief devices are incorporated along with shutdown devices in redundant systems.

AUTOMATIC SHUTDOWN DEVICES

Automatic shutdown devices are used to protect equipment and/or personnel when process variables exceed or drop below preset levels. Automatic shutdown devices may be activated by variables such as the following:

- Temperature
- Pressure
- Flow
- Level
- Composition (on-line analyzers)

High alarms (shown in Figure 23-8) are triggered when a process variable, such as fluid in a tank, rises above a pre-determined high level. Low level alarms (shown in Figure 23-9) are triggered when a process variable drops below a pre-determined low level.

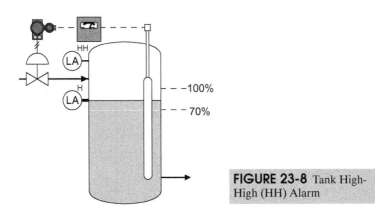

FIGURE 23-8 Tank High-High (HH) Alarm

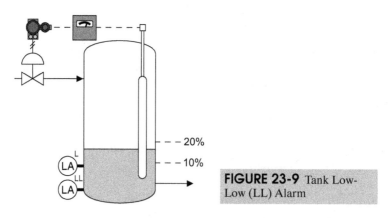

FIGURE 23-9 Tank Low-Low (LL) Alarm

When levels exceed preset thresholds, shutdown devices may automatically close valves in order to reduce the flow or shutdown the system (e.g., shutting down a pump to prevent the tank from overflowing). When levels drop below preset levels, a shutdown device may automatically open valves to increase flow or shutdown the system (e.g., automatically shutting down a pump to prevent the pump from running dry).

INTERLOCKS

Interlocks prevent an action unless a certain condition within the process is satisfied. For example, an interlock could shut down or start up equipment to prevent equipment damage or serious process upset.

There are usually two types of interlocks used in the process industries:

• Safety—this type of interlock circumvents a process. If a condition or requirement is not satisfied, then the entire process is shut down (e.g., an interlock switch on the lid of a washing machine prevents the machine from agitating if the lid is open).

• Process—this type of interlock is built into a process, meaning that if certain requirements in the process are not fulfilled, the process will not continue to the next step (e.g., a washing machine will not begin the agitation step until the wash water has reached a preset level).

Process Upset Controls

There are a number of **process upset controls** or devices or systems used in the process industries to control upsets. These include the following:

• Flares
• Pressure-relief devices
• Deluge systems
• Explosive-suppression systems

Did You Know?

Process technicians often use the slang term "pop valve" or "pop-off valve" to refer to a safety valve.

That is because they open quickly or "pop off" once the pressure threshold has been exceeded (as opposed to relief valves that open slowly).

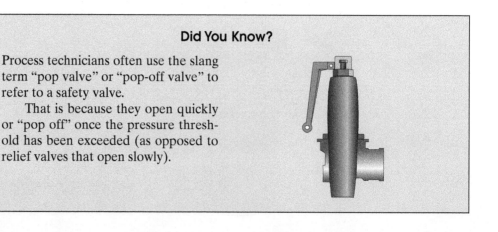

- Explosive-gas alarms and detectors
- Spill
- Containment

FLARES

Flares are environmentally approved devices which burn waste gases that have been collected from various process sources via a piping system. During process upsets, the process may generate pressure. Gases are released to reduce pressure, but instead of releasing to the atmosphere, the gases are piped into a flare. The gases are burned and released into the atmosphere as carbon dioxide (or carbon dioxide and water), thereby controlling the process upset.

A flare on a process unit is normally elevated on a tower several hundred feet in the air due to the heat that can be generated. Flares must be designed to provide smokeless burning of the gases, which is usually accomplished with the use of steam addition during the combustion process.

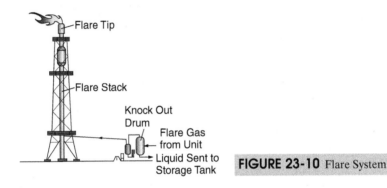

FIGURE 23-10 Flare System

PRESSURE-RELIEF DEVICES

Pressure-relief valves are valves that automatically open at a set pressure to protect vessels or piping from over-pressuring. Pressure-relief devices do not rely on external sensing devices. Instead, they rely on spring tension settings adjusted to a set pressure.

Pressure-relief devices are inherently very reliable. For example, the spring tension of a relief valve may be set at 50 psi. If the pressure in the process exceeds 50 psi, the valve will automatically open to relieve the pressure. Once the pressure drops below 50 psi, the valve will close and the material being released may be vented to a flare system or to the atmosphere, depending on the type of material being vented.

There are two types of pressure-relief devices: **relief valves** and safety valves. (Figure 23-11) Relief and safety valves are both similar in their design. The difference, however, is that relief valves are designed to open slowly if the pressure of a liquid exceeds a preset level, while safety valves are designed to open quickly and are used to vent gases.

Rupture disks (sometime referred to as burst disks) are another type of pressure-relief device. Rupture disks are metal disks, held in place with a vented plug, that are designed to rupture (pop out of their holder) if tank pressure is greatly exceeded.

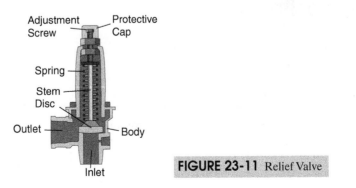

FIGURE 23-11 Relief Valve

DELUGE SYSTEMS

Deluge systems dump or spray large volumes of water over a short period of time for the purpose of extinguishing a fire or suppressing a toxic release or a hydrocarbon spill. Deluge systems may be automatic or require manual activation.

FIGURE 23-12 Deluge System

EXPLOSION-SUPPRESSION SYSTEMS

An **explosion-suppression barrier** includes walls erected to contain an explosion. The walls may be erected around a particular process area, or the walls of a control room may be built for explosion suppression.

An example of an explosion-suppression barrier is a concrete blast wall around a high-pressure reactor. There may also be "safe havens" within a process facility (e.g., a control room) that are designed to withstand the force of an explosion.

EXPLOSIVE-GAS ALARMS AND DETECTORS

Explosive-gas alarms and detectors operate similarly to toxic-gas alarms and detectors, but are designed to detect explosive gas. They will activate an alarm and may also automaticly activate a deluge system.

SPILL CONTAINMENT

Spill containment is designed to contain spilled materials in the immediate area of the spill, prevent spread and contamination, and facilitate disposal.

Dikes are used around storage tanks and are sized to contain more than the full contents of the tanks within the enclosure. **Curbs** are used to provide containment for process units. The containment area is sloped to a process sewer collection system. Other containment devices (e.g., sandbags to protect storm sewers) may also be used.

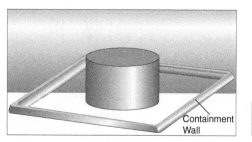
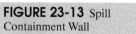
Containment Wall

FIGURE 23-13 Spill Containment Wall

Industrial-Hygiene Monitoring

Industrial-hygiene monitoring, which is conducted by an industrial hygienist, involves sampling the working environment to determine if any hazardous agents are present at unacceptable OSHA (Occupational Safety and Health Administration) levels. Industrial hygienists use a variety of tools to capture data on workplace hazards, and then analyze these data and compare them to published standards. Some of the data industrial hygienists monitor are taken from personal monitoring equipment (e.g., badge and pump dosimeters). Hygienists use these data to determine how much of a substance workers are exposed to.

Common industrial hygiene sampling activities include the following:

- Noise monitoring
- Toxic substances sampling
- Ergonomic studies

Fugitive-Emissions Monitoring

Fugitive-emissions monitoring requires that samples be taken, analyzed, and compared to government or company standards. Fugitive emissions can occur at many scattered points (e.g., valve packing, pump seal, or a gasket on a flange) that are not caught by a capture system. However, emissions monitoring involves sampling of the atmosphere to determine if any hazardous agents are present at unacceptable EPA (Environmental Protection Agency) levels. Also, emissions monitoring can be performed either by in-line sensors or by people with detectors. You may be asked to participate in fugitive-emissions monitoring by being trained to operate a gas-detection device and collect data or collect readings from in-line sensors.

Government Regulations and Industry Guidelines

On-site air monitoring is required by OSHA's air contaminant standard and standards specific to some chemicals (such as benzene, ethylene oxide, asbestos) to identify and quantify levels of hazardous substances. This monitoring is performed periodically to ensure the proper protective equipment is used on site. For more information on the OSHA standards, refer to their Web site at www.osha.gov.

The EPA requires monitoring of fugitive emissions through gas-detection devices. Companies are required to monitor these emissions to ensure that hazardous agents are detected and quantified to ensure they are within the acceptable levels. For more information on the EPA requirements, refer to their Web site at www.epa.gov.

Summary

The very nature of the process industries involves the use of hazardous materials. These materials must be monitored and controlled to ensure physical and environmental safety. Monitoring equipment is used to identify unacceptable hazards so that these hazards can be mitigated or controlled and employees can be alerted.

There are many different types of monitoring equipment and systems. These types include Lower Explosive Limit (LEL) meters, O_2 meters, gas detection equipment, personal monitoring devices, alarm and indicator systems, automatic shutdown devices, and process upset controls.

Lower Explosive Limit (LEL) meters measure the combustible content of atmosphere. O_2 meters measure the oxygen content of the atmosphere. Gas-detection equipment detects the components within a gaseous atmosphere. Personal monitoring devices (dosimeters) are used to measure an individual's exposure to airborne contaminants. Alarm and indicator systems (e.g., fire and toxic gas alarms) are designed to alert everyone in the facility if something goes wrong and notify those in control that a potentially dangerous situation exists. Automatic shutdown devices are used to protect equipment and/or personnel when process variables exceed or drop below preset levels. Automatic shutdown devices may be activated by variables such as temperature, pressure, flow, level, and composition (on-line analyzers). Process-upset controls are devices and systems used in industry to respond to and control upsets.

Process monitoring is essential for health and safety. Monitoring of the process and its environment (industrial-hygiene monitoring) is performed to ensure the workplace is safe for employees and for the surrounding communities.

Checking Your Knowledge

1. Define the following terms:
   - a. Alarms and indicators
   - b. Automatic shutdown device
   - c. Deluge system
   - d. Explosion suppression barrier
   - e. Explosive-gas alarms and detectors
   - f. Flares
   - g. Interlocks
   - h. Personal monitoring device
   - i. Pressure-relief valve
   - j. Shutdown device

2. Explain the purpose of Lower Explosive Limit (LEL) oxygen meters.

3. List and describe the two most commonly used oxygen detector cell types.

4. Explain the purpose of a personal monitoring device (dosimeter) and why you would want a personal monitoring device (i.e., what kinds of things could be monitored).

5. Which of the following fire suppression substances might be used for fires involving computer systems and other sensitive electronics?
   - a. Halon
   - b. Water
   - c. Carbon dioxide
   - d. A and C
   - e. All of the above

6. Redundant (repeated or duplicated) equipment is often incorporated into critical systems like those used for fire detection. Explain why this redundancy is important.

Match the process upset control device to the appropriate description.

Process Control Device	Description
7. Deluge system	a. A building or walls designed to contain the impact of an explosion
8. Explosion suppression system	b. A flame-containing device used to burn waste gases during process upsets
9. Explosive gas detector and alarm	c. A safety valve that automatically opens if the pressure of a vessel exceeds a preset threshold
10. Flare	d. A system that delivers a high volume of water in order to extinguish a fire or suppress a toxic gas leak
11. Pressure relief device	e. Dikes or curbs used to contain a spill and prevent it from spreading
12. Spill containment device	f. Used to detect explosive gases and may also trigger the activation of a deluge system

13. Compare and contrast industrial-hygiene monitoring and fugitive-emissions monitoring.

Student Activities

1. Given a Lower Explosive Limit (LEL) monitor, determine the LEL or gas contents of the atmosphere around you.

2. On most washing machines, there is a switch that will only allow the tub to spin if the lid is closed. This is an example of a safety interlock. Examine the world around you and identify five additional examples of commonly used interlocks. In a one-page report, list and describe the function of each of these interlocks.

3. Write a one to two-page report explaining the purpose and function of flares.

4. Given a bottle of ammonia and an ammonia-detector tube, sample the air near the mouth of the ammonia bottle and then read the detector tube to determine the concentration of ammonia in the air.

5. Given an O_2 meter, determine the oxygen level in the air around the exhaust of a running car (Caution: carbon monoxide is extremely hazardous; do not stand in or near the exhaust!).

6. Observe how confined spaces can become oxygen-deficient environments by placing a piece of iron scrap metal and some leaves in a gallon jar. Add enough water to barely cover the leaves and the metal. Seal the jar and leave it sealed for one week. At the end of that week, use an O_2 meter to monitor the oxygen level inside the jar.

Fire, Rescue, and Emergency Response Equipment

Objectives

This chapter provides an overview of fire, rescue, and emergency response equipment found in the process industries.

After completing this chapter, you will be able to do the following:

- Indicate the function and purpose of fire, rescue, and emergency response equipment typically found in the process industries:

  Emergency response

  Fires (different types)

  Spills

  Fire and rescue retrieval

  Escape

  Chemical exposure

- Demonstrate the correct use of a safety shower and eyewash station.
- Demonstrate the proper selection and use of fire extinguishers.
- Participate in a tabletop drill in preparation for response to a fire, release, or spill.
- Describe government regulations and industry guidelines that address fire protection and emergency response.

Key Terms

Ambulance—a standard vehicle of emergency medical service departments used to treat injury victims at remote sites and transport them to other medical facilities as needed.

Backboard—a board used to immobilize injured victims to prevent damage to the spinal cord while they are being moved.

Chemical protective clothing (CPC)—clothing that provides protection to those who work around chemicals. Four classifications of CPC allow employees to access the proper level of protection for the chemicals to which they may be exposed.

Class A fire—the classification given to ordinary combustible materials, such as wood, paper, cloth, and rubber.

Class B fire—the classification given to flammable liquids, such as greases, tars, oils, and lacquers.

Class C fire—the classification given to electrical equipment fires.

Class D fire—the classification given to combustible metals, such as magnesium, lithium, and sodium.

Escape mask—protective gear that provides respiratory protection for a limited period of time in the event of a suddenly occurring emergency.

Extinguishing agent—the material in a fire extinguisher (e.g., water, dry chemical, and foam); varies depending on the fire classification being fought.

Eyewash station—used to flush the eyes of contamination from harmful chemicals.

Fire brigade—a private fire department or industrial fire department composed of a group of employees who are knowledgeable, trained, and skilled in basic firefighting operations.

Fire engine—a vehicle that carries and pumps water.

Fire hose—used to deliver large quantities of water when the fire is too large to be controlled by a fire extinguisher or other means.

Fire hydrant—a point of water supply for fire hoses; mounted on trucks or freestanding and connected to a water main.

Fire truck—an emergency response vehicle that does not carry or pump water.

Handheld fire extinguisher—used to extinguish small fires; must be mounted, located, and identified so it is easily accessible during an emergency.

Incipient fire—a small fire that requires immediate attention to prevent it from becoming larger and more serious.

OSHA fire protection standard (29 CFR 1910.155)—intended to prevent and or minimize the consequences of fire within the workplace.

Ropes and mechanical retrieval devices—equipment used to rescue workers from a hazardous situation. Ropes are primarily used to lower rescuers into confined spaces and to hoist injury victims out of those spaces. Mechanical retrieval devices are set up on a tripod and use a winch or crank to lower and raise individuals with the rescue rope.

Safety shower—located throughout the plant; provides the ability to rinse the body off in the event of a large exposure to a chemical spill.

SCBA—an apparatus used to provide breathing air when the wearer needs to enter an area with an atmosphere that might be oxygen-deficient or harmful if inhaled.

Standpipe system—a system that provides hose stations with water supplied by a main, fire pump, or tank; standpipes can be either wet (connected to a permanent water supply and constantly filled with water) or dry (not connected to a permanent water supply).

Stretcher—a bed with wheels used to transport injured victims from one place to another.

Introduction

Regardless of the safety systems and equipment used in the process industries and the amount of training that is provided for employees, emergencies sometimes occur. The response to those emergencies can often mean the difference between life and death. Therefore, the response to emergencies must be efficient and timely.

There are several types of emergencies that require an efficient and timely response in the process industries. They include fires, leaks, spills, and releases.

Fires

Incipient fires can be classified as small fires that need immediate attention before they get larger and more serious. Fires are classified by the types of materials or equipment involved. The four main classes of fires are A, B, C, and D. However, there is also a Class K, which is specific to the food service and food preparation industries.

- **Class A** — ordinary combustibles (e.g., wood, paper, cloth, and rubber)
- **Class B** — flammable liquids (e.g., greases, tars, oils, and lacquers)
- **Class C** — electrical equipment
- **Class D** — combustible metals (e.g., magnesium, lithium, and sodium)
- **Class K** — cooking oil, fat, grease, or other kitchen fires

FIGURE 24-1 Five Classes of Fire

Class "A"   Class "B"   Class "C"   Class "D"   Class "K"

Leaks, Spills, and Releases

When process fluids are not handled properly, they can create dangerous situations for the process technician. When toxic process fluids manage to escape the confines of their containers, other dangerous situations are created.

Leaks, spills, and releases all involve the uncontrolled discharge of hazardous process fluids. A leak occurs when a container is compromised and the compromised area allows a small amount of liquid to escape. A spill is similar to a leak in that it involves the uncontrolled liquid discharge, but a spill usually involves more volume than a leak. A release refers to the uncontrolled discharge of process materials into the environment.

Leaks, spills, and releases can cause adverse consequences for both the worker and the environment. Workers can accidentally inhale toxic vapors, absorb toxic chemicals into their body, or have other types of acute exposure to lethal doses of toxic substances. The environment can be harmed if toxic materials are released into the air,

FIGURE 24-2 Soil Pollution Caused by a Spill

leak, spill, or seep into the soil offsite or into groundwater. Because of this, leaks, spills, and releases must be reported. The RQ is the Reportable Quantity listed for any substance having an EHS RQ or a CERCLA RQ. This includes a loss to the environment of 1 lb or more from a RECRA storage tank and associated piping within a 24-hour period. States also have their own individual reporting requirements.

Fire and Rescue

Sometimes, fires become too large to be controlled by one or two individuals. When this is the case, more experienced and highly trained firefighters are needed. Process industry plants may rely on the city fire departments, internal fire brigades, or a combination of both.

FIRE BRIGADES

Fire brigades are private fire departments or industrial fire departments composed of an "organized group of employees who are knowledgeable, trained, and skilled in at least basic firefighting operations." (29 CFR 1910.155).

Employers who choose to form fire brigades are responsible for ensuring they select competent members, train them, and provide them with the equipment and protective clothing necessary to fight fires. Employers must select employees who are capable of performing the firefighting duties to which they are assigned. They may not select employees with heart disease, epilepsy, emphysema, or other serious medical conditions to participate in fire brigades. Once selected, employers must provide training commensurate with the assigned duties of each fire brigade member. Training for general fire brigade members must be provided annually, and employees may be required to pass a physical fitness test. Anyone assigned to fight fires within a structure must receive quarterly training.

Employers are required to inform fire brigade members about any special hazards to which they may be exposed when responding to emergencies. The brigade members must be trained on special actions required during emergencies to minimize hazards.

Employers must also provide, maintain, and inspect the firefighting equipment onsite. Protective clothing must include protection for the feet, legs, hands, body, eyes, face, and head.

To help fight fires, fire brigades may use the following types of equipment:

- Fire hoses
- Standpipe systems
- Fire hydrants
- Emergency vehicles

FIGURE 24-3 Fire Brigades Consist of Employees Trained in Basic Fire-Fighting Operations

Fire Hoses

Fire hoses are one of the most common firefighting devices used to deliver large quantities of water when the fire is too large to be controlled by other means (e.g., fire extinguishers).

FIGURE 24-4 Fire Hose

Fire hoses come in different sizes. The larger the hose, the greater the volume and force the hose will deliver. Because large fire hoses deliver such huge amounts, they can be very difficult to maneuver and control, and may require several people to operate them.

Standpipe Systems

Standpipe systems provide hose stations with water supplied by a main, fire pump, or tank. There are three classifications of standpipes.

- Class 1 are for the largest hoses (2 ½-inch outlets) and are used by trained firefighters.
- Class 2 are equipped with 1½-inch outlets and are used primarily by the occupants of the plant until firefighters arrive.
- Class 3 provide a combination of Class 1 and 2 standpipes.

There are two types of standpipe systems:

- Wet standpipes are connected to a permanent water supply and are constantly filled with water.
- Dry standpipes are not connected to a permanent water supply.

Fire Hydrants

Fire hydrants serve as a point of water supply for fire hoses. Hydrants can be mounted on trucks or be freestanding and connected to a water main.

Fire monitors are strategically located spray nozzles, attached to a hydrant or some other water supply. Fire monitors can be directed at a fire and locked in place so a technician can exit the area while still fighting the fire and limiting its damage. Figure 24-5 shows an example of a fire monitor.

FIGURE 24-5 Fire Monitor Attached to a Fire Hydrant

Emergency Vehicles

There are several types of emergency vehicles used in plants. These include fire engines and trucks, and ambulances.

FIGURE 24-6 Ambulance

Fire brigades and fire departments, for the most part, are equipped with fire engines and fire trucks. While they are usually customized to the needs of the specific fire brigade or department, there are a variety of fire rescue vehicles that are utilized by fire brigades. The most common fire rescue vehicles are pumper and ladder trucks. A **fire engine** is a vehicle that carries and pumps water, while a **fire truck** does not carry or pump water.

Ambulances are the standard vehicles for emergency medical service departments. They function as mobile medical units to help treat injury victims at remote sites and transport them to other medical facilities if needed. Ambulances are staffed by trained emergency medical personnel and are well-equipped to aid injured parties in the event of an emergency.

FIRE EXTINGUISHERS

Handheld fire extinguishers are primarily used to extinguish small incipient fires. These must be regularly inspected, tested, and maintained, and employers are required to train employees in their proper use. All portable fire extinguishers must be mounted, located, and identified so that employees may easily access them when needed. They must be maintained in a fully charged and operable condition and remain in their designated place at all times except when in use.

Employees are required to have annual training in the use of fire extinguishers. The extinguishers must also undergo an annual maintenance check. Any abnormalities must be corrected, or the extinguisher must be replaced. While extinguishers are being checked and/or replaced, employers are required to provide alternate equivalent protection.

FIGURE 24-7 Handheld Fire Extinguisher

Extinguishing Agents

Extinguishers come equipped with different **extinguishing agents** depending upon the fire classification. A fire extinguisher is required to be clearly marked as to what type and size of fire it can extinguish.

Several extinguishing agents are used, depending on the type and size of fire. The following chart (Table 24-1) identifies the types of agents used for the various fire classifications.

Process technicians must be aware that some extinguishing agents have hazards (e.g., halon can create an oxygen-deficient environment, while CO_2 extinguishers can displace oxygen and produce static electricity — another potential ignition source).

TABLE 24-1	Fire Extinguishing Agents and the Fire Classifications They are Used For
Extinguishing Agent	*Fire Class(es)*
Water	A, B
Dry Chemical	A, B, C, D
Foam	A, B
Halon	A, B, C
Carbon Dioxide	A, B, C
Wet-Chemical	A

Fire Extinguisher Use

When using a fire extinguisher, it is imperative that employees follow general rules to ensure their own safety, as well as the ability to extinguish the fire as quickly and efficiently as possible. Whenever a fire extinguisher is required, technicians should contact the emergency response team at the facility to ensure that help is on the way in the event the fire is greater than what a technician can extinguish.

RESCUE EQUIPMENT

In the process industries, various types of emergencies can occur. During these emergency situations, it is sometimes necessary to rescue workers from a hazardous situation. To help aid in retrieval, rescue workers use various types of equipment.

Ropes and Mechanical Retrieval Devices

Ropes come in a variety of materials and strengths. In rescue situations they are primarily used to lower rescue workers in confined spaces to help injured victims and/or to hoist victims out of those spaces.

To accomplish vertical retrieval, ropes may be used in conjunction with rigging mechanisms (e.g., harnesses, bolts, bolt hangers, karabiners, and other types of clips) and mechanized retrieval devices.

Mechanized retrieval devices are usually set up on a sturdy tripod and have a wench or crank with which to lower and raise individuals into vertical confined spaces via a rescue rope.

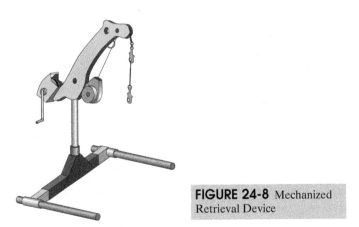

FIGURE 24-8 Mechanized Retrieval Device

Stretchers and Backboards

Backboards are used to immobilize injured victims to prevent damage to the spinal cord while they are being moved. They usually consist of a rigid backboard of some type and restraints to hold the patient's head, torso, and limbs in place while in transit. **Stretchers** are beds with wheels that are used to transport patients from one place to another.

Did You Know?

The following mnemonic is an easy way to remember the classes of fire extinguishers and what they are used for:

Type:	Used for:
A (ashes)	Solids
B (bucket)	Liquids
C (current)	Electrical equipment
D (difficult)	Metals
K (kitchen)	Grease/cooking oils

Breathing Apparatuses

As discussed in the chapter dealing with personal protective equipment (PPE), respirators and breathing apparatuses come in different types and forms.

Most rescue teams use **SCBA** (self-contained breathing apparatus) to enter into confined spaces or fires. **Escape masks** serve as respiratory protection and allow potential victims to quickly escape in the event of a suddenly occurring emergency. Some escape masks (or hoods) can be used in conjunction with air supplies that allow for five to 15 minutes of breathing time.

FIGURE 24-9 Person Donning a Self-contained Breathing Apparatus (SCBA)

CHEMICAL EXPOSURE EQUIPMENT

Exposure to hazardous chemicals is an emergency that needs immediate attention. To help respond to such emergencies, plants have various types of safety equipment that are easily accessible. Some of the equipment used to attend to such emergencies is listed below.

Eyewash Stations

Eyewash stations are used to flush the eyes of contamination (e.g., harmful chemicals that have entered the eyes as a result of a splash). This helps reduce the risk of injury associated with chemical burns.

Eyewash stations can either be personal portable units or permanent fixtures. Eyewash stations should be placed in strategic locations that are easily accessible to all employees working around chemical agents.

FIGURE 24-10 Eyewash Fountain

Process technicians should always note the location of the eye-wash station when entering a work area, since it may be difficult to find the station once something has been sprayed in the eyes.

Safety Showers

Safety showers are used to rinse the body off in the event of a large exposure to a chemical spill. Like eyewash stations, safety showers are strategically located within the plant so they can be easily accessed by employees.

Safety showers may require technicians to pull a chain or step on a grate or footplate to activate them. Once activated, the safety shower releases water and may activate an alarm in the control room.

Process technicians should always note the location of the safety shower station when entering a work area, so they can get to it quickly in the event of prolonged exposure to hazardous agents.

Safety showers and eyewash stations must be checked periodically to ensure proper flow and maintained so they are free of debris.

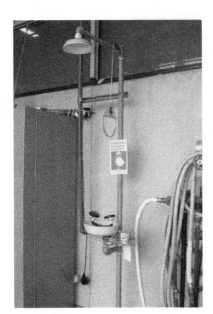

FIGURE 24-11 Safety Shower with Pull Chain

Decontamination

During a chemical spill, equipment can also become contaminated, so it is standard practice to wash the chemical(s) off the equipment to decontaminate it. Some chemicals may need other treatments to complete the decontamination process. What treatments need to be applied depends upon the chemical agent and how it reacts to water or the surrounding atmosphere (e.g., some chemicals, if mixed with water or air, become highly volatile).

Chemical Protective Clothing (CPC)

As discussed in the chapter on PPE, **chemical protective clothing (CPC)** is essential to the health and safety of those who work around chemicals. CPC are classified based on the level of protection they provide.

• Level A—gives the highest level of protection for the skin, eyes, and respiratory system. It is used when there is a high concentration of vapors, gases, or particulates and a high potential for splash, immersion, or exposure to unexpected or unknown toxic chemicals.

• Level B—gives the highest level of respiratory protection but less skin protection. It is used when IDLH (Immediately Dangerous to Life and Health) concentrations of substances do not represent a skin hazard and levels do not meet the criteria for air-purifying respirators, but a high level of respiratory protection is still needed (i.e., the atmosphere contains less than 19.5% oxygen or levels of substances may not be completely determined, but there is no apparent threat of harmful skin contact).

• Level C—gives the same amount of protection for skin as Level B and requires less respiratory equipment than Level A. It should be used when direct contact with atmospheric and chemical contaminants will not adversely affect the skin and levels meet criteria for air-purifying respirators, meaning the material and its airborne concentration are known and the air-purifying equipment provides adequate protection.

• Level D—gives minimal skin and respiratory protection. It should be used when the atmosphere contains no known hazards and the work situation presents no immediate threat from chemical splashes or immersion or unexpected contact or inhalation of hazardous substances.

Technicians must always select the CPC that is approved for the chemical(s) they are working with, and be aware that CPCs cannot be used for firefighting.

FIGURE 24-12 Person in a Chemical Protective Suit

Government Regulations and Industry Guidelines

Government regulations and industry guidelines affect how fire, rescue, and emergency responses are handled in the process industries. These include the following:

FIRE PROTECTION STANDARD

The **OSHA fire protection standard (29 CFR 1910.155)** is intended to prevent and/or minimize the consequences of fire within the workplace. When hazardous chemicals and other materials are present within the workplace, the potential consequences are even more dangerous than in non-hazardous locations.

Equipment and alarms must be in place to protect employees from fire. Unless employees are members of a plant's fire brigade or emergency response team, they are not typically expected to fight fires. The fire protection standard requires the use of the following:

• Fire brigades
• Portable fire extinguishers

- Sprinkler systems
- Fixed extinguishing systems
- Fire detection systems
- Alarm systems

HAZMAT EMERGENCY RESPONSE

The Department of Transportation (DOT) requires that plants transporting and/or loading or unloading hazardous materials follow specific procedures to ensure the safety of employees and of the general public. These requirements, identified in 49 CFR 172.600, list the following emergency response requirements:

- Identification of the hazardous materials
- Identification of the health hazards and risks of fire and explosion
- Documentation of the precautions to be taken in the event of an accident
- Means or methods for handling fires and spills
- Plans for first-aid measures

Summary

Despite the safety systems and equipment used in the process industries and the amount of training provided for employees, emergencies sometimes occur. The response to those emergencies can often mean the difference between life and death, so the response to emergencies must be efficient and timely.

Emergency response procedures should always be in place, and emergency response equipment should be available when it is needed. Employees should be trained on these procedures and equipment so they can respond to an emergency until professional help arrives.

Technicians involved in emergency response may use a variety of equipment. For example, during fire situations, technicians may use fire hoses, fire hydrants, fire extinguishers, ropes and mechanical retrieval devices, stretchers, backboards, and self-contained breathing apparatus. During hazardous chemical exposure, technicians may use equipment such as eyewash stations, safety showers, and chemical protective clothing (CPC).

It is important for technicians to be familiar with the different types of emergency response equipment and how to use them, as well as government and company rules and regulations governing emergency response.

Checking Your Knowledge

1. Define the following terms:
   a. Chemical protective clothing
   b. Eyewash station
   c. Fire brigade
   d. Incipient fire
   e. Safety shower
   f. SCBA
   g. Standpipe
2. List the five classes of fire extinguishers and the substances for which each class is used.
3. Which of the following involves the discharge of gas into the atmosphere?
   a. Leak
   b. Spill
   c. Release
4. *(True or False)* Fire trucks carry and pump water.
5. Explain the similarities and differences between Level A, B, C, and D chemical protective clothing.

Student Activities

1. Given a fire extinguisher, practice extinguishing a fire (or simulated fire) using proper extinguisher technique.
2. Practice proper use of a safety shower and an eyewash station.

Appendix A: Government/Regulatory and Industry Resources

American Chemical Council

www.americanchemistry.com

1300 Wilson Blvd.
Arlington, VA 22209
Phone: 703-741-5000

**American Conference of Governmental
Industrial Hygienists (ACGIH®)**

www.acgih.org

1330 Kemper Meadow Drive
Cincinnati, OH 45240
Phone: 513-742-6163

**American National Standards
Institute (ANSI)**

www.ansi.org

1819 L Street, NW
Suite 600
Washington, DC 20036
Phone: 202-293-8020

American Petroleum Institute (API)

www.api.org

1220 L Street, NW
Washington, DC 20005-4070
Phone: 202-682-8000

American Red Cross

www.redcross.org

2025 E. Street, NW
Washington, DC 20006
Phone: 202-303-5000

American Society of Mechanical Engineers

www.asme.org

Three Park Avenue
New York, NY 10016-5990
Phone: 800-843-2763

American Society for Testing & Materials

www.astm.org

100 Barr Harbor Drive, PO Box C700
West Conshohocken, PA 19428-2959
Phone: 610-832-9585

**Centers for Disease Control
and Prevention (CDC)**

www.cdc.gov

1600 Clifton Rd
Atlanta, GA 30333
Phone: 404-639-3311
Public Inquiries: 800-311-3435

Coast Guard

www.uscg.mil

2100 Second Street, SW
Washington, DC 20593

Department of Energy

www.doe.gov

1000 Independence Ave, SW
Washington, DC 20585
Phone: 800-342-5363 (DIAL DOE)

Department of Health and Human Services

www.hhs.gov

200 Independence Avenue, SW
Washington, DC 20201
Phone: 877-696-6775

Department of Labor

www.dol.gov

Frances Perkins Building
200 Constitution Ave, NW
Washington, DC 20210
Phone: 866-487-2365 (4-USA DOL)

Department of Transportation

www.dot.gov

1200 New Jersey Ave, SE
Washington DC 20590
Phone: 202-366-4000

Environmental Protection Agency (EPA)

www.epa.gov

Ariel Rios Building
1200 Pennsylvania Avenue, NW
Washington, DC 20460
Phone: 202-272-0167

Federal Aviation Administration (FAA)

www.faa.gov

Federal Aviation Administration
800 Independence Avenue, SW
Washington, DC 20591
Phone: 866-835-5322 (TELL-FAA)

Federal Railroad Administration

www.fra.dot.gov

1200 New Jersey Avenue, SE
Washington, DC 20590
Phone: 202-493-6000

Homeland Security

www.dhs.gov

U.S. Department of Homeland Security
Washington, DC 20528
Phone: 202-282-8000

International Organization for Standardization (ISO)

www.iso.org

1, ch. de la Voie-Creuse, Case postale 56
CH-1211 Geneva 20, Switzerland
Phone: +41 22 749 01 11

Mine Safety and Health Administration (MSHA)

www.msha.gov

100 Wilson Blvd, 21st Floor
Arlington, VA 22209-3939
Phone: 202-693-9400

Minerals Management Service (MMS)

www.mms.gov

1849 C Street, NW
Washington, DC 20240
Phone: 202-208-3985

National Fire Protection Association (NFPA)

www.nfpa.org

1 Batterymarch Park
Quincy, MA 02169-7471
Phone: 617-770-3000

National Institutes of Health (NIH)

www.nih.gov

9000 Rockville Pike
Bethesda, MD 20892
Phone: 301-496-4000

National Institute for Occupational and Safety Health (NIOSH)

www.cdc.gov/niosh/

Centers for Disease Control
and Prevention
1600 Clifton Rd
Atlanta, GA 30333
Phone: 800-356-4674 (35-NIOSH)

National Institute of Standards and Technology

www.nist.gov

100 Bureau Drive, Stop 1070
Gaithersburg, MD 20899-1070
Phone: 301-975-6478 (NIST)

National Oceanic & Atmospheric Association (NOAA)

www.noaa.gov

1401 Constitution Avenue, NW
Room 5128
Washington, DC 20230
Phone: 202-482-6090

National Paint and Coating Association

www.paint.org

1500 Rhode Island Ave, NW
Washington, DC 20005
Phone: 202-462-6272

National Safety Council (NSC)

www.nsc.org

1121 Spring Lake Drive
Itasca, IL 60143-3201
Phone: 630-285-1121

National Transportation Safety Board (NTSB)

www.ntsb.gov

490 L'Enfant Plaza, SW
Washington, DC 20594
Phone: 202-314-6000

National Weather Service

www.nws.noaa.gov

1325 East West Highway
Silver Spring, MD 20910

Nuclear Regulatory Commission (NRC)

www.nrc.gov

Office of Public Affairs (OPA)
Washington, DC 20555-0001
Phone: 800-368-5642

Occupational Safety & Health Administration (OSHA)

www.osha.gov

200 Constitution Avenue, NW
Washington, DC 20210
Phone: 800-321-6742 (OSHA)

Pipeline and Hazardous Materials Safety Administration (PHMSA)

www.phmsa.dot.gov

East Building, 2nd Floor
Mail Stop: E21-317
1200 New Jersey Ave., SE
Washington, DC 20590
Phone: 202-366-4433

U.S. Government's Official Web Portal

www.firstgov.gov

U.S. General Services Administration
1800 F Street, NW
Washington, DC 20405
Phone: 800-333-4636 (FED-INFO)

Glossary

4-to-1 rule a safety rule for using straight or extension ladders.

ABC model a model used to provide observation and feedback on safety performance to workers: events or activators (A) in an environment often direct performance or behavior (B), which is motivated because of the consequences (C) that people expect to avoid or receive.

Absorption the complete uptake of a contaminant into a liquid or solid.

Acid a substance with a pH less than 7.0.

Acute short-term health effects.

Administrative controls policies, procedures, programs, training, and supervision to establish rules and guidelines for workers to follow in order to reduce the risk of exposure to a hazard.

Adsorption the adhesion of a contaminant to the outer surface of a solid or liquid.

Affected and other employees employees who are responsible for recognizing when the energy control-procedure is used and understand the purpose of the procedure and importance of not starting up locked/tagged equipment.

Agreement a plan for coordinating activities between different organizations.

Air a layer of gases surrounding Earth; composed mainly of nitrogen (79%) and oxygen (21%).

Air pollution the contamination of the atmosphere, especially by industrial waste gases, fuel exhausts, smoke, or particulate matter (finely divided solids).

Air-purifying respirator a type of PPE that usually covers a wearer's nose and mouth, using a filter or cartridge to remove any contaminants before they enter the wearer's lungs.

Air-supplying respirator a type of personal protective equipment (PPE) that covers a wearer's face with a mask, providing breathable air through a hose that connects the mask to a clean-air source (usually a compressed-air tank or compressor).

Alarm a signal that indicates the existence of an unusual or potentially hazardous situation.

Algae simple, plantlike organisms that grow in water, contain chlorophyll, and obtain their energy from the sun and their carbon from carbon dioxide (through photosynthesis).

Allergen a substance that causes an allergic reaction or unhealthy response by the body's immune system.

Ambulance a standard vehicle of emergency medical service departments used to treat injury victims at remote sites and transport them to other medical facilities as needed.

Ampere (amp) a unit of measure of the electrical current flow in a wire; similar to "gallons of water" flow in a pipe.

Amplitude the measurement used to describe the intensity of sound.

Anaphylaxis a life-threatening, allergic reaction that can result in shock, respiratory, failure, cardiac failure, or death if left untreated.

Annunciator a device that displays alarm conditions through the use of flashing and continuously lit panels.

Arachnid a class of arthropod that has four pairs of segmented legs (i.e., the legs are made up of sections) and includes scorpions, spiders, and ticks.

Arc a spark that occurs when current flows between two points (contacts) that are not intentionally connected. See *spark*.

Arc welding a welding process that uses an electrical arc produced between two electrodes to generate heat.

Arthropod a type of animal that has jointed limbs and a body made up of segments, such as crustaceans (crabs), arachnids (spiders), and insects (mosquitoes).

Atmosphere the air space or environment in which the process technician is working.

Atmospheric pressure the pressure at Earth's surface (14.7 psi at sea level).

Attitude a state of mind or feeling with regard to some issue or event.

Audible alarm an alarm that uses sound to warn workers of a particular condition or hazard.

Audit a review, typically conducted by people from the company, a hired third party, regulatory agencies, or a combination of these groups, to determine if a particular facility is complying with established safety, health, and/or environmental programs.

Authorized employees employees who are responsible for implementing the energy-control procedures and performing service or maintenance work. They receive the most detailed training on procedures.

Backboard a board used to immobilize injured victims to prevent damage to the spinal cord while they are being moved.

Bacteria single-celled, microscopic organisms that lack chlorophyll and are the most diverse group of all living organisms.

Base a substance with a pH greater than 7.0; also referred to as alkaline.

Behavior an observable action or reaction of a person under certain circumstances.

Biological hazard a living or once-living organism, such as a virus, a mosquito, or a snake, that poses a threat to human health

Bloodborne pathogens pathogenic microorganisms that are present in human blood and can cause disease in humans. These pathogens include, but are not limited to, hepatitis B virus (HBV) and human immunodeficiency virus (HIV).

Bonding a system that connects conductive equipment together, keeping all bonded objects at the same electrical potential to eliminate static sparking.

Carcinogen a cancer-causing substance.

Ceiling a concentration that should not be exceeded by workers during an exposure period. Also called the Maximum Acceptable Ceiling.

Chain reaction a series of reactions in which each reaction is initiated by the energy produced in the preceding reaction.

Chemical hazard any hazard that comes from a solid, liquid, or gas element, compound, or mixture that could cause health problems or pollution.

Chemical protective clothing (CPC) clothing that provides protection to those who work around chemicals. Four classifications of CPC allow employees to access the proper level of protection for the chemicals to which they may be exposed.

Chronic long-term health effects.

Class A fire the classification given to ordinary combustible materials, such as wood, paper, cloth, and rubber.

Class B fire the classification given to flammable liquids, such as greases, tars, oils, and lacquers.

Class C fire the classification given to electrical equipment fires.

Class D fire the classification given to combustible metals, such as magnesium, lithium, and sodium.

Closed-environment drain system a system of devices such as pumps, piping, and scrubbers to prevent the release of liquids, gases, and vapors to the atmosphere.

Cold work any work performed in an area that contains bulk quantities of combustible or flammable liquids or gases or bulk quantities of liquids, gases, or solids that are toxic, corrosive, or irritating.

Combustible liquid any liquid that has a flashpoint at or above 100 degrees F but below 200 degrees F.

Combustion the process by which substances (fuel) combine with oxygen to release heat energy, through the act of burning (oxidation).

Combustion point the ignition temperature at which a fuel can catch on fire.

Conduction the transfer of heat through matter via vibrational motion.

Conductor a substance or body that allows a current of electricity to pass continuously along it.

Confined space a work area, not designed for continuous employee occupancy, that restricts the activities of employees who enter, work inside, and exit the area, and provides a limited means of egress.

Contaminant a substance not naturally present in the atmosphere or present in unnaturally high concentrations; also called an impurity. Can be a physical, chemical, biological, or radiological substance.

Convection the transfer of heat through the circulation or movement of a liquid or a gas.

Corrosion the eating away of materials by a chemical process (e.g., iron rusting).

CPR cardiopulmonary resuscitation, which is an emergency method to assist a victim whose heart has stopped beating properly.

Critical lift any lift that could result in death, injury, health impacts, property damage, or project delay if there is an accident.

Curb a method of spill containment for process units. It is sloped to a sewer collection system.

Cyber relating to computers and computing items, such as data, the Internet, and computer networks.

Cyber attack an attack against information, computers, and communication systems, to cause harm, steal information, disrupt productivity, or take control of a computer system.

Cybersecurity security measures intended to protect information and information technology from unauthorized access or use.

Decibel (dB) the measurement of the intensity of a sound, based on the human ear's perception. A unit that is used to measure sound-level intensity (how loud a sound is).

Deflagration a process of subsonic combustion that usually propagates through thermal conductivity (i.e., hot burning material heats the next layer of colder material and ignites it).

Deluge system a system of dumping or spraying water to extinguish a fire or suppress a toxic release or hydrocarbon spill.

Detection device equipment designed to sense a particular condition (e.g., smoke, vapors, flame) and send a signal to an alarm system if the condition exceeds a pre-set limit.

Detonation a violent explosion that generates a supersonic shock wave and propagates through shock compression.

Dike a trenchlike structure built around storage tanks to hold the contents of the tanks in the event of a leak.

Dockboard a temporary platform used during the loading operations of cargo vehicles.

Dose the amount of a substance taken into or absorbed by the body.

Dose-response relationship the connection between the amount (dose) and the effect (response) that a substance can have on the body.

DOT U.S. Department of Transportation; a U.S. government agency with a mission of developing and coordinating policies to provide efficient and economical national transportation system, taking into account need, the environment, and national defense.

Earthquake a shaking and moving of the earth resulting from a sudden shift of rock beneath the surface (geologic stress).

Effluent liquid wastewater discharge from a process facility.

Electric tool a tool operated by electrical means (either AC or DC).

Electricity a flow of electrons from one point to another along a pathway, called a conductor.

Endothermic a chemical reaction that requires the addition or absorption of energy.

Energy the ability to do work.

Engineering controls the use of technological and engineering improvements to isolate, diminish, or remove a hazard from the workplace.

Engulfment the state of being surrounded or completely covered by materials or products within a confined space.

EPA Environmental Protection Agency; a federal agency charged with authority to make and enforce national environmental policy.

Ergonomic hazard a hazard that can create physical and psychological stresses because of forceful or repetitive work, improper work techniques, or poorly designed tools and workspaces.

Ergonomics the study of how people interact with their work environment.

Erosion the wearing away (abrading) of materials by a physical process (e.g., sandblasting).

Escape mask protective gear that provides respiratory protection for a limited period of time in the event of a suddenly occurring emergency.

Exothermic a chemical reaction that releases energy.

Explosion a sudden increase in heat energy, released in a violent burst.

Explosion-suppression barrier a wall or other device built around a specific process area or the control room to suppress explosions.

Explosive a substance that causes a sudden, almost instantaneous release of pressure, gas, and heat when subjected to sudden shock, pressure, or high temperature.

Exposure incident an incident involving the contact of blood (or other potentially infectious materials) with an eye, mouth, other mucous membrane, or non-intact skin, which results from the performance of an employee's duties.

Extinguishing agent the material in a fire extinguisher (e.g., water, dry chemical, and foam); varies depending on the fire classification being fought.

Eyewash station used to flush the eyes of contamination from harmful chemicals.

Facility also called a plant. Something that is built or installed to serve a specific purpose.

Fall protection a system designed to minimize injury from falling when the work height is six feet or greater (above or below grade).

Fire a type of combustion, resulting from a self-sustaining chemical reaction.

Fire brigade a private fire department or industrial fire department composed of a group of employees who are knowledgeable, trained, and skilled in basic fire-fighting operations.

Fire engine a vehicle that carries and pumps water.

Fire hose used to deliver large quantities of water when the fire is too large to be controlled by a fire extinguisher or other means.

Fire hydrant a point of water supply for fire hoses; mounted on trucks or free-standing and connected to a water main.

Fire point the temperature at which burning is self-sustaining after removal of an ignition source.

Fire tetrahedron the elements of a fire triangle (fuel, oxygen, and heat) combined with a fourth element, a chain reaction that keeps the fire burning.

Fire triangle the three elements (fuel, oxygen, and heat) that must be present for a fire to start.

Fire truck an emergency response vehicle that does not carry or pump water.

Fire watch a trained employee who monitors the conditions of an area for a specified time during and after hot work to ensure that no fire danger is present.

First-degree burn a burn that affects only the outer layer of skin and causes pain, redness, and swelling.

Flammable (inflammable) the ability (inability) of a material to ignite and burn readily.

Flammable gas any gas that, at ambient temperature and atmospheric pressure, forms a flammable mixture with air at a concentration of 10% or less by volume.

Flammable liquid any liquid that has a flashpoint below 100 degrees F.

Flammable solid any solid other than a blasting agent or explosive that is liable to cause fire through friction, absorption of moisture, spontaneous chemical change, or retained heat from manufacturing or processing. It can be ignited readily, and when ignited burns so vigorously and persistently that it creates a serious hazard.

Flare an environmentally approved device that burns waste gases collected from various process sources to reduce pressure.

Flashpoint the minimum temperature at which a liquid gives off a vapor in sufficient concentration to ignite.

Flood the rising of water to cover normally dry land.

Frequency the number of sound vibrations per second (peaks of pressure in a sound wave).

Fuel any material that burns; can be a solid, liquid or gas.

Fugitive emission an intentional or unintentional release of a gas.

Fugitive-emissions monitoring measures emissions from sources (e.g., flange connection around packing) to determine if any hazardous agents are present at unacceptable EPA levels.

Fungi plantlike organisms that obtain nutrients by breaking down decaying matter and absorbing the substances into their cells. They are similar to algae but do not contain chlorophyll.

Gas-detection equipment equipment that detects the components within a gaseous atmosphere.

Gas-detector tube an instrument that provides instant measurement of a specific gas, usually indicated by a color change within the tube.

Ground-Fault Circuit Interrupter (GFCI) a device that protects personnel from the possibility of electrical shock by shutting off the power to electric tools when a small amount of current-to-ground is detected.

Grounding connecting an object to the earth using metal in order to provide a path for electricity to travel and dissipate harmlessly into the ground.

Guardrail a rail secured to uprights and erected along the exposed sides and ends of platforms (OSHA).

Hand tool a tool that is manually powered.

Handheld fire extinguisher used to extinguish small fires; must be mounted, located, and identified so it is easily accessible during an emergency.

Handrail a single bar or pipe supported on brackets from a wall or partition, as on a stairway or ramp, to furnish persons with a handhold in case of tripping (OSHA).

Hazard a substance (such as a chemical or disease) or action (such as lifting a heavy object or hearing a loud noise) that can cause a harmful effect.

Hazard control the recognition, evaluation, and elimination (or minimization) of hazards in the workplace.

Hazardous agent a substance, method, or action by which damage or destruction can happen to personnel, equipment, or the environment.

Hazardous atmosphere an atmosphere that can cause death, illness, or injury if people are exposed to it. Examples of hazardous atmospheres are flammable, oxygen deficient/enriched, toxic, or irritating/corrosive environments.

HAZCOM OSHA 29 C.F.R. 1910.1200 Hazard Communication/Employee Right-to-Know, a standard to ensure that employees are aware of the chemicals they are exposed to in the workplace and the measures to take to protect themselves from such hazards.

HAZWOPER the acronym for the OSHA standard for Hazardous Waste Operations and Emergency Response.

Heat the transfer of energy from one object to another as a result of a temperature difference between the two objects.

Hertz (Hz) a measurement used to describe frequency. One hertz is one cycle per second.

Highly toxic a substance that requires only a small amount of exposure to be lethal.

Host an organism whose body provides nourishment and shelter for another, smaller organism.

Hostile nation-state a country that poses a threat to other countries.

Hot work any fire or spark-producing operation (e.g., welding, burning, riveting).

Hurricane an intense, low-pressure tropical (warm area) weather system with sustained winds of 74 miles per hour or more. Hurricanes can rotate clockwise or counter-clockwise, depending on their location of origin.

HVAC ventilation systems used to control workplace environmental factors such as temperature, humidity, and odors. The acronym is short for heating, ventilating, and air conditioning.

Hydraulic the use of liquid (hydraulic fluid) as the power source.

Hydraulic tool a tool that is powered using hydraulic (liquid) pressure.

Immediately Dangerous to Life and Health (IDLH) any condition that presents an immediate threat to a person's life or causes permanent health problems. This usually refers to an airborne concentration that is immediately dangerous to life and health, or can impair a person's ability to escape the atmosphere.

Incipient fire a small fire that requires immediate attention to prevent it from becoming larger and more serious.

Indicator device a generic term for a type of equipment that indicates process variables; may be visual (e.g., light), audible (e.g., horn), or both.

Industrial-hygiene monitoring monitoring the health and well-being of workers exposed to chemical and physical agents in their work environment.

Infectious capable of infecting or spreading disease.

Inorganic compound a chemical compound that does not contain carbon chains.

Insider a person inside a company who causes harm, either intentionally or unintentionally.

Inspection a proactive activity conducted prior to a need for action by plant personnel to ensure that safety, health, and environmental programs are being followed.

Insulator a device made from a material that will not conduct electricity; the device is normally used to give mechanical support or to shield electrical wire or electronic components.

Intensity the loudness of a sound (pressure-peak intensity of a sound wave).

Interlock a type of hardware or software that does not allow an action to occur if certain conditions are not met.

Ionizing radiation radiation that has enough energy to cause atoms to lose electrons and become ions; an extreme health hazard that is usually confined to restricted areas and requires employees to follow strict guidelines to keep exposure as low as possible.

ISO 14000 an international standard that addresses how to incorporate environmental aspects into operations and product standards.

Job-safety analysis a method of analyzing how a job is performed in order to identify and correct undesirable conditions.

Kinetic energy energy associated with mass in motion.

Leak a condition that occurs when a container or equipment is compromised, allowing a material to escape.

LEL meter Lower Explosive Limit (LEL) meter; measures the combustible content of an atmosphere.

Lockout device a device placed on an energy source, in accordance with an established procedure, that ensures the energy is isolated and the equipment cannot be operated until the lockout device is removed.

Lockout/tagout a procedure used in process industries to isolate energy sources from a piece of equipment.

Machine guard a barrier that prevents a machine operator's hands or fingers from entering into the point of operation.

Malware a computer program developed to cause intentional harm.

Material Safety Data Sheet (MSDS) a document that provides key safety, health, and environmental information about a chemical; required by OSHA's HAZCOM standard.

Means of egress an exit or way to evacuate a building or facility during an emergency.

Microorganism a very small form of life, often viewable only through through a microscope, that includes viruses, bacteria, algae, and fungi.

Monitor a process used to gather data to evaluate the work environment using specialized equipment.

Musculoskeletal Disorder (MSD) a health condition characterized by damage to muscles, nerves, tendons, ligaments, joints, etc.

Mutagen a chemical suspected to have properties that change or alter a living cell's genetic structure; mutagens can lead to cancer or birth defects if the egg or sperm is affected.

Near miss an unsafe act that does not result in an incident or accident.

Network two or more computers linked together for sharing data, programs, and resources such as printers and scanners.

Neurotoxins a poison that affects the nervous system.

Nip point a dangerous area where contact is made between two points on the equipment (e.g., a belt meeting a pulley or two gears intermeshing); also called a pinch point.

Noise any unwanted or excessive sound.

Non-ionizing radiation low-frequency radiation such as radio and television waves.

NRC Nuclear Regulatory Commission; a U.S. government agency that protects public health and safety through regulation of nuclear power and the civilian use of nuclear materials.

Ohm a measurement of resistance in electrical circuits.

Opening/blinding permits a permit used used to help ensure that accidental leaking from pipes does not occur.

Organic compound a chemical compounds that contains carbon chains.

Organic peroxide an organic compound that contains the bivalent -O-O- structure and which can be considered to be a structural derivative of hydrogen peroxide, where one or both of the hydrogen atoms has been replaced by an organic radical.

OSHA Occupational Safety and Health Administration (OSHA); a U.S. government agency created to establish and enforce workplace safety and health standards, conduct workplace inspections and propose penalties for noncompliance, and investigate serious workplace incidents.

OSHA fire protection standard (29 CFR 1910.155) intended to prevent and or minimize the consequences of fire within the workplace.

Oxidizer a chemical that can initiate or promote combustion in other materials.

Oxyacetylene welding a welding process that burns a blend of oxygen and acetylene to generate heat.

Oxygen-deficient an atmosphere in which the oxygen concentration is less than 19.5%.

Oxygen-enriched an atmosphere in which the oxygen concentration is greater than 23.5%.

Pathogen a specific cause of a disease, such as bacteria or a virus.

Permissible Exposure Limit (PEL) an OSHA limit representing the maximum acceptable exposure of workers to a hazard over a specific period of time.

Permissive a type of interlock that does not allow a process or equipment to start up unless certain conditions are met.

Personal monitoring device an instrument that measures an individual's exposure to airborne contaminants.

Personal protective equipment (PPE) specialized gear that provides a barrier between hazards and the worker using the PPE.

pH a measure of the amount of hydrogen ions in a solution that can react and indicate if the substance is an acid or a base.

Physical hazard any hazard that comes from environmental factors such as excessive levels of noise, temperature, pressure, vibration, radiation, electricity, or rotating equipment.

Physical security security measures intended to counter physical threats from a person or group seeking to intentionally harm other people or vital assets.

Plan a method, prepared in advance, for carrying out an action.

Pneumatic the use of air pressure or a gas as a power source.

Pneumatic tool a tool that is powered using pneumatic (air or gas) pressure.

Point of operation the area where the equipment actually performs its intended task (e.g., cutting, rotating, stamping).

Policy a guiding principle.

Positive pressure control a system used to keep external air, which may contain airborne toxic substances, from entering the building.

Potential energy stored energy; energy that has the potential to become kinetic.

Powder-activated tool a tool that is powered using a small explosive charge (e.g., nail gun).

Power tool a tool that is powered by a source such as electricity, pneumatics, hydraulics, or powder-activation.

Powered industrial truck the American Society of Mechanical Engineers (ASME) defines a powered industrial truck as a mobile, power-propelled truck used to carry, push, pull, lift, stack, or tier materials. Forklifts and other similar vehicles are considered to be powered industrial trucks.

Powered platform equipment designed to lift personnel on a platform to work at heights.

Pressure the force exerted on a surface divided by its area.

Pressure-relief valve a safety valve that automatically opens at a set pressure to protect process vessels or piping from excessive pressure.

Pressure switch a mechanical device that uses electrical contacts to complete an electric circuit and generate an alarm signal.

Principle a set of rules or standards.

Proactive a preventive activity conducted prior to a need for action.

Procedure a set of step-by-step instructions for accomplishing a task.

Process fluid any material that flows; it can be either liquid or gas. When under pressure, both gases and liquids transmit force equally. Process gases are compressible and liquids are not.

Process Hazard Analysis (PHA) an organized, systematic process used to identify potential hazards that could result in accidents causing injury, death, property damage, or environmental damage.

Process industries a broad term for industries that convert raw materials, using a series of actions or operations, into products for consumers.

Process technician a worker in a process facility who monitors and controls mechanical, physical, and/or chemical changes, throughout many processes, to produce either a final product or an intermediate product, made from raw materials.

Process technology a controlled and monitored series of operations, steps, or tasks that converts raw materials into a product.

Process upset controls devices and systems used in the process industries to respond to and control upsets.

Program computer software (sometimes called an application).

Protein allergen an allergen caused by substances produced by vertebrate animals, including blood, feces, hair, and dead skin.

Pyrophoric a chemical that will ignite spontaneously in air at a temperature of 130 degrees F or below.

Rad radiation absorbed dose; the unit of ionizing radiation absorbed by a material, such as human tissue.

Radiation the transfer of heat energy through electromagnetic waves.

Reactive a corrective activity conducted in response to a need.

Redundant system a system that provides a backup in the event the primary system fails.

Release a controlled or uncontrolled discharge of process materials into the environment.

Relief valve a safety device designed to open if the pressure of a liquid in a closed vessel exceeds a preset level.

Rem roentgen equivalent man; a unit of measure of the dose of radiation deposited in body tissue, averaged over the body.

Repetitive-motion injury (RMI) an injury caused by repeating the same motion.

Resistance welding a welding process that uses electricity generated through the material to be welded combined with pressure at the weld point, to create the weld.

Respiration the bodily process of taking oxygen from air breathed in (inhalation) and giving off carbon dioxide (exhalation); also called breathing.

Risk a combination of vulnerabilities and threats.

Ropes and mechanical retrieval devices equipment used to rescue workers from a hazardous situation. Ropes are primarily used to lower rescuers into confined spaces and to hoist injury victims out of those spaces. Mechanical retrieval devices are set up on a tripod and use a winch or crank to lower and raise individuals with the rescue rope.

Routes of entry the ways in which a hazardous substance can enter the body, such as inhalation through the nose, absorption through the skin or eyes, accidental ingestion, or through injection.

Rule a statement describing how to do something or what may or may not be done.

Runaway reaction a reaction that is out of control; can be either endothermic or exothermic.

Safe-work permit a permit used to ensure that the area is safe for work to be performed and for communicating that information.

Safety shower located throughout a plant to provide the ability to rinse the body off in the event of a large exposure to a chemical spill.

Scaffold tag a tag used to clearly label the status of a scaffold and communicate its approved use.

SCBA an apparatus used to provide breathing air when the wearer needs to enter an area with an atmosphere that might be oxygen-deficient or harmful if inhaled.

Second-degree burn a burn that affects both the outer and the underlying layer of skin.

Security hazard a hazard or threat from a person or group seeking to intentionally harm people, computer resources, or other vital assets.

Sensitizer an agent that can cause an allergic reaction.

SHE Safety, Health, and the Environment. Also referred to as HSE or EHS.

Short circuit a short circuit occurs when electrons in a current flow find additional unwanted paths outside of the intended circuit or conductor, and flow to it.

Short-Term Exposure Limit (STEL) a concentration to which workers can be exposed for a short term (e.g., 15 minutes) before suffering any harm.

Shutdown device equipment that ensures that processes and/or equipment are secured from failures that may result in hazardous conditions.

Soil pollution the accidental or intentional discharge of any harmful substance into the soil.

Sound a form of vibrational energy conducted through a medium (e.g., solid, liquid or gas) that creates an audible sensation that can be detected by the ear.

Sound wave a pressure wave that moves through the air and is audible to the human ear.

Spark a single burst of electrical energy. See *arc*.

Spill an uncontrolled discharge of a liquid; usually involves more volume than a leak.

Spill containment a combination of procedures and devices designed to contain spilled materials in the immediate area of the spill and minimize the effects of a spill on the environment.

Standpipe system a system that provides hose stations with water supplied by a main, fire pump, or tank; standpipes can be either wet (connected to a permanent water supply and constantly filled with water) or dry (not connected to a permanent water supply).

Static electricity electricity "at rest"; an electrical charge caused by friction between two dissimilar materials.

Storm surge one of the greatest dangers created by hurricanes and one that pushes water toward the shoreline by the hurricane-force winds swirling around the body of the storm. As a storm surge advances on a shoreline, it can create a wall of water 15 feet or more above normal water levels. Storm surges pound coastlines with water that weighs 1,700 pounds per cubic yard, thereby having the capability of destroying many structures in its path.

Stretcher a bed with wheels used to transport injured victims from one place to another.

Superfund a monetary fund that comes from tax dollars paid by the chemical industry to pay for the cleanup of abandoned waste sites in the event no responsible party can be found.

System an organized, interdependent set of related principles or rules.

Temperature the degree of hotness or coldness that can be measured by a thermometer and a definite scale.

Teratogen a substance believed to have an adverse effect on human fetus development.

Terrorist a radical person who uses terror as a weapon to control others.

Third-degree burn a burn that affects deeper tissues and causes white or blackened, charred skin that may be numb.

Threat a perceived or implied feeling or communication; an individual or a group that will harm people or property.

Three-point contact a safety practice in which both feet and at least one hand are used when ascending or descending stairs.

Threshold Limit Value (TLV) a limit set by the American Conference of Government Industrial Hygienists, representing the maximum acceptable exposure of workers to a hazard over a specific period of time.

Time-Weighted Average (TWA) an average concentration of a chemical or a noise to which an employee can be exposed over an eight-hour period, or 40 hours a week.

Toeboard a vertical barrier at floor level erected along exposed edges of a floor opening, wall opening, platform, runway, or ramp to prevent falls of materials (OSHA).

Tornado a destructive, localized windstorm. Tornadoes are produced by severe thunderstorms. A funnel cloud may or may not occur during a tornado.

Toxic material a substance determined to have an adverse health impact.

Toxin a poisonous substance that can harm living organisms.

Unit an integrated group of process equipment used to produce a specific product or products. All equipment contained in a department.

Universal precautions an approach to infection control. According to the concept of Universal Precautions, all human blood and certain human body fluids should be treated as if known to be infectious for HIV, HBV, and other bloodborne pathogens.

Vacuum any pressure below atmospheric pressure (14.7 psi).

Venom a poisonous substance created by some animals (such as snakes and spiders) transmitted to prey or an enemy by biting or stinging.

Vertigo a sensation or illusion of movement in which a person feels as if revolving in space (called subjective vertigo) or senses the surrounding environment to be spinning (called objective vertigo).

Vibration a rapid, back-and-forth, periodic movement of an object along its radial or horizontal axis.

Visible alarm an alarm that uses visual means (e.g., lights, motion, color) to warn workers of a particular condition or hazard.

Volatility the ability of a material to evaporate.

Volt one volt is the electromotive force that will establish a current of one amp through a resistance of one ohm.

Voltage the driving force needed to keep electrons flowing in a circuit.

Voluntary Protection Program (VPP) an OSHA program designed to recognize and promote effective safety and health management.

Vulnerability a weakness, or "hole" in a defense system.

Walking and working surfaces how OSHA refers to floors, walkways, passageways, corridors, platforms, and other similar surfaces.

Warning a weather advisory issued when certain weather conditions (e.g., thunderstorm, hurricane, flash flood) are expected in the specified area.

Watch a weather advisory issued when certain weather conditions (e.g., thunderstorm, hurricane, flash flood) are possible in the specified area.

Water pollution the introduction, into a body of water or the water table, any EPA listed potential pollutant that affects the water's chemical, physical, or biological integrity.

Water-reactive any chemical that reacts with water to release a gas that is either flammable or presents a health hazard.

Wavelength the distance between successive points of equal amplitude on a sound wave.

Working-load limit the maximum weight that can be lifted and should not be exceeded when working with a load.

Index

Chemical Hazards

Chemical protective suit
Courtesy of Brazosport College Industry Training Class

Chemical protective suit with breathing apparatus
Courtesy of Brazosport College Industry Training Class

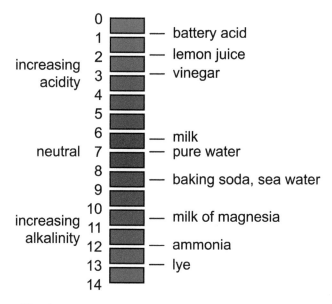

pH scale

Biological Hazards

Ants

Scorpion

Black widow spider

Brown recluse spider

Deer tick

Bees

Wasp

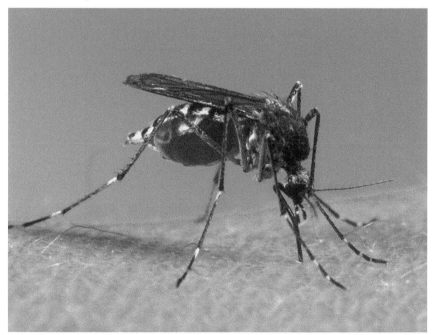

Mosquito

Copperhead snake

Coral snake

Cottonmouth snake

Rattle snake

Poison sumac

Poison oak

Biohazard container

Equipment Hazards

Breaker Box open

Breaker

Tag secured with cable tie
Courtesy of Eastman Chemical Company

Technician applying a field lock
to valve
Courtesy of Eastman Chemical Company

Grounding wire to the ground

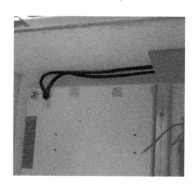

Grounding wire from breaker box

A wire that has shorted out

Hot, neutral, and ground wires

Fire and Explosion Hazards

Truck displaying flammable solution placard

Container for flammable materials

Self contained breathing apparatus and chemical suit

NFPA 704 color diamond on truck

Pressure, Temp, and Radiation Hazards

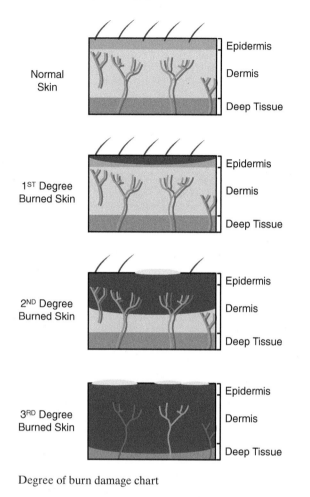

Normal Skin

Epidermis

Dermis

Deep Tissue

1ST Degree Burned Skin

Epidermis

Dermis

Deep Tissue

2ND Degree Burned Skin

Epidermis

Dermis

Deep Tissue

3RD Degree Burned Skin

Epidermis

Dermis

Deep Tissue

Degree of burn damage chart

Hazardous Atmospheres and Respirator Hazards

Self contained breathing apparatus

Working Area and Height Hazards

Guardrail

Caged ladder

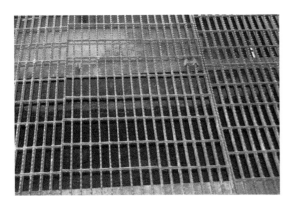

Textured walking surface

Vehicle and Transportation Hazards

Pipeline

Rail storage tank

Crane

Man lift

Helicopter

Gasoline truck filling the gasoline storage tanks of a service station

Oil tanker

Natural Disasters and Inclement Hazards

Lightning

Land slide

Avalanche

Earthquake damage

Tornado

Tornado damage

Hurricane damage

Hurricane Rita in the summer of 2005

Flooding

Flood damage

Hurricane damage

Engineering Controls

Flare

Relief valve
Courtesy of Bayport Training and Technical Center

Administrative Controls

Inspection of equipment and piping

Monitoring of processes

Fire, Rescue, and Emergency Response Equipment

Fire extinguisher

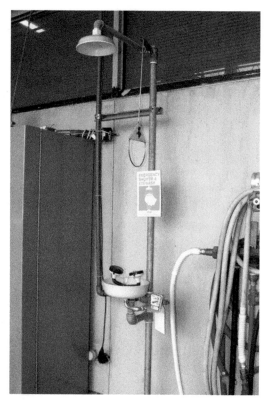

Safety shower and eye wash station

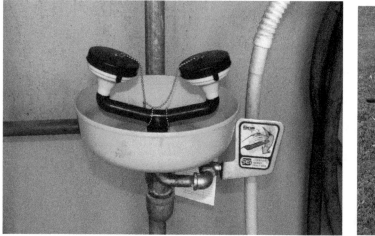

Eye wash station close up

Fire monitor mounted on a fire hydrant